3rd edition

POLICING IN AMERICA
A BALANCE OF FORCES

ROBERT H. LANGWORTHY
University of Alaska at Anchorage

LAWRENCE F. TRAVIS III
University of Cincinnati

Prentice
Hall

Upper Saddle River, New Jersey 07458

Library of Congress Cataloging-in-Publication Data

Langworthy, Robert H.
    Policing in America : a balance of forces / Robert H. Langworthy, Lawrence F. Travis III.—3rd ed.
        p. cm.
    Includes bibliographical references and index.
    ISBN 0-13-092624-8 (alk.paper)
    1. Police—United States. 2. Law enforcement—United States. I. Travis, Lawrence F. II. Title.

HV8138 .L277 2002
363.2'0973—dc21

2002020605

Publisher: Jeff Johnston
Executive Editor: Kim Davies
Assistant Editor: Sarah Holle
Production Editor: Emily Bush, Carlisle Publishers Services
Production Liaison: Barbara Marttine Cappuccio
Director of Production & Manufacturing: Bruce Johnson
Managing Editor: Mary Carnis
Manufacturing Manager: Cathleen Petersen
Creative Director: Cheryl Asherman
Cover Design Coordinator: Miguel Ortiz
Cover Designer: Scott Garrison
Cover Image: Thierry Dosogne/ The Image Bank
Marketing Manager: Jessica Pfaff
Editorial Assistant: Korrine Dorsey
Composition: Carlisle Communications, Ltd.
Printing and Binding: Phoenix Book Tech Park

Pearson Education LTD., *London*
Pearson Education Australia PTY. Limited, *Sydney*
Pearson Education Singapore, Pte. Ltd.
Pearson Education North Asia Ltd., *Hong Kong*
Pearson Education Canada, Ltd., *Toronto*
Pearson Educacíon de Mexico, S.A. de C.V.
Pearson Education-Japan, *Tokyo*
Pearson Education Malaysia, Pte. Ltd.

10 9 8 7 6 5 4 3 2 1
ISBN 0-13-092624-8

Dedicated, with thanks to:

Jack and Betsy Langworthy
Sharon Langworthy
Larry and Peggy Travis
Sergeant Francis Phelan

CONTENTS

3

THE ENGLISH ROOTS OF AMERICAN POLICING 54

4

THE EVOLUTION OF POLICING IN AMERICA 79

PART II

THE LAW-ENFORCEMENT INDUSTRY IN AMERICA 105

5

FEDERAL AND STATE POLICE 107

6

PRIVATE AND SPECIAL-PURPOSE POLICE 134

7

MUNICIPAL AND LOCAL POLICE 158

PART III

CORRELATES OF POLICING: ORGANIZATIONS, OFFICERS, AND COMMUNITIES 185

8

POLICE ORGANIZATIONS 187

9

INDIVIDUALS IN POLICING: OFFICERS AND SUPERVISORS 212

10

POLICE OFFICERS 233

11

POLICE AND COMMUNITY 265

PART IV

THE FUNCTIONS OF POLICING IN AMERICA 289

12

LAW ENFORCEMENT AND THE POLICE 291

13

SERVICE AND THE POLICE 325

14

ORDER AND THE POLICE 355

PART V

DILEMMAS IN POLICING 385

15

COMMUNITY POLICING: TYING IT ALL TOGETHER 387

16

CONTROLLING THE POLICE 410

17

CURRENT TRENDS AND FUTURE ISSUES IN POLICING 444

PREFACE

Writing the third edition of *Policing in America* was a challenge. Reactions to the first two editions from colleagues and our students were gratifying. Our original goal was to produce a book that could both serve as an introductory textbook and spur the interest and thinking of both our students and our colleagues. The comments we have received indicate that we have been successful. Instructors and students were satisfied with the level and depth of coverage, and the style of presentation. Our challenge in this edition was twofold. First, how might we improve on our earlier success? Second, how could we best incorporate the explosion of knowledge about the police?

The third edition retains the best of the earlier editions and includes improvements suggested by colleagues and the latest findings from the continually expanding body of policing knowledge. We continue to use a "conversational" tone, writing in "plain English." Where we need to use precise or jargonistic terms, we define them in the text. Our purpose remains that of communicating ideas, and we still think that is best done simply. The ideas are complex, but the reading is clear. We still want readers to wrestle with the ideas, not the vocabulary or sentence structure.

FRAMEWORK

The third edition follows the same framework we used in the second. To organize the large and diverse body of information, we provide a conceptual framework within which we hope to understand the police. We believe that policing in practice—what the police do on the street—is a product of a number of factors or forces. We use a balance-of-forces metaphor for understanding the police and devote chapters to identifying the important forces and for illustrating ways in which different balances are reached. We recognize not only that differences exist among police agencies in the United States but also that these differences are purposeful. What works in one community may not work in another for very legitimate and understandable reasons.

We don't use words such as *cause* or *determinant* when discussing factors that appear to be linked to police practice. Instead, we focus on *correlates* of policing—factors that may not explain any particular police action, but that do seem to explain police practices in general. In combination with our balance-of-forces metaphor, we make recurrent reference to correlates of policing in substantive chapters. This recurring topic provides unity and continuity to our examination of police practice. Our colleagues and students tell us that this framework encourages readers to develop their own integrative skills.

PEDAGOGICAL FEATURES

As a learning tool, this textbook is designed to assist students in learning about the police. Each chapter begins with a detailed outline of the topics included within it. As we introduce new words, we define them in the text so that readers do not have to flip through the book. To assist readers further, we have included review questions at the end of each chapter, called the "Chapter Checkup." Probably the best way to use these questions is to read them first, then read the chapter knowing what we believe readers should gain from the chapter. Upon finishing the chapter, readers should take a few moments to answer each of the questions to be sure that they have understood the material. Each chapter also contains a summary, sometimes under the heading "Correlates," and sometimes simply titled "Conclusion." We have included a detailed and exhaustive index to make it easier for readers to find specific topics. Finally, each chapter is extensively referenced and the list of references provides a solid bibliography for readers who wish to begin an independent study of any of the topics discussed in the book.

We believe that the use of a recurrent theme and writing in plain English makes the book "reader friendly." Where appropriate, we have included boxes and insets to supplement discussion in the text. We try to present information in a visual fashion, using illustrations, graphs, and figures to help "visual learners" grasp the meaning of the words and data included in the chapters. These insets and graphics include biographies of people important to the development of policing, photographs to serve as visual cues, and analytic schemes to illustrate how important factors are correlated with policing. Our goal is to have the insets communicate ideas, and we agree with the saying, "A picture paints a thousand words." Whenever possible, we try to use pictures with, or in place of, words.

ORGANIZATION

This book is divided into five parts, and each part contributes to a global understanding of policing in America. Our own perspective is that policing is organic, that the history, structure, organization, functions, and issues in policing are all related. Nonetheless, we chose these five sections as a means to organize the presentation of information. Part I analyzes the history of policing and assesses the social, political, and historical forces that are correlated with both the rise of formal policing and the variety of shapes such policing has taken. Part II examines the police industry in the United States. It describes federal, state, special-purpose, private, and local policing agencies; their history, and their current status. Part III describes what we identify as the major correlates of policing; organizations, officers, and communities. Part IV examines the basic functions of police in American society. Part V applies the lessons learned to an analysis of the development of community policing, an assessment of police misconduct and control, and the likely future of policing in America.

In the preface of the previous editions we wrote that we wanted this book to do "double duty." Novices can learn enough to be sufficiently grounded in existing theory and knowledge of the police to pursue further study. Beyond this, readers will develop the habit of integrating the available theory and research on policing—in short, of looking at the big picture. In addition, instructors will be spurred to reconsider the research and theory on policing as a result of the way in which we organize and present these materials in the book. We wanted to integrate materials as well as communicate them in a way that would prove stimulating to instructors, advanced students, and introductory students alike. Eight years later, we are satisfied that we have accomplished this goal. This edition is designed to continue the tradition.

CHANGES FOR THE THIRD EDITION

As with the second edition, when we set about the task of preparing the third edition we believed that the basic book was solid and well received. Following the old adage, "If it ain't broke, don't fix it," we made only those changes we felt were essential. A quick comparison of the tables of contents for this edition and the second edition will reveal no major changes. What it will not show is the subtle fine tuning we have attempted. Throughout the book we have taken great pains to improve the wording, where possible. We have also been careful to build on earlier chapters as we revised later ones. Although each chapter can, we believe, stand on its own, we have been careful to reintroduce and employ the theoretical concepts developed in the descriptive chapters as we present information in the analytic chapters.

When the second edition went to press in late 1998, the empirical literature about policing was entering a period of rapid growth. Since then, literally hundreds of federally supported policing studies have been released, the federal Office of Community Oriented Policing Services reached its stride, new policing journals were launched, and our knowledge of and understanding about the police has grown tremendously. In that same time period, concern about international organized crime has grown—crimes, especially violent crimes in schools—have emerged as a major issue, and the question of racial bias in police actions has retaken center stage on the national agenda. At the same time, serious crime has been in decline, and the role of policing in crime control has been debated.

The major challenge we faced in this edition was how best to integrate and examine these emerging and re-emerging topics. In the end, our own analysis led us to conclude that the balance-of-forces metaphor applies as well to understanding the growth and development of these "new" issues as it does to a general understanding of policing in America. Thus, the most important change for the third edition involved integrating emerging issues into the existing analytic framework. Another change in this edition reflects the fact that we now have some empirical research addressing several of the questions and issues highlighted in the chapters. As a result, we have been able to incorporate more graphs and tables to present this information to readers. Each chapter is a bit longer than the previous edition, reflecting the availability of pertinent information. Finally, the reference section for each chapter is also a bit longer, incorporating the latest information about relevant topics.

We have retained our focus on police discretion throughout the book, and we have moved coverage of domestic violence from our examination of police service to the discussion of order maintenance, where we explore the relatively recent movement to control police discretion by mandating police take certain actions, such as mandatory or preferred arrest. Finally, we completed the necessary updating of previous information. We hope we have kept the contents of the book current while retaining a focus on a broader understanding of policing in the United States over its entire history and into the future.

ACKNOWLEDGMENTS

We would like to take this opportunity to express our gratitude to the many people who helped in the preparation of this book. Although a listing of all the people who have contributed to our ability to write this book would be too long, several individuals deserve special recognition. We would like to thank again, Christine Cardone for her efforts with the first edition, and Kim Davies and especially Sarah Holle of Prentice-Hall who were immensely helpful in this most recent effort.

Several knowledgeable educators and researchers reviewed and commented on the first and second editions of the book. Their insights and suggestions, while not always welcomed, were first rate and helpful. The critical appraisal of this book by respected colleagues was often a source of inspiration and direction to us. For their willingness to serve, and for the job they did, we want to thank Terry C. Cox, Eastern Kentucky University; Armand P. Hernandez, Arizona State University; H. Bruce Pierce, John Jay College of Criminal Justice; Allen E. Wagner, University of Missouri, St. Louis; James D. Stinchcomb, formerly of Miami-Dade Community College; Edward Maguire, George Mason University; Steven A. Egger, University of Illinois, Springfield; Alex del Carmen, University of Texas, Arlington, TX; Marty Gruher, Rogue Community College, Grants Pass, OR; Jim Newman, Rio Hondo College, Whittier, CA; Ronald Burns, Texas Christian University, Fort Worth, TX; and Steven Brandl, University of Wisconsin, Milwaukee, WI. Additional suggestions were provided by colleagues and students, Darrell Cook, John Crank, Jim Frank, Colleen Kadleck, Stephen Holmes, Bill King, John Liederbach, Ken Novak, Beth Sanders, Brad Smith, and Bill Walsh.

Finally, Patricia Travis and RoseMarie Langworthy, our spouses, want to thank Prentice-Hall for keeping us both busy and off the streets for the past year.

Robert H. Langworthy
Anchorage, Alaska

Lawrence F. Travis III
Cincinnati, OH

PART ONE

DEVELOPING A PERSPECTIVE

The first part of this book introduces the subject of our study, policing in America. It is a fascinating and complicated topic that requires careful analysis and deep thought. The very idea of hiring people to direct and control the behavior of citizens may be confusing to Americans who prize liberty and generally distrust governmental interference. The number and variety of police officers, organizations, and functions also often lead to confusion. It is difficult to answer questions such as, "Who are the police?" and "What do they do?" It is, we think, impossible to give a short, simple answer to such questions.

The task of studying and understanding the police is made difficult by the number and variety of people, organizations, and jobs that compose policing. It is made more difficult by the volume and diversity of knowledge about the police that exist in the literature of several academic disciplines. Our first requirement is to bring order to this confusion.

Chapter 1 defines our topic and briefly describes the role and structure of policing in the United States. This chapter examines types of police and police functions and presents this topic in social, organizational, and political contexts. It then identifies sets of factors that appear to explain differences in police organization and practice. The chapter concludes with a perspective for the study of the police. This perspective maintains that policing in practice is the product of a unique balance among the people, organizations, and communities involved.

Because a primary purpose of understanding the police is to develop the capacity to change or control police practice, the remainder of Part I is dedicated to a description of the evolution and development of policing. The history of policing tells us where we have been and indicates the direction in which contemporary policing appears to be moving. Further, an analytic history, one that seeks to explain and understand change over time, identifies factors that may influence police practice.

Chapter 2 presents a model of the evolution of policing as a product of the characteristics of the society in which it develops. By applying the model to an analysis of the development of policing in continental European and Asian settings, we are able to see how theory (the model) can guide analysis. We are also able to identify social characteristics that should be important to a general understanding of the evolution of policing, and of American policing in particular.

Chapter 3 presents a history of police development in England, in the American colonies, and during earlier times in the United States. As the most direct ancestor of American policing, the English experience is important as a guide to the creation of police in the United States. Equally important, however, is how the early American police differed from their English predecessors.

The final chapter in this part of the book, Chapter 4, examines the more recent history of policing in America. Covering the period between the Civil War era and the present, this chapter identifies the people and forces that produced contemporary American policing. It concludes with the observation that the lessons of history can be instructive to those seeking to understand the present and to change the future.

With the conclusion of Chapter 4, we are up to date and ready to begin an analysis of policing in America today. The remaining parts of the book examine important segments of this complicated topic in detail. Part I provides the background and foundation for the work that follows.

UNDERSTANDING THE POLICE

CHAPTER OUTLINE

If you are asked to think about the police, what images and emotions come to mind? Do you picture a uniformed patrol officer, a police cruiser, an undercover detective, or something else? Is your emotional reaction to this image positive, negative, or neutral? How well do you think your image truly reflects the police in America?

All of us carry images of the police in our minds. Probably none of our individual perceptions of the police are wholly accurate or adequate. Rather, we tend to form impressions of the police based on a few personal experiences, casual observations, media portrayals, and things we are told about the police from others. That is, while everyone has a view on the police, these views usually stem from a narrow, personal focus.

The purpose of this book is to provide a broader perspective on policing in America. Our goal is to foster an understanding of the variety of forces that shape the practice of policing in our society. Instead of limiting ourselves to seeing the police through our personal and therefore narrow lens, or even sharing our views, we hope to build a "broad-based lens" that will give us a "wide-angle" view of the police.

DEFINING THE POLICE: THE CENTRALITY OF FORCE

Before we can begin to broaden our view of the police, it is important to define our subject more clearly. Thus far we have been talking about "the police" as if that title meant something specific. Beginning with the observation that we are likely to have different visions of the police means that we expect "the police" to mean different things to different people at different times. What, then, are the police?

George Rush (1977:271), in his *Dictionary of Criminal Justice,* defines the **police** as "an organized body of municipal, county, or state officers engaged in maintaining public order, peace, safety, and in investigating and arresting persons suspected or formally accused of crime." Rush distinguishes between the police and law enforcement; he includes the variety of other agencies in the latter category. Police officials have various titles, such as police officer, deputy (marshall or sheriff), peace officer, constable, or ranger (Conser and Russell, 2000:5–6).

Although a good start, this definition is too narrow for our purposes. It is too narrow because it excludes many organizations that are typically covered under the term police, such as the Federal Bureau of Investigation, military police, and U.S. Postal Inspectors. It is also too narrow because it limits us to looking at "organized bodies" and does not permit us to examine individual police officers. Finally, it is too narrow because it focuses exclusively on the "ends" of policing—what it is that we expect from the police, such as order maintenance, public safety, and law enforcement.

Carl Klockars (1985:9) has observed that definitions of the police that focus on the goals, or ends, of policing tell us more about the person making the definition than they do about the police. He argues that police are police whether or not they are in the process of maintaining public order. Rather than looking at what the police are supposed to do, he suggests we focus on the meaning of *police*. The meaning is not found in what is done but in how it is done. Policing is done by coercive force.

The importance of coercive force to the definition and function of the police was first identified by Egon Bittner (1970:36–47). He suggested that the capacity to use coercive force is the core of the police role in society. The reason we have police and the reasons for which we call on police are based on a belief that force may be necessary. Police are authorized to use force to resolve all sorts of social problems. Calling the cops is calling for force.

A variety of people, organizations, and functions make up the police in America.
(Hazel Hankin/Stock Boston)

For example, police are commonly called to the scene of automobile
wrecks. Why do we call the police? To be sure, the responding officer can
provide emergency medical care, and usually conducts an investigation of
the accident for insurance and other purposes, but these tasks do not require a police officer. Why do we not simply call a life squad and an insurance adjuster?

Perhaps, you might think, the reason is because the maintenance of an
orderly flow of traffic and, by extension, public safety on the roadway require that someone direct traffic. The question remains, however, why must
that someone be the police? Could not just any concerned citizen, or even an
official "traffic director," be assigned this task? Why do we call the police?

One reason is convenience—we call the police because we can (i.e., if
we call, an officer will come). But the more compelling reason for calling the
police is that force may be necessary. Especially in grisly wrecks involving
injuries, many drivers wish to "rubberneck" and view the carnage. Their cu-
riosity can create a traffic jam and pose a danger to other drivers and emergency personnel at the scene. These drivers might not listen to the average
citizen or even an appointed traffic director telling them to move along.
How could a nonpolice traffic director force drivers to obey? The director
would have to either call the police or be the police. Only the police are authorized to force obedience.

We can return to our definition. As Klockars (1985:12) concludes, "Police are institutions or individuals given the general right to use coercive force by the state within the state's domestic territory." This definition of *police* is perhaps the best available, and so it is the one we will use.

Defining the police in this way allows us to consider both police organizations and individual police officers. It lets us include federal, private, and special-jurisdiction agencies and officers. It focuses our attention on the process of policing and does not limit us to looking only at the goals of policing. With this definition, we can proceed to develop an understanding of policing in America.

POLICING IN AMERICA: FREEDOM VERSUS ORDER

Now that we have an idea of what is meant by police, we must turn our attention to the matter of policing in America. How is it that we have police, and what role do they play in our society? In many ways, the very idea of police can be viewed as un-American.

As Herman Goldstein (1977:1), a noted police scholar, remarks, "The police, by the very nature of their function, are an anomaly in a free society." This comment recognizes the core conflict that characterizes policing in America. On the one hand, we have a society and system of government that values and protects individual freedom. On the other, we have police who apply their capacity to use coercive force to make people behave in certain ways. The police serve to limit individual freedom.

The fact is, for all of us to exercise and enjoy our individual rights and freedoms, we each must be careful not to infringe on the rights of others. Richard Lundman (1980:22–30) identifies this dilemma as "a dynamic tension" between liberty and civility. **Liberty** refers to individual freedom to act as one chooses, while **civility** refers to the need for order in social relations. In all societies, the two interests, freedom and order, compete. The "price" for more freedom is likely to be less order; and the price for more order is likely to be less freedom.

Efforts to regulate cigarette smoking illustrate this tension. Smokers wish to be able to enjoy their cigarettes wherever they please. Nonsmokers wish to avoid the health hazards and related negative consequences of cigarette smoke wherever they are. In a situation of totally unrestrained freedom, we are likely to witness considerable incivility as smokers and nonsmokers fight over smoking in spaces they share. One solution is to designate certain areas for smoking. Then smokers are free to enjoy cigarettes in those areas, and nonsmokers can avoid them. The freedoms of both smokers (to smoke in nonsmoking areas) and nonsmokers (to be free from smoke in smoking areas) are constrained. To achieve a level of order, each group sacrifices some freedom.

The police must not only restrict the freedom of individuals, but also protect that freedom. (Bob Daemmrich/The Image Works)

As we shall see later in our discussion of the development of police, policing serves to balance the tension between freedom and order in society. Especially in large, complex societies such as the United States, the need arises for formal rules to control the exercise of individual freedoms. The police enforce those rules. Still, a society that places a high value on individual freedom constrains its police.

The powers of the police to use coercive force are controlled by procedural requirements (McEwen, 1997). Police are not allowed to indiscriminately arrest, detain, or apply force to individuals. Police actions must be justified to be considered appropriate. The police are required not only to enforce the laws but to obey them. This often means that the options available to police are limited. This limitation, too, is a component of the dynamic tension between freedom and order.

Policing a free society requires a balance between police powers and the rights of citizens. To ensure the orderly workings of society, the public behavior of citizens must be somewhat predictable and controlled. Motorists must obey traffic laws, or else driving would be impossible. On the other hand, to protect individual freedom, the power of government to coerce citizens must also be predictable and controlled. Allowing the police to forcibly enter homes and arrest citizens at whim would be intolerable (Davis, 1997). To meet these conflicting needs for order and for the protection of freedom, procedural laws are enacted that set out (1) the conditions under which the police may intervene with citizens, and (2) the obligations that the government, through the police, owe individual citizens (Vizzard, 1995).

In short, the police must not only maintain order, but they must also protect individual rights. The protection of individual rights requires that the police balance competing interests of people in conflict, in addition to controlling their own use of power. Our societal concern with limiting the powers of the police has resulted in a peculiar organizational structure. There is not one police force in America, but rather a variety of American police and policing.

VARIETIES OF POLICE AND POLICING

Policing in America is conducted by hundreds of thousands of individuals employed by thousands of distinct agencies and organizations. This staggering array of police is one of the reasons it is so difficult to define what is meant by the word *police*. There are public and private police, large and small organizations, general and special-purpose departments, village, town, city, county, state, and federal agencies and agents. Policing in America is a large, diverse, and complex topic.

Our primary focus in this book will be on general-purpose, municipal police agencies and their officers. These are the police we typically encounter in our daily lives, and they are the most numerous of all the various police in America. But we will also periodically examine other types of police such as federal agents, transit authority, state police, and highway patrols. We will review the full range of police and police activities in our attempt to understand police in America.

Types of Police

The types and structures of U.S. police organizations are discussed further in later chapters, but for now it is necessary to briefly describe them as a prelude to our continued overview in this chapter. Unlike most other nations of the world, the United States does not have a national police force. Rather, in keeping with our beliefs in federalism and local autonomy, American police have evolved largely in response to local needs and pressures. This evolutionary process has produced a uniquely American police structure that includes municipal, state, federal, and private police agencies.

Municipal Police. The majority of American police agencies are operated at the municipal, or local, level. These agencies include village, township, city, and county police departments, sheriffs' departments, and a variety of special-purpose agencies such as transit-authority and housing-authority police. These police departments employ the most officers. They range in size from departments consisting of one officer to departments having thousands of personnel. Police in these municipalities typically have general policing duties for their respective jurisdictions.

General police duties include peacekeeping, law enforcement, and service delivery. Thus, officers employed by these agencies are expected to handle a variety of problems within their local communities. They can be called to investigate crimes, settle disputes, provide emergency medical service, regulate traffic, protect visiting dignitaries, and for a host of other tasks. They are generally responsible for maintaining order and protecting life and property.

State Police.
In addition to municipal police, almost every state also has one or more state-level policing agency. Perhaps the most familiar of these state agencies are highway patrols, which typically are given the task of regulating traffic and maintaining order and safety on state and federal highways. In many states, the highway patrol function is a part of a broader mandate given the state police that includes the provision of general police services to people living in unincorporated areas. Thus, the state police are expected to patrol and protect people and property in areas not constituted as official municipalities. These areas are typically rural and, absent the county sheriff, have no independent police department.

In addition to these types of state police agencies, many states also have special investigative agencies that concentrate on statewide law enforcement. Such agencies may conduct undercover investigations, operate crime laboratories, and provide assistance to local (municipal) police departments. State governments often include other police agencies such as fish and game wardens and state park rangers or police.

Federal Police.
At the national level, the federal government operates a number of police agencies. Most of these have limited responsibilities and do not provide general police services to citizens. The Federal Bureau of Investigation (FBI) is an example. The FBI investigates violations of federal law, such as robberies of federally insured financial institutions. It also manages national crime data, operates a national crime laboratory, keeps fingerprint records, runs a national academy for police administrator training, and provides a variety of other assistance to state and local police.

Other similar federal law enforcement agencies exist, such as the Bureau of Alcohol, Tobacco and Firearms, U.S. Postal Inspection Service, and U.S. Customs Service, to name a few. In addition to these types of federal police, there are agencies that have more general police duties, but these duties are typically constrained to specific locations. For example, the federal government maintains separate police forces for the U.S. Capitol, U.S. Supreme Court, National Art Gallery, National Park Service, and even the U.S. Mint. To this list of multipurpose federal agencies can be added the various military police units that have jurisdiction on military bases and over military personnel.

Private Police. In addition to those police agencies operated by different levels of government, a variety of privately run policing and security services exist. Traditionally, private police have been seen as less qualified and less important than public police (Becker, 1989). Recently, though, the distinction between private and public police has lessened. It is more and more common for private concerns to contract with public police officers or departments for police services (Reiss, 1988). In many instances, such contracts allow private entities to secure additional police protection by employing sworn public police officers who work for the private contractor when they are not officially on duty for their public employer.

The private police have a long tradition, and as we will see in the next chapter; current public police evolved from this tradition of private security. Through the 1980s, the private security industry grew more rapidly than public police, with the result that there are more private police officers in America than public ones. However, given that many officers working as private police do so as off-duty public police officers, it is difficult to gauge the relative sizes of public and private police personnel pools (Reiss, 1988:2).

Private police provide added police protection for the areas in which they are employed. For example, if a sports arena retains its own police force, the demand on the public police for crowd control at sporting events is diminished. Further, the overlap between public and private police represented by the hiring of off-duty public police officers similarly augments the level of policing. As Reiss (1988:32) observes, "Most departments regard their officers as responsible for enforcing the law whether or not they are on duty. Consequently, officers on paid details are obliged to deal with police matters that come to their attention." Clearly, then, it is important to include private police in any consideration of policing in America (Palmiotto, 1989).

Police Functions

The large number and variety of police agencies and officers in the United States serve several purposes. Thus, not only are the structure and organization of policing complex, so, too, is the role of the police. This is particularly true of the local, general-purpose police agencies that are the focus of our attention.

James Wilson (1968:4–5) identifies the three primary functions of local police as service delivery, law enforcement, and order maintenance. Though often related, these are distinct sets of activities that characterize the police role. Wilson roughly classifies police work according to (1) situations where the police assist citizens without evidence of law violations and (2) situations where laws have, or may have, been broken. The first set of tasks are service-delivery functions. The latter set includes law enforcement (when the police base their actions on the law) and order maintenance (when the police do not rely on the law as the primary justification for action).

Service Delivery.   Many people consider the police to be the "agency of last resort" for people in trouble (Goldstein, 1977; Wilson, 1968). This is because the police are always available to citizens, only a telephone call away. In large part because of this availability, citizens frequently call on the police for assistance with all sorts of problems that do not directly involve either law enforcement or the maintenance of order. The traffic accident example we discussed earlier illustrates this point. The fact is that the police can and will do "something" about most problems, so, lacking any alternatives, people tend to "call the cops." To sum up, there are three reasons citizens frequently call on the police for assistance with problems involving neither law enforcement nor the maintenance of order:

1. The police are always available to the public.
2. They have the authority to take charge and will usually do something about most problems.
3. They are only a telephone call away.

The service duties of the police include helping stranded motorists, giving directions, locating missing children, checking residences for absent owners, providing first aid and transportation for the ill and injured, and a variety of other tasks. The service function of the police has evolved over time, in part because there is no other resource, especially outside business hours.

Suppose you attend a concert and on returning to your car discover that you have locked your keys inside. What do you do? Often the easiest solution is to call the police, so that an officer equipped with a "slim jim" can unlock the door and allow you to drive home. Similarly, what should you do with found property? The logical response is to notify the police.

The full-time availability of the police, coupled with their ability to use force when necessary to resolve problems, makes them an ideal "catch-all" agency for a variety of problems. No examination of the police in America can be complete without an appreciation of their important role as a social-services provider. Although many observers, including Wilson (1968), question the wisdom of using police for social services, the fact remains that service activities are an important part of policing in America.

Law Enforcement.   When we think of the police, we most commonly consider them in terms of their law-enforcement role. Portrayals of the police in the entertainment and news media, descriptions of police by elected officials, and the police themselves stress their law-enforcement and crime-control functions (Manning, 1978). The police are expressly authorized and required to enforce the criminal law. Whereas we may disagree over whether the police should routinely be called to traffic accidents, nearly everyone would agree that the police should be concerned with an armed robbery, burglary, and other crimes.

Crime control through law-enforcement actions is the manifest, or stated, purpose of the police in America. That is, the simple answer to the question of why do we have police is that we need them to enforce the laws. Police officers are hired, trained, and deployed to prevent crime (by deterring violators) and to detect and apprehend those who break the law. A basic measure of police effectiveness is the amount of crime that occurs in an area. Yet most police time is not spent on crime control or law enforcement. Even though we have the police principally for the control of crime (in theory at least), in operation we require them to perform a variety of other functions. As long as the police are available in their crime-fighting role, the reasoning goes, why not use them to meet service needs of the citizenry when they are not actively engaged in law enforcement?

Law-enforcement activities are those in which the police perform in a mostly ministerial capacity. That is, law enforcement takes place when there is little doubt that a crime has occurred, and the police follow a fairly routine procedure to identify and arrest the violator. The problem the police must handle is identified as a crime, and the appropriate steps to be taken are known by the police. Typical outcomes in a law-enforcement situation are the issuance of a citation or the arrest of a citizen. Many times, however, the police face problems where it is unclear if the law has been broken. In these cases police are usually engaged in order-maintenance activities.

Order Maintenance.   Perhaps the primary obligation of the police is the maintenance of order in society. Sometimes order-maintenance activities are also called *peacekeeping,* a more descriptive term. Above all, the police are expected to maintain the public peace by ensuring that everyone behaves in an orderly fashion. In order maintenance, the police establish the balance between freedom and order. They serve to define the limits of individual liberty caused by the need for predictability in society.

Order maintenance involves the absence of disorder and is achieved by preventing or settling conflicts in a peaceable manner. In other words, the police maintain order in two basic ways: by preventing disorder and by restoring order when it has been disrupted. Police intervention in neighbor disputes, traffic regulation, and crowd control are examples of order-maintenance activities. For example, the police prevent disorder by directing traffic at an intersection where the signal light is broken. They restore order by dispersing a crowd gathered at the site of an accident.

Legitimate crime-control interests are served through these peace-keeping actions, such as preventing fights, traffic violations, and riots. The police do not wait for crimes to occur. Rather, order maintenance is often a preemptive action in which the police intervene to preserve order. Because it often involves preventive action, order maintenance is possibly the most troublesome police function.

In most peacekeeping situations, the police act without evidence of a law violation. Indeed, the point of order maintenance is to prevent such a violation. Thus, the officers involved operate in a "gray area" lacking specific legal justification for intervention. They must choose when to intervene and how best to maintain the peace (Eck and Spelman, 1987). Because they work to balance freedom and order in these activities, it is frequently the case that some, if not all, of the parties involved are dissatisfied. For example, if the police disband a group of people "hanging out" on a street corner, those people are likely to feel oppressed.

Wilson (1968) suggests that order maintenance is the most important, as well as the most troublesome, police function. He distinguishes types of local policing based on how officers typically accomplish order maintenance. Sometimes the police act primarily to restore order. In other settings, the police rely on the criminal law to define disorderly situations as needing their attention. In still other settings, the police reflect community definitions of order, which they support without recourse to the criminal law. In peacekeeping, then, the police are often on unsteady legal footing, may antagonize citizens, and are sometimes unsure of what to do. For these reasons, peacekeeping is troublesome.

Order maintenance is the most important police function, because at bottom, what the police do is social control. They are the agency most clearly charged with the task of ensuring civil relations in society. Order-maintenance activities are those in which the police attempt to resolve problems without recourse to the law (although arrest or the threat of arrest may be an important tool in order maintenance). The police have a range of choices about how to act in maintaining order.

Social Control.   In combination, these three major functions of the American police—service delivery, law enforcement, and order maintenance—represent the purpose of the police: social control. Whether by the delivery of services, the enforcement of the criminal law, or the prevention of disorder and restoration of order, the police are an agency of social control. Specifically, the American police are the government agents of social control, empowered to use coercive force when necessary.

In order to function, the members of all social groups need to be able to rely on the predictability of behavior by others (Cohen, 1966:3). Predictability is achieved through a shared understanding of "rules" for action. As Cohen observes, "If the actions of many people are to be fitted together, there must be understandings about who is supposed to do what and under which circumstances." But understandings alone are not sufficient. Not only must members of social groups understand the rules, they must follow them.

Social control is the mechanism by which social entities ensure predictable behavior. It involves teaching the rules to members and then sanctioning those rules by punishing violators and rewarding rule followers. Talcott Parsons (1966) suggests that social control is a critical component of social groups, without which there is no "society." In this vein, Olson (1968:118) notes, "Because all social organization is perpetuated through this control process, *social control is inherent in organized social life*" (emphasis in the original).

In society, social control is typically achieved through the three related processes of training, structures, and coercion (Travis, 2001:6–7). That is, members of the group are trained, or taught the rules of what constitutes proper and improper behavior, through what is called *socialization*. Further, opportunities for certain behaviors are controlled through the social structure (Cullen, 1983). For example, if one is never placed in a position of trust, such as a bank teller, account executive, or other financial officer, one has little opportunity to engage in embezzlement. Finally, various pressures are brought to bear on individuals to force them to obey the rules through punishment (coercion) or to encourage obedience through rewards.

Social controls are of two basic types: formal, or external, and informal, or internal. **Informal control** exists when the individual chooses to abide by the rules. **Formal control** exists when the group imposes limits on the individual's behavior. To the degree that informal, or internal, controls are sufficient, there is no need to exercise formal, or external, controls. Informal controls work best when individuals share common activities and values. They work less well when individuals lack common interests and values. Generally, the larger and more complex the social group is, the less the mutual understanding and sharing of values among members. Consequently, the bigger and more diverse groups often require greater levels of formal, or external, controls.

If everyone in your neighborhood works at the same place and must rise early to go to work, chances are that the neighborhood will be quiet in the evening. Suppose, however, that one neighbor works a late shift. This neighbor may hold a loud party during the night that disrupts the sleep of everyone else. Probably this offending neighbor will learn through informal channels that the loud party was unacceptable. If the neighbor is concerned about how others view him or her, there will be no more parties. If not, the neighbor may consider everyone else to be "fuddy-duddies" and schedule another party. In the event of another party, informal mechanisms will have failed to work. What can the early risers do? It is likely that some of them will "call the cops" and thus impose an external control on late-night parties in the neighborhood.

Informal controls operate through processes such as ridicule and ostracism, or "the cold shoulder." By treating offenders rudely, we hope to convince them to behave. Formal controls, on the other hand, tend to operate through the exercise of coercive power. Both types try to limit or prevent **deviance,** which is any violation of the social rules. Complex societies develop social institutions to control deviance.

All sorts of acts can be considered to be deviant, from mass murder to using the dessert fork to eat the salad. Different mechanisms and levels of social control are applied to different types of deviance within society. The more serious and disruptive the deviant act, the more likely it is that formal institutions of external social control will be mobilized. Two formal social-control systems that exist in American society illustrate this point: the civil justice system and the criminal justice system.

Most deviance is of a minor nature, and control is attempted informally. Deviance that is not controllable in this way is defined as the appropriate target for formal controls. Some deviance may be particularly disruptive, yet be considered beyond the control of the offender. These acts are often the products of mental illness and are controlled through the mental health system. Other acts are considered to be willful, or at least controllable by the offender, and are the subjects of the legal system. The legal system consists of two formal social-control components: civil justice and criminal justice. The **civil justice system** is concerned with deviance that harms individuals, while the **criminal justice system** concentrates on deviance that harms the group. As Roscoe Pound (1929:4) observes, "Law does but a part of this whole task of social control; and the criminal law does but a part of that portion which belongs to the law."

The Police and Social Control

We stated earlier that order maintenance is the primary function of the police. It should be clear now that this function is one of social control. The police are a formal organization charged with a broad social-control mission. As a general human services agency, the police provide an immediate response to citizen problems. In the words of Herman Goldstein (1977:41), "The police function, if viewed in its broadest context, consists of making a diagnostic decision of sorts as to which alternative might be most appropriate in a given case."

In choosing alternatives, the police select from a variety of possible actions, including the possibility of doing nothing at all (Davis, 1969). When the problem facing the police involves some form of deviance, the police can select the appropriate mechanism of social control. For example, responding officers can encourage disputing citizens to resolve a situation informally. Similarly, they can suggest that one or all of the parties involved pursue a civil justice action. In cases involving inebriates, addicts, or the mentally ill, the police can turn the case over to mental health officials. The police provide an important social service by assigning specific cases of deviance to particular forms of social control.

The criminal justice system is always an option for the police, at least for the immediate solution of problems. For example, a boisterous drunk can be turned over to a detoxification center or arrested for public intoxication. A mentally ill person can be taken to a hospital or arrested for disturbing the peace. Absent workable alternatives, the police can employ the criminal process (Finn and Sullivan, 1988).

POLICING IN THE COMMUNITY

At first glance, the multiple functions of the police and the variety of police structures and organizations seem to defy study. How are we to understand the police when there is so little agreement about what they are and what they do? Yet, it is precisely because the police in America are so complex that we need to develop that understanding.

Recognition of the complexity of the topic is the first step toward understanding it. Our goal is to understand how policing came to have such variety in the first place. If this amorphous mix of structures and functions is the product of a random evolutionary process, it may not be possible to understand policing in America. On the other hand, if, as we suspect, the complexity of contemporary policing represents different products of the same developmental process, our task is to understand how the differences occurred. We need to learn why we have so many structures and functions, and how they relate to each other.

We can begin to understand policing by placing it in context. The first important contextual characteristic of the police is the community. That is, American police must be viewed within the context of the communities they serve and in which they operate. As Judge George Edwards (1968:36) writes, "When all is said and done, the police function in the United States is primarily a local function, governed at the municipal level." He went on to say that as a social institution the police do not exist in a vacuum, but are shaped by the values and attitudes of the society in which they operate. Thus, according to Judge Edwards, the police reflect the communities in which they exist.

It is also important to consider the police as an institution with a broad social service mandate. One of the problems we encounter in studying the police comes from trying to decipher which of their functions is dominant. That is, are the police mostly crime fighters, or are they mostly service providers? Daniel Kennedy (1983) makes a persuasive argument that the police today are best understood as a human services agency. With this perspective, enforcement of the laws is only one service available through the police. Recognition of the police as a general service agency clarifies the reason for them having multiple functions.

As institutions with a broad service mandate, police agencies are still comparable with one another. Although any given department may emphasize one function over the others, all police agencies serve all functions. As McIver and Parks (1983) observe, knowing that the police serve a variety of functions does not render assessment of the police impossible. Instead, this knowledge is useful because it prepares us to recognize differences in agencies.

In his book *Policing a Free Society,* Goldstein (1977:33–34) argues that it is most appropriate to view the police as a multipurpose agency of municipal government. Doing so, he says, recognizes the many obligations of the police that are not a part of law enforcement. Further, "viewing the police primarily as an agency of municipal government is a way of emphasizing the fact that each community has the opportunity to make its own judgments as to what its police force should do." In other words, seeing the police as a generic government service agency recognizes that police organizations will differ from each other to the degree that communities and their local governments differ.

Wilson (1968) echoes this conceptualization of the police in *Varieties of Police Behavior.* He contends that the culture of a community (specifically, its "political culture") exerts an influence on police operations. By this he means that local police are sensitive to, and swayed by, community values and attitudes. However, Wilson rejects the idea that communities or city governments typically determine police practices in a direct fashion.

Rather than exercising direct control over police practices, Wilson suggests that the structure of interests and balance of political power in a community shape the nature and definition of police functions. He identifies three types of police agencies—the watchman style, service style, and legalistic style—which we examine in more detail in Part IV. Each of these styles is related to the characteristics of the population and government of the eight communities he studied. The important point is that different types of communities are served by different types of policing.

In all of the communities Wilson studied, the police engaged in service-delivery, order-maintenance, and law-enforcement activities. They differed in terms of the formality of their actions (reliance on the criminal law) and the frequency with which they intervened with citizens. Thus, differences in police organization and practice can be understood, in part, as a reflection of differences in communities.

POLICING IN THE CRIMINAL JUSTICE SYSTEM

A second important contextual characteristic of the police in America is their role in the criminal justice system. Placing the police within the context of the broader criminal justice system serves two important purposes. First, it makes us aware of the interaction between police and other social-control institutions, particularly courts and corrections. Second, it invites us to examine the structure of social control in society as it relates to the police. This latter effect helps us understand how social or cultural values serve to define and structure police operations.

Although the police operate as a multipurpose human services agency within society, the role of the police in the criminal justice system is central to an understanding of policing in America. The law-enforcement function of the police is what supports their capacity to use force in society. That is, we grant police the right to exercise coercive force so that they can control criminal deviance. To the extent that the capacity to use force is central to the definition of the police, then, their role in the criminal justice process is the source of their central role in social control.

Beyond the justification for force, there are two other reasons why it is important to understand the police as a component of the criminal justice process. First, the common perception of the police is that of crime fighters. Both the police themselves and the general public tend to think of the police as agents of criminal justice. Second, the criminal justice process is a primary resource of the police. In dealing with all sorts of social problems, the potential of invoking the criminal process underlies police actions. One of the reasons citizens obey the police is because failure to do so may result in criminal charges.

The **criminal justice system** is the formal social institution charged with the control of that deviance labeled criminal in American society. The President's Commission on Crime and Administration of Justice (1967) described the justice process as being composed of three parts: the police, the courts, and corrections. The commission prepared a model of the justice system, presented here as Figure 1.1. In this model, they show the criminal justice process as a series of decisions. The police are responsible for the detection and investigation of crime and for the arrest of criminals. These decisions come at the "front end" of the justice process.

Detection involves the determination that a crime has occurred. It is a decision by the police as to whether a citizen complaint, for example, involves criminality or whether evidence of a crime exists. **Investigation** involves the collection of evidence to establish that a crime has occurred and to identify a suspect. Investigations vary in terms of intensity and scope, and the determination of whether and how to proceed with a criminal investigation is another police decision. **Arrest** involves taking a suspected criminal offender into custody. In arrest, the police must decide whether a

This figure seeks to present a simple yet comprehensive view of the movement of cases through the criminal justice system. Procedures in individual jurisdictions may vary from the pattern shown here. The differing weights of line indicate the relative volumes of cases disposed of at various points in the system, but this is only suggestive because no nationwide data of this sort exist.

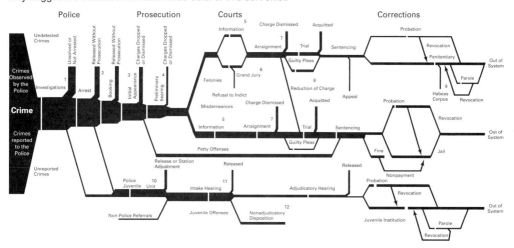

1 May continue until trial.

2 Administrative record of arrest. First step at which temporary release on bail may be available.

3 Before magistrate, commissioner, or justice of peace. Formal notice of charge, advice of rights. Bail set. Summary trials for petty offenses usually conducted here without further processing.

4 Preliminary testing of evidence against defendant. Charge may be reduced. No separate preliminary hearing for misdemeanors in some systems.

5 Charge filed by prosecutor on basis of information submitted by police or citizens. Alternative to grand jury indictment; often used in felonies, almost always in misdemeanors.

6 Reviews whether government evidence sufficient to justify trial. Some states have no grand jury system; others seldom use it.

7 Appearance for plea; defendant elects trial by judge or jury (if available); counsel for indigent usually appointed here in felonies. Often not at all in other cases.

8 Charge may be reduced at any time prior to trial in return for plea of guilty or for other reasons.

9 Challenge on constitutional grounds to legality of detention. May be sought at any point in process.

10 Police often hold informal hearings, dismiss or adjust many cases without further processing.

11 Probation officer decides desirability of further court action.

12 Welfare agency, social services, counseling, medical care, etc., for cases where adjudicatory handling not needed.

FIGURE 1.1 A general view of the criminal justice system

suspect probably committed a crime, and if so, whether to take the suspect into custody. These decisions occur at the "front end" of the justice process, representing the intake of cases into the criminal justice system.

The police are the gatekeepers of the criminal justice system (Manning and Van Maanen, 1978). For the most part, the criminal process concerns itself with cases identified by the police. Thus, the courts decide the fate of persons arrested by the police. Those who are under correctional supervision were first arrested by the police. In short, absent police action, most cases do not enter the justice process. Police decisions, then, are important factors in the work of the courts and correctional agencies.

Similarly, the decisions of prosecutors, judges, and correctional officials affect the police. For example, if prosecutors continually refuse to file formal charges, or if courts fail to convict or punish, persons arrested for the possession of small quantities of marijuana, the police may eventually decide not to arrest such violators. Further, correctional officials or prosecutors may request the police to investigate or arrest certain criminals. Finally, every arrest made by the police is subject to a judicial review.

As an agency of criminal justice, the police are evaluated in large part for their ability to prevent and solve crimes. Therefore, despite the range of police functions, every police agency will devote considerable attention to crime, and criminal matters will generally take precedence over other activities of the police. The position of the police in the criminal justice system explains the importance of crime control to police agencies. That the police decide to invoke the criminal process is equally important.

Police Discretion

The decisional aspect of criminal justice at the police level means that the police not only serve the law, they use it as well. If the police have choices about invoking the criminal law, we must strive to understand how those choices are made. When the police can choose whether to arrest, and how, if at all, to investigate potential crimes, they are said to have **discretion** (Davis, 1969). It would be a mistake to assume that the police enforce all laws equally in all circumstances. Rather, the police frequently choose to use the law just as they might choose other alternatives such as referral to the mental health system or civil courts. Thus, sometimes the police arrest not for criminal law purposes but for police purposes.

It is the fact that police exercise discretion in the performance of the job that makes understanding policing so important. What happens in a particular instance is up to the officer(s) involved. If a citizen is stopped for violating the speed limit, the outcome of the encounter may be that the citizen receives a citation. Alternatively, the citizen may receive only a warning from the officer. Or, indeed, the citizen may be arrested for something more serious than speeding, discovered when she or he was stopped for speeding.

In his classic treatise on police discretion, Kenneth Davis (1975:164–165) concludes, "Enforcement policy is not based on studies, and specialized staffs do not contribute to it. It is made primarily by patrolmen, the least qualified in the organization." Officers, Davis argues, determine law-enforcement policy by their personal discretionary decisions about when and whether to enforce laws.

The police are responsible for both the detection and investigation of crime.
(Erin N. Calmes)

Michael Feldberg (1995:207) observes, "Discretion is not only proper, but it is a necessary part of police work." The ability to choose among different courses of action—how to handle a particular problem—is a core component of policing. Officers dealing with the public must be able to choose which police goals to pursue (peacekeeping, law enforcement, or service delivery). They must also decide how best to achieve those goals (Bryett, 1997). Few today argue that police discretion is inappropriate or unnecessary. Indeed, the current movement toward community-oriented or problem-oriented policing explicitly recognizes the wide discretionary power each officer exercises.

Discretionary authority allows officers and police agencies to set priorities for policing and to make the best use of limited resources. Discretion enables the police to individualize justice and to avoid problems with the overreach of the criminal law. Further, discretionary authority is a hallmark of professionalism, and the exercise of discretion provides the officer with a challenging job. Finally, the need to balance interests in individual liberty and community safety requires that officers be able to "fine tune" reactions to specific problems.

On the other hand, the exercise of discretion by police officers poses problems for police officers, administrators, and the communities they serve. Discretion contributes to police corruption, supports discriminatory practices, and allows police to avoid their duty to protect the public.

Discretion makes discipline of officers difficult and can expose the police agency to civil liability. The exercise of discretion can produce "bad feelings" between the police and the public.

It is no secret that many African Americans in the United States suspect that police target them, and other minorities, for special attention. Concern about **racial profiling,** the selection of people for police attention based on race, reflects a recognition of police discretion (Kennedy, 1997; Smith and Petrocelli, 2001). It is likely that the race of the officer does not matter. In a study of residents of Washington, DC, Ronald Weitzer (2000) found that citizen attitudes towards the police were not strongly influenced by the race of the officer. In many ways, police are seen as a separate group, and the actions of police officers are viewed as representative of all police. Comparing African American, Hispanic, and white citizen attitudes toward the police, Cheurprakobkit (2000) found that citizen attitudes were more greatly affected by type of police contact than by race. Citizens who were least satisfied with the police were those with whom the police initiated contact. The most satisfied citizens were those who contacted the police on their own.

The existence of discretionary authority enables the police to use the criminal sanction as a tool. Research into the exercise of discretion by police officers indicates wide variation in how officers deal with citizens. This research has found that a large number of factors explain police officer decisions, especially those involving arrest and use of force. It is precisely this variation in police behavior that is the topic of our investigation.

We will return to this and related issues later in the book. For now it is sufficient to realize that the police in America both shape the criminal justice system and are shaped by it. Still, as important as their role in the criminal justice system of social control may be, their role in society is broader than that of law enforcement and no less discretionary.

CORRELATES OF POLICING

Up to this point we have been primarily concerned with defining the subject of our study and placing it into a context that will enable us to understand it. We have seen that the police in America represent a complex topic that must be considered within a social or environmental context. Thus, the police are a multipurpose agency of government that exists in and reflects characteristics of communities. Further, the police occupy an important position within the criminal justice system of formal social control. They both influence and are influenced by the other components of criminal justice in America.

The time has come to identify those factors or forces that may explain the variety of structures, functions, and activities involved in policing in

America. The remaining chapters of the book will explore these factors as correlates of what we will learn about the police. In simple terms, **correlates** are things that go together in ways that allow you to guess correctly about one, based on knowledge about the other. For example, clouds are a correlate of rain. If the sky is cloudy, rain is more likely than if the sky is clear.

We need to be very careful, however, in our use of correlates. The fact that things seem to go together does not necessarily mean that one causes the other. Sometimes cloudy skies produce snow, not rain. Sometimes they produce neither. Or "sun showers" may occur in which it rains while the sun is clearly visible. Cloudy skies are correlated with rain, but do not fully explain it. Although we would like to be able to fully explain policing, we must settle for identifying its many correlates.

Lawrence Sherman (1980) reviewed and summarized the existing quantitative research on police work including crime detection, arrest decisions, the use of force, and service activities. His review indicates that five sets of correlates had been reported. These were individual officer characteristics and situational, organizational, community, and legal factors. Later, Riksheim and Chermak (1993) extended this review, using the same basic set of correlates. For our purposes, this list of correlates can be reduced to three: people, organization, and community.

People

At base, policing is an interactive process. Most police work is done at an individual or small-group level and consists of one or a few officers interacting with one or a few citizens. The problems for which the police are called to intervene are usually "people" problems. Therefore, a major influence on police practice is the people involved in police–citizen encounters, both police officers and citizens.

Manning and Van Maanen (1978:215–219) identify the five salient concerns in every police–citizen encounter as

1. Questions of authority
2. Context
3. Components of the interaction
4. Outcomes expected by involved parties
5. Demeanor of those involved

Most policing involves negotiation between the officers and citizens involved. Arrest or the use of coercive force are police strategies of last resort and typically occur when negotiations have failed. Thus, the behavior of police officers reflects the interpersonal interaction between them and the citizens with whom they deal.

Questions of authority involve perceptions of the justification of police intervention. The *context of the encounter* relates to the time and place in which the police meet citizens. *Components of the interaction* refer to the roles assumed by those involved. As it implies, *expected outcomes* are what the police and citizens expect will be the product of the interaction. The *demeanor of the parties involved* relates to how the police and citizens treat each other. The combination of these factors determines in large part not only what the actual outcome of the interaction will be, but also how satisfied both the police and the citizen will be with the interaction.

For example, if the citizen approaches a police officer seeking directions to some location, a number of outcomes are possible. If the citizen asks politely, in the middle of the day, and the officer responds politely, the interaction is likely to end there. On the other hand, if the citizen approaches the officer at night, the officer may be suspicious, fearing an attack. The cautious officer may "pat down" the citizen. The citizen, feeling the officer is overstepping his or her authority, may become quarrelsome. In response, the officer may become curt and discourteous. The situation may deteriorate, and the citizen may end up arrested for some violation such as disturbing the peace. Depending upon the time of day, the nature of the request, officer and citizen expectations, and how the officer and citizen behave toward each other, simply asking for directions can result in an arrest, although this is very unlikely.

Similarly, a motorist may have been speeding but not receive a citation if the officer decides that citing the motorist is inappropriate in the circumstances. A respectful and apologetic motorist who cooperates with the officer may receive only a warning if the officer feels that the warning is sufficient in the situation. However, another motorist speeding at the same place in the same way may get a ticket if she or he is disrespectful to the officer. Sometimes the outcome of a police–citizen encounter is largely the product of the interaction between them.

Students and observers of the police have long recognized the discretionary power of officers (Goldstein, 1960; La Fave, 1965; Wilson, 1968; Davis, 1969; Davis, 1975; Goldstein, 1977; Manning and Van Maanen, 1978; Bordner, 1983). In their everyday dealings with citizens, individual officers are able to choose whether and how to intervene. One of the factors that helps explain these choices by individual officers is the personal characteristics of the officer. Another important factor is the personal characteristics of the citizens with whom the officers must deal. To the degree that policing represents an interpersonal interaction, the characteristics of the people involved are important to outcomes. However, personal characteristics of both the police and the people involved in policing, while very important, do not completely explain policing.

Organization

Police officers work within police organizations, and the organizations themselves help to explain officer behavior. The characteristics of the police organization shape the officer's exercise of discretion more or less directly. For example, if the officer is expected to issue a certain number of citations each month, a motorist who might normally receive only a warning can get a ticket if the officer is behind on his or her "quota."

The policies, procedures, and structures of police organizations influence both the officer's choices and the available opportunities to make certain choices. For example, most police agencies that have traffic-control responsibilities also have what are called **tolerance limits.** These tolerances define a threshold for the routine writing of speeding citations to be somewhat higher than the legal maximum. For example, a five-mph tolerance limit instructs officers to cite motorists for speeding only when they have exceeded the legal speed limit by at least five miles per hour. The effect of such a tolerance limit is to encourage officers not to intervene with drivers who are traveling within five miles per hour of the limit, or if they intervene with such drivers for some other reason, not to issue a speeding citation.

The effect of organizational policies such as quotas and tolerance limits is to structure the exercise of discretion by officers. These policies define when and how officers should intervene with motorists. Similar organizational effects can be seen in the assignment of officers. For example, officers assigned to foot patrol are not expected, nor are typically able, to cite speeding motorists. The formal and informal expectations of officers are expressed in departmental policy and accrue to various assignments, thus influencing individual officer decisions.

The link between the police organization and the activities of individual police officers was established by Wilson (1968) in his study of police organizations. Wilson focused on the exercise of discretion by patrol officers, but he sought organizational explanations of those decisions. That he was able to characterize police organizational styles by looking at officer behavior indicates that the police organization is an important determinant of policing in America. As Wilson (1968:11) explains, "The principal limit on managing the discretionary powers of patrolmen arises not from the particular personal qualities or technical skill of these officers, but from the *organizational* and legal definition of the patrolman's task" (emphasis added).

So a second important factor in understanding how policing is accomplished is the police organization. Like the personal characteristics of the individuals involved, however, the characteristics of the police organization are not a sufficient explanation.

Community

We have established that policing represents to some degree the outcome of interpersonal interactions between police and citizens, and that these interactions are at least partly moderated by the organizational characteristics of police agencies. In analyzing police organizations, though, Wilson (1968) observes that they are the product of the *political culture* of the communities in which they are found. In a replication of Wilson's study, Langworthy (1985) finds support for Wilson's theory that police department activities reflected differences in the political organization of the communities in which they operated. However, Langworthy's (1985:97–98) data reveal considerable variety in police activity within types of political organization.

That is, while it is possible to classify police agencies by style as legalistic, service, or watchman, within any one style, the activities of specific agencies show important differences. The political culture of a community may lead its police department to generally adopt a certain style of enforcement, but policing within communities of similar political cultures may still vary greatly. For example, watchman-style agencies differ from legalistic ones in significant ways, yet these watchman agencies also differ from each other.

These findings raise the possibility that there are factors beyond political culture that influence police organizations and, by extension, police officer behavior. We can refer to these other factors as *community values*. For example, in the introduction to his second edition of *Varieties of Police Behavior,* Wilson (1978:vii–viii) reports that a study in Boston discovered different policing styles in two neighborhoods in that city. Thus, the one city police department exhibited at least two styles of policing depending on neighborhood characteristics. Wilson explains the differences by referring to the values of the neighborhoods involved.

In a review of police organizational structures, Langworthy (1990) notes that all forms of police organizations "work." The question, according to Langworthy, is not whether or not certain structures are generally better ways of organizing the police in a community. Rather, the question is normative, because the best method of organizing a police agency depends on how the goals of that agency are defined in the community. The goals of the police are products of community values, and the structure of the police agency will reflect those values.

Richard Ericson and Kevin Haggerty (1997) developed an explanation of policing as *risk management*. They suggest that the role of the police is increasingly structured by the interactions between police organizations and other formal social institutions like schools, businesses, medical care facilities, and the like. To the degree that these other institutions are concerned about risks and require information about risks, they place demands on the police to provide this information. In turn, the police agency develops rules and forms for reporting information. Ericson and Haggerty (1997:429) argue that it is the report requirements that structure officer behavior: "Police officers' decisions in the field are not governed primarily by the informal rules prevalent in the local occupational subculture, but they are constituted by the communication rules embedded in reporting formats."

In sum, the demands for information placed on the police by other organizations (including the courts) work to define both what kinds of things are "police business" and what sorts of police actions are proper. This is an innovative concept of policing, one worthy of further examination. Nonetheless, the police, and the police occupational subculture, exert an influence, at a minimum, in negotiating with other institutional audiences over what kinds of information ought to be the responsibility of the police to gather.

As we saw earlier, policing, like all the rest of the criminal justice process in America, involves a conflict between the value citizens place on the competing interests of freedom and order. Herbert Packer (1968) suggests two models of the criminal justice system as being due process and crime control. The **due-process model** emphasizes the protection of individual freedom. The **crime-control model** emphasizes order maintenance. These models suggest that the agencies of the criminal justice process, including the police, adapt to changes in the importance we, as citizens, place on freedom versus order. Thus, the police reflect societal definitions of the appropriate balance of important social values. These social values are supported by the police organization (Langworthy, 1990), through which they are translated into practice by the actions of police officers. Neighborhood or community characteristics are correlated with citizen ratings of satisfaction with police service (Kurtz, Koons, and Taylor, 1998; Weitzer, 1999).

Throughout the 1990s police departments across the United States adopted a reform model of policing known as community policing. This model requires, among other things, that the police form a partnership with the community to identify appropriate police functions and strategies. To the extent that police agencies are actually taking direction from communities, the importance of community context to an understanding of policing will increase.

PERSPECTIVE: POLICING AS A BALANCE OF FORCES

The purpose of this book is to develop an understanding of policing in the United States. With all of the foregoing serving to set the stage for our investigation of American policing, we are now ready to proceed with this complex and fascinating topic. First let us describe our approach to understanding the police.

The view we take of the police is that in practice, policing represents the balancing of competing and often conflicting forces in society. These forces exist on community, organizational, and individual levels, and the balance among them is often seen in the decisions of officers in their day-to-day activities. To understand the police, we must be able to broadly identify the issues involved in police decisions and recognize the community values, organizational constraints, and individual preferences that correlate with those decisions.

This concept of policing recognizes that although patterns in decision making can be expected to emerge over time, each decision is an independent event. In any given instance, an officer may decide to make an arrest of a citizen regardless of organizational constraints or social values. In the aggregate, however, the decisions of officers will reflect community values as they are moderated through the police organization structure. Accordingly, in some communities police are more likely to intervene with persons holding loud parties than in others. Within those communities where the police are likely to intervene, the officers from some police departments are more likely to arrest or cite violators than others. Yet, in any instance of a loud party, in any community, it is possible that the responding officer will invoke the criminal law.

The choices of officers are structured by community values, which are expressed in terms of which situations should require police intervention. Communities that place a higher value on freedom and individual liberty narrowly define what kinds of behaviors warrant police attention (Black, 1973). The policies and procedures of the police agency further instruct the officer as to how and when to intervene with citizens. Finally, the personal attributes of the officer and the citizen with whom the officer deals shape the exact resolution of the problem.

The remainder of Part I of this book is devoted to a historical review of the development of policing in America. The purpose of this review is to clarify and describe the social values reflected in policing so as to better understand how these values are balanced in contemporary police decisions. Later sections examine the organization of police agencies as resolutions of value conflicts, including the social and psychological forces that shape decision making by individual officers. In combination, these sections serve to identify how policing reflects a balance of competing forces.

Chapter Checkup

1. Define *police*.
2. What is the "core conflict" between freedom and order?
3. What does it mean to say the police must not only enforce the law, but obey it as well?
4. What are the four major types of police in America?
5. What are the three principal police functions?

6. What is social control, and how is it achieved?
7. What are Wilson's three styles of policing?
8. What is the criminal justice system?
9. What is the role of the police in the criminal justice system?
10. Define *discretion* in a policing context.
11. What are three correlates of policing?
12. How can policing be understood as a product of the balance of forces?

References

Becker, D. C. (1989) "Private Security," in W. G. Bailey (ed.) *The encyclopedia of police science.* (New York: Garland Publishing): 519–526.

Bittner, E. (1970) *The functions of police in modern society.* (Washington, DC: National Institute of Mental Health).

Black, D. J. (1973) "The mobilization of law," *Journal of Legal Studies* 2(1):125–149.

Bordner, D. C. (1983) "Routine policing, discretion, and the definition of law, order and justice in society," *Criminology* 21(2):294–304.

Bryett, K. (1997) "The dual face of policing: An Australian scenario," *Policing* 20(1):160–174.

Cheurprakobkit, S. (2000) "Police-citizen contact and police performance: Attitudinal differences between Hispanics and non-Hispanics," *Journal of Criminal Justice* 28(4):325–336.

Cohen, A. K. (1966) *Deviance and control.* (New York: Prentice-Hall).

Conser, J. A. and G. D. Russell (2000) *Law Enforcement in the United States.* (Gaithersburg, MD: Aspen).

Cullen, F. T. (1983) *Rethinking crime and deviance theory: The emergence of a structuring tradition.* (Totowa, NJ: Rowman & Allenheld).

Davis, K. C. (1969) *Discretionary justice.* (Baton Rouge, LA: Louisiana State University Press).

Davis, K. C. (1975) *Police discretion.* (St. Paul, MN: West).

Davis, R. (1997) "What Fourth Amendment? HR 666 and the satanic expansion of the good faith exception," *Policing* 20(1):101–112.

Eck, J. and W. Spelman (1987) "Who you gonna call? The police as problem busters," *Crime and Delinquency* 33(1):31–52.

Edwards, G. (1968) *The police on the urban frontier: A guide to community understanding.* (New York: Institute of Human Relations Press).

Ericson R. and K. Haggerty (1997) *Policing the risk society.* (Toronto: University of Toronto Press).

Feldberg, M. (1995) "Police discretion," in W. Bailey (ed.) *The encyclopedia of police science,* 2nd ed. (New York: Garland Publishing): 207–211.

Finn, P. E. and M. Sullivan (1988) "Police respond to special populations" *NIJ Reports* (May/June):2–8.

Goldstein, H. (1977) *Policing a free society.* (Cambridge, MA: Ballinger).

Goldstein, J. (1960) "Police discretion not to invoke the criminal process: Low visibility decisions in the administration of justice," *Yale Law Journal* 69(March):543–594.

Kennedy, D. B. (1983) "Toward a clarification of the police role as a human services agency," *Criminal Justice Review* 8(2):41–45.

Kennedy, R. (1997) *Race, crime, and the law.* (New York: Pantheon).

Klockars, C. B. (1985) *The idea of police.* (Beverly Hills, CA: Sage).

Kurtz, E. M., B. A. Koons, and R. B. Taylor (1998) "Land use, physical deterioration, resident-based control, and calls for service on urban streetblocks," *Justice Quarterly* 15(1):121–149.

La Fave, W. R. (1965) *Arrest.* (Boston: Little, Brown).

Langworthy, R. H. (1985) "Wilson's theory of police behavior: A replication of the constraint theory," *Justice Quarterly* 2(1):89–93.

Langworthy, R. H. (1990) "Police organizational structures: Which work?" in G. Cordner and D. Hale (eds.) *What works in policing? Operations and administration examined.* (Cincinnati, OH: Anderson):87–105.

Lundman, R. (1980) *Police and policing: An introduction.* (New York: Holt, Rinehart & Winston).

Manning, P. K. (1978) "The police: Mandate, strategies, and appearances," in P. K. Manning and J. Van Maanen (eds.) *Policing: A view from the street.* (Santa Monica, CA: Goodyear):7–31.

Manning, P. K. and J. Van Maanen (eds.) (1978) *Policing: A view from the street.* (Santa Monica, CA: Goodyear).

McEwen, T. (1997) "Policies on less-than-lethal force in law enforcement agencies," *Policing* 20(1):39–59.

McIver, J. P. and R. B. Parks (1983) "Evaluating police performance: Identification of effective and ineffective police actions," in R. R. Bennett (ed.) *Police at work: Policy issues and analysis.* (Beverly Hills, CA: Sage):21–44.

Olson, M. E. (1968) *The process of social organization.* (New York: Holt, Rinehart & Winston).

Packer, H. L. (1968) *The limits of the criminal sanction.* (Englewood Cliffs, NJ: Prentice-Hall).

Palmiotto, M. J. (1989) "The law enforcement-security connection: Equal status for crime prevention and control," *Journal of Security Administration* 12(1):37–48.

Parsons, T. (1966) *Societies: Evolutionary and comparative perspectives.* (New York: Prentice-Hall).

Pound, R. (1929) *Criminal justice in America.* (New York: Henry Holt & Company).

President's Commission on Law Enforcement and Administration of Justice (1967) *The challenge of crime in a free society.* (Washington, DC: U.S. Government Printing Office).

Reiss, A. J. (1988) *The private employment of public police.* (Washington, DC: U.S. Department of Justice).

Riksheim, E. and S. Chermak (1993) "Causes of police behavior revisited," *Journal of Criminal Justice* 21(4):353–382.

Rush, G. E. (1977) *The dictionary of criminal justice.* (Boston: Holbrook Press).

Sherman, L. W. (1980) "Causes of police behavior: The current state of quantitative research," *Journal of Research in Crime and Delinquency* (January): 69–100.

Smith, M. R. and M. Petrocelli (2001) "Racial profiling? A multivariate analysis of police traffic STOP data," *Police Quarterly* 4(1):4–27.

Travis, L. F. (2001) *Introduction to criminal justice,* 4th ed. (Cincinnati, OH: Anderson).

Vizzard, W. (1995) "Reassessing Bittner's thesis: Understanding coercion and the police in light of Waco and the Los Angeles riots," *Police Studies* 18(3–4):1–18.

Weitzer, R. (1999) "Citizens' perceptions of police misconduct: Race and neighborhood context," *Justice Quarterly* 16(4):819–846.

Weitzer, R. (2000) "White, black, or blue cops? Race and citizen assessments of police officers," *Journal of Criminal Justice* 28(4):313–324.

Wilson, J. Q. (1968) *Varieties of police behavior: The management of law and order in eight communities.* (Cambridge, MA: Harvard University Press).

Wilson, J. Q. (1978) *Varieties of police behavior,* 2nd ed. (Cambridge, MA: Harvard University Press).

THE EARLY HISTORY OF POLICING

This chapter traces the early history of policing from ancient times to the nineteenth century. In doing so, it demonstrates that policing in America is in many ways unlike anything that has come before, yet it has a long tradition. Contemporary policing can be understood as a product of the balance of social forces that arose over time. What we hope to learn is how America came to have a variety of formal organizations with paid employees charged with serving the functions of police.

*   A historical perspective adds to our understanding in several important ways. Walker (1983:2) suggests that a historical perspective identifies enduring aspects of policing, allows us to evaluate the effects of prior reforms, and enables us to anticipate future developments. Johnson (1988:4) similarly urges a historical approach because learning the history of social*

institutions prepares us to recognize changes that take place over a long pe-riod and prevents us from "reinventing the wheel." An added benefit is that historical analysis requires us to be both careful and critical in our thinking about the police.

Victor Kappeler (1996) has critiqued recent histories of the police. He correctly notes that contemporary histories are a product of modern views of what happened. That is, they are modern interpretations of what forces in-fluenced the development of policing. The writing of a history of the police is an exercise in interpreting the past, not simply a recording of what has come before. This is true of all histories (Rawlings, 1999).

If a review of the history of the police is to be useful, it needs to be more than a description of events over time. Rather, we must take an analytic per-spective that directs our attention to events and processes that explain the development of modern policing. In the words of Carl Klockars (1983:56), our purpose should be to present "not the history of what happened in any period of time, but one way of explaining certain features of it [emphasis in the original]." Therefore, we are more concerned with explaining why (or how) the police developed than with when they appeared.

A DEVELOPMENTAL MODEL OF THE POLICE

Richard Lundman (1980) identifies three developmental stages of police or-ganization from informal to formal policing. In **informal policing,** every member of the group shares equal responsibility for meeting the police function of social control. In **formal policing,** specific members are charged with policing responsibility as their primary occupation and role within the society.

Between these two extremes, Lundman identifies a middle-ground or-ganization he calls **transitional policing,** in which the police function is as-sumed by (or assigned to) group members on a voluntary and part-time basis.

Lundman argues that the history of the police, or at least of English and American police, has been a shift from informal, through transitional, to for-mal policing. He explains this evolution with reference to three factors: the form of social organization, the interests of social elites, and the rates and images of crime within the society. Changes in these three factors are related to developments in policing and, he suggests, explain the shift from informal to formal policing. Thus we can see that the evolution of formal police corre-lates with certain changes in the society in which the police develop.

Klockars (1985) distinguishes between what he terms avocational and vocational policing. **Avocational policing** involves the exercise of police authority or the performance of the police function by individuals who are not primarily police officers. In contrast, **vocational policing** resembles formal policing in that it is done by individuals whose primary occupation is police work. Therefore, one important shift in the evolution of the police is the move to policing as a vocation. This involves the definition of a specific organization as the repository of police power, and the identification of a set of people as "police" officials.

At some point in their histories, all industrialized societies have created a police vocation. A separate police occupation develops, in part, as a logical extension of general specialization of social functions. That is, as societies develop distinct institutions or organizations for economic, religious, welfare, health, and other activities, they also develop them for social control. Additionally, the creation of a police vocation entails a recognition that the problems of social order are no longer adequately solved by "episodic" reactions but that a more permanent and continuous control response is needed. Finally, a police vocation allows control over those who enforce the rules of society because they are now on the "payroll" and hence can be fired.

A formal police figure represents a significant change in social relationships. Informal and transitional policing are largely voluntary and short lived. The posse, or vigilante, groups arise in response to a critical problem of social order, deal with that problem, and then disband. Members of these types of "police" serve in a part-time capacity, and they are generally beyond the control of government authorities. Formal police are permanent in that they continue to exist whether or not a crisis in social order is recognized. Further, because officers must be paid, an organization develops to recruit, deploy, and manage the police. The formal police thus become a continuing and more or less controllable force for social order within society. Lundman developed a model of the evolution of police (Figure 2.1).

CORRELATES OF POLICE EVOLUTION

Social Organization

As Lundman (1980) notes, the type of police found in a society reflects the organization of that society. Societies are organized in one of two basic ways: mechanical or organic. A **mechanical organization** is one in which the members of the group share similar beliefs and activities. Thus, the society emerges from, and depends on, a community of interests that has been called a *collective conscience* (Durkheim, 1933). An **organic organization** is one where the members of the group are dependent on one another because the tasks of society are specialized. In such a society, members are no longer self-sufficient, and the role of a collective conscience is not as great in creating a social order.

As Katherine Newman (1983:15–17) explains Durkheim's theory, "All societies share a collective conscience, since all individuals within them share at least some experiences and some aspects of a common world view." The difference between a mechanical and organic social organization relates

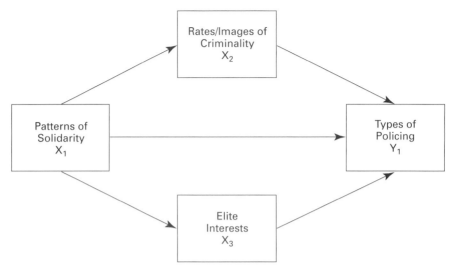

FIGURE **2.1** A Model of Police Development. (From Lundman, R. J. (1980). *Police and policing: An introduction.* New York: Holt, Rinehart & Winston, p. 15. Copyright © 1980 by Richard J. Lundman. Reprinted by permission of Richard J. Lundman.)

to specialization and differentiation in society. In mechanical social organizations, individuals are independent, and thus the collective conscience is the basis of social cohesion, or group identity. In organic social organizations, interdependence for meeting basic needs is the basis of social cohesion.

Mechanically solidary societies (those bound by communal interests) are often described as being *primitive, underdeveloped,* or *preindustrial.* In such social groups, members are typically independent of each other in terms of meeting basic needs. The members of the group may farm, hunt and gather, or fish for their subsistence, but they do so independently. Those who need food hunt; those who need shelter or furniture build it themselves. What makes these independent individuals form a social whole is a set of shared experiences (everyone hunts and builds for himself or herself), and shared beliefs about the nature of humanity, life, and the supernatural. Social control is designed to protect those beliefs, so there is little distinction between religious or taboo behaviors and what we might call secular crime.

In contrast, organically solidary societies are often described as *advanced, developed,* or *industrial.* In these instances, people are engaged in specific functions. Some farm, some build, and some perform service duties. As a result, individuals do not share common experiences but must rely on others for basic needs. Farmers rely on builders for shelter and furniture, builders rely on farmers for food, and so forth. Social control is designed to ensure that the necessary relationships among individuals are predictable and orderly. While there may be considerable agreement about some behaviors, distinctions are made between religious and secular offenses, and social control is often applied differently to members of different social classes. As with other activities, social control becomes more complex and specialized in organically solidary societies.

A formal, specialized police organization does not make sense in a mechanically organized society. There is not the need, ability, or process for role specialization in such a social order. However, an organically solidary society virtually requires a specialized, formal police organization. Such a society needs police to ensure order, assigns its members to a policing role, and develops processes of specialization in other areas of social function that can be applied to social control. For these reasons, Lundman (1980) maintains that the pattern of social organization correlates with the development of formal police.

If the history of societies can be understood as a general movement toward "increasing complexity" and differentiation (Applebaum, 1970:9–10), the emergence of formal police can be understood as a part of that general process. As societies change from informal, mechanical organizations to more formal, organic organizations, social functions and roles become specialized. The evolution of the police will similarly show a shift from unspecialized to specialized policing within those societies.

Olivero (1990) notes that few scholars have applied a cross-cultural, anthropological approach to understanding the evolution of police. The few studies done tend to agree that the police develop in response to increasing differentiation or specialization in the social structure. Robinson and Scaglion (1987) illustrate that the police emerge in the transition from a kinship to a class-dominated society. Newman (1983) similarly argues that the police, and formal law as a whole, develop in response to economic changes in society. She found that as the means of production became more specialized and industrial, access to scarce production resources in society becomes more limited, and the likelihood to find formal law and formal police within a society increases. Olivero (1990) noted that the more specialized the political organization of a society, the more specialized are its police.

This body of scholarship illustrates the link between general social organization and the evolution of police. As societies become more stratified and specialized, they create specialized police. As Lundman's (1980) theory for explaining types of police anticipates, the general level of societal complexity is related to the types of police we find in different societies.

Elite Interests

Newman (1983) observes that the purpose of law in a society is the settlement of disputes. She quotes Fried (1967), who observed that the purpose of the state is social control, typically through the exercise of power. He wrote, "In the final analysis the power of a state can be manifested in a real physical force, an army, a militia, a police force . . . and other paraphernalia of social control. It is the task of maintaining general social order that stands at the heart of the state" (Newman, 1983:95).

The wording of this comment is important. The purpose of the state, and by extension the purpose of the police, is not to change or even to achieve social order. Rather, the purpose is to *maintain* that order. Thus, the police, as an instrument of social control, are required to protect the social status quo. In other words, the function of the police is maintenance of the social order. The police serve the interests of the elites within a society.

Elites are those persons at the upper levels of society by virtue of their wealth, prestige, or other power. One outcome of the development of organic social organization is the creation of social classes, or stratification. In mechanically solidary society, individuals exist in a state of relative equality largely because each is self-sufficient. In more complex societies, certain members acquire wealth, status, and power. While everyone may get some benefits from a specialized society, these powerful people receive the most. It is in their interests to ensure that the social structure does not change. It is also in their power to create laws and structures for law enforcement. They do so to protect their economic and political interests.

Lundman (1980:35) states, "In an organically solidary society, only elites have the power and resources to bring about change." The nonelites, or relatively powerless people in society, do not possess the ability to create and maintain the police. Thus, the police evolve in response to elite perceptions of the need for formal social-control organizations, which are created for the express purpose of protecting the interests of social elites through maintenance of the status quo.

Returning to Lundman's conception of liberty versus civility (or, as we described it in Chapter 1, freedom versus order), elites are moved to establish police when disorder or incivility becomes intolerable. When those with power in a society perceive a threat to their social positions from the existence of disorder in society, they move to restore order by establishing a police force. To paraphrase Lundman (1980:34), disorder does not cause the creation of police; elite fear of disorder does.

Thus, the type of social organization in a society may allow a formal, specialized police organization to develop but does not necessarily require one. Coupled with general social specialization is the creation of social elites. With elites come conflicts between the powerful and powerless in society. When the interests of the elites of an organically solidary society are threatened by disorder, the elites move to create police to protect those interests.

Perceptions of Crime and Disorder

The third factor Lundman identified as important to understanding the evolution of police was the rate and images of crime. This factor relates directly to the interests of social elites, because it is their interpretation of the problem of crime as an indicator of intolerable levels of disorder that results in formal police. Formal police are created when the elites in an organically solidary society define crime and disorder as a threat to their position of power.

A certain level of crime and disorder is natural in any society and particularly so in cities. It is precisely these types of events that are the targets of episodic and avocational forms of transitional policing. As long as disorder and crime remain relatively disorganized and intermittent, there is generally no call for a more formal policing organization. To the degree that the elites can insulate themselves and their social positions from the threat of disorder, they do not move to create police forces. However, when the level of crime and disorder increases in either frequency or seriousness to the point that it threatens to disrupt social functioning, formal police are instituted.

In England and the United States, social elites called for the creation of formal police when crime and disorder were rampant and when the cause of crime was identified as the "dangerous classes." As long as crime and disorder were seen as the work of a few, isolated wrongdoers, there was little cause for concern. However, as crime rates increased, and particularly in the wake of large-scale destructive and bloody riots, the elites came to fear for the existence of society itself. Charles Loring Brace wrote in 1872 of the growth of a dangerous class in New York and the need for control, stating "Let but the Law lift its hand from them for a season, or let the civilizing influences of American life fail to reach them, and, if the opportunity offered, we should see an explosion from this class which might leave this city in ashes and blood" (1872:21).

What the social elites feared was not so much crime and disorder but chaos and revolution. The mission of the police was not simply the control of crime, but the larger task of controlling the dangerous classes. For example, a proponent of formal police in England, Patrick Colquhoun, suggested that, in addition to crime control and order maintenance, the police should also be responsible for the "correction of public manners." As Silver (1965) suggested, the formal police in London served to provide an example of proper, middle-class behavior and also created a "capacity for awareness" that let them anticipate and control disorder.

Thus, these three factors—social organization, the interests of elites in society, and perceptions of the crime and disorder problem—correlate with the development of formal police organizations. In combination, they create the conditions under which formal policing can be instituted and in which the benefits of a formal police organization are perceived to outweigh their possible negative effects.

If this conception of the forces that explain the evolution of policing from unspecialized to specialized organization is correct, it should serve equally well to explain types of policing found in various places at different times (Bayley, 1996). Before we apply this evolutionary framework to the study of policing in America, we will examine its usefulness in other settings.

The Evolution of Policing in Non-English Settings

Thus far we have suggested that the creation of a formal police organization correlates with changes in the social organization that allow the emergence of an elite society. This social elite, in turn, institutes a formal police system in response to perceived threats to social order. The emergence of a formal police organization should correlate with increasing specialization in society, including separation of economic, political, religious, and other institutions. Further, a product of this specialization is the identification of "leaders," or elites, who control social institutions and who seek to preserve their status. The evolution of the police within a society will correlate with efforts of social elites to protect and expand their power.

Policing and Social Organization

As Wormser (1962) indicates, in its earliest forms, social control was the product of "self-help." The redress of wrongs was a personal matter. If someone stole, attacked, or trespassed on the lands of another, the injured person was responsible for seeking justice. Most often, this justice came in the form of personal vengeance or revenge. If someone stole from you, you would be justified in demanding repayment from the offender, and you might personally use, or threaten to use, force to secure payment. If, however, the victim did not seek compensation, no one else would become involved. With rare exceptions (usually violations of taboos or other religious regulations), the settlement of disputes was not the concern of the group.

One unfortunate result of this system of personal vengeance was a tendency toward excess and resulting feuds. That is, if someone stole from you, you might demand repayment, but if repayment were not made, you might assault or kill the offender. In response, the offender or her or his kin might seek vengeance against you for your actions. In short order a situation would develop in which you and your kin would be "at war" with the offender and her or his kin. Responsibility for harm could quickly spread from the individual offender to include the entire kinship group. Like the Hatfields and McCoys of American fame, mere membership in the family would then render one an acceptable target of revenge for members of the other family.

Mutual responsibility for wrongs done by individual members of a kinship group develops over time. In the interests of safety and peace, family members begin to "police" each other. In this way, while the greater society may become increasingly complex, social control is largely mechanistic and authorized along kinship lines. Social problems and disorder arise when kinship lines become blurred (families grow too large) or when disputes occur between families. Eventually a need arises for regulations to control relations among families.

For example, suppose a member of one family throws a stone and strikes a member of another family. The injured party seeks revenge, asking that the offender be stoned. The offender's family refuses this request, offering to pay damages instead, but the victim is not satisfied. What will happen? The questions at hand are: (1) the right of the victim to seek revenge, and (2) the limits on revenge. Without some outside arbitration, there is likely to be a feud between the families.

As Wormser sums up:

> About this time [in history] we find the beginnings of a judicial system, for there had to be some way of determining how much revenge you were entitled to. Then chiefs and headmen began to make an effort to reduce the slaughter within the group and prevent manhunts and clan feuds. They would get the disputants together and try to induce a compromise, a voluntary form of arbitration in which the matters at issue were the right to get personal revenge, and how that revenge should be satisfied. (Wormser, 1962:15–16)

In its early development, the law and the "courts" that administered it were peacemakers. They did not determine facts and assign penalties as much as negotiate settlements. "Court rulings" in these cases might still be ignored, but it became increasingly difficult to do so. As the chief or headman who administered the court became more and more powerful, the decisions of the courts were more likely to be accepted. In time, the court was the mechanism of deciding and maintaining the "king's peace."

Policing and the Consolidation of Elite Power

Eventually one family comes to dominate in an area, and the leader of that family becomes a political power in the region. Because social control is organized on kinship, this political leader initially relies upon the cooperation of the leaders of other families to maintain the peace. Over time, this leader begins to consolidate authority so that he or she can become independent of the other family heads. To be successful, it is necessary to realign the social organization from individual loyalty to the family to individual loyalty to the political leader.

One way of doing this is to redefine individual wrongs so that they become "public wrongs." This can be done by defining certain behaviors as "breaches of the king's peace." The victim of a theft, for example, is no longer the individual whose property was stolen, but the king whose peace was broken. If the king is now the wronged party, it is the king who reacts to "crimes" and the king to whom the individual is responsible. To maintain the king's peace, the monarch must establish some sort of police. The police, in turn, represent the king and serve not only to maintain the peace but to maintain the power of the king. Crime and disorder are no longer simply individual wrongs, but rather represent challenges of the monarch's authority.

In this way, the rise of policing is intricately tied in social evolution to the rise of the state as a unit of government. As political power is consolidated into a state government entity, formal police are created to enforce and administer the laws. Being an extension of the state, the police derive their power from the power of the state. The organization of the state correlates with the structure and authority of the police. In cases where the state, or monarch, acquires a monopoly on power, the police tend to be centrally organized and are given a high degree of power over many aspects of life. In cases where the state is granted authority by kinship or other local groups, the police are less centrally structured and their powers more limited.

Policing systems can be distinguished in terms of the basis for their authority. Since all police systems rely on state authority, the source of state power ultimately represents the basis of police authority as well. As Fosdick (1915:14–18) observes, there are two basic models of the police. He distinguishes between a *continental model* (referring to policing systems on the European continent) and a *British model.* (Later we will discuss another author's use of the term "continental model" and a parallel to the British model called an "Anglo-Saxon model.") Fosdick notes, "The great safeguards to personal liberty established in England by Magna Carta and the Bill of Rights and sustained by centuries of judicial interpretation are almost entirely lacking on the Continent."

 Raymond Blaine Fosdick (1883–1972)

Raymond Fosdick was born in Buffalo, New York. He earned a law degree from New York University in 1908 and began work for the City of New York. At the request of John D. Rockefeller, Fosdick undertook a study of European police systems. In 1915, he published *European Police Systems.* This project was followed five years later with *American Police Systems.* Fosdick published fourteen books during his career and was perhaps the first to attempt to document and explain differences in police structures and development. (Culver Pictures)

We will return to this distinction in our discussion of the evolution of policing in English settings. For now it is enough to note that different societies arrive at different balances between freedom and order. In societies where the state acquires power, and therefore dominance, over lower social groups, the police have more pervasive powers, and order outweighs liberty. In those societies where the state is given power by lower social groups, as in the English tradition, the police are more limited because freedom interests are protected.

Louise Shelley (1994) identifies four models of policing in connection with her examination of the origins and development of policing in the Soviet Union. The continental model existed in places where the central government had gained dominance. The result was police that were centralized, armed, and charged with crime control, with an emphasis on political and administrative control of the people. In the colonies of most major powers, the colonial model of police developed. Here police were centralized and armed, relied on colonial authority, and emphasized political and administrative control. In the Anglo-Saxon model (what Fosdick called the "British model") the police were local and emphasized both crime control and order maintenance. In the Soviet Union, a communist model of policing developed. The police were similar to both the continental and colonial police in terms of their structure and duties, but they had the added dimension of supporting the communist ideology. This resulted in a militia that exercised wide control over all manner of civil functions and also used censorship and undercover investigations to ensure that citizens followed official doctrine.

With this introduction, we can now turn to an examination of the evolution of policing in societies where the state achieved a power dominance. Here we see the role of the police in terms of supporting consolidated political power; that is, the police as a tool of state authority.

Policing in Imperial Rome

Robert Nisbet (1964) explains the development of law in ancient Rome as a result of a general social and political movement toward centralized authority. Traditional Roman society was organized on strong kinship lines. Romans drew a rigid distinction between domestic and civil responsibilities, and the Roman citizen was first and foremost a family member. The Roman family was patriarchical, headed by the *pater familias,* or head of the family. The family father had sole possession of all property and absolute authority over all members of his household. Each Roman owed her or his first allegiance to the family, and the government would not intervene in family matters. For example, if a Roman was convicted of a crime, the punishment was imposed within the family, by relatives of the offender, and not by agents of the state (Nisbet, 1964:266).

Roman society and government began to change during the reign of Augustus, beginning in 27 B.C. As Nisbet noted, Augustus started a movement that continued for some 500 years in which the state, through the emperor, came to dominate all aspects of Roman life. By a series of decrees and laws on marriage, property, and crime, Augustus and later emperors reduced the power of the family and created an individual responsibility of the citizen to the emperor. By breaking the bonds of kinship, the Roman emperors created a monopoly on power in the Roman state.

Among other actions, Augustus created the *Praetorian Guard,* consisting of some 7,000 Roman soldiers within the city of Rome. The Praetorian Guard was assigned the task of protecting the emperor and his property and maintaining the peace of the city. Prior to Julius Caesar's march into Rome at the head of his army, Roman law did not permit the maintenance of armies within the city. With the creation of the Praetorian Guard, Augustus created a precedent for the maintenance of a standing army in Rome for police purposes.

Other officials in Rome were also responsible for law-enforcement and policing duties. These included **quaestores,** who worked for magistrates and had the authority to arrest those accused of crimes. There were also **vigiles,** who served as watchmen and were responsible for both maintaining the peace and fighting fires (Germann, Day, and Gallati, 1988:46–47).

Members of the Roman Praetorian Guard, the police of Ancient Rome. (The Bettmann Archive)

But until the reign of Augustus, the powers of these officials were quite limited. They all worked under the control and direction of magistrates or judicial officers.

Augustus and the emperors who followed him developed the power of Roman "police." The tradition of "self-help" was increasingly replaced by official law enforcement through agents of the state. Thus, while the individual Roman citizen could invoke the law by reporting offenses to judicial officers, "police" agents also increasingly began to impose the laws of the state on their own intiative. The ability of the emperor, through the actions of his agents, to regulate individual behavior served to enhance his power and consolidate his control over the public. The task of the police became the maintenance of public order more than the management of individual disputes.

Policing in Nineteenth-Century Italy

The Roman Empire collapsed late in the fifth century, and Europe entered what has been called the Dark Ages. Little is known about law enforcement in this era as the European continent was in a state of turmoil with constant warfare (Germann, Day, and Gallati, 1988). During this period, *feudalism* developed as a pattern of social organization. Under feudalism three classes, or "estates," existed in society: the nobility, the clergy, and the common people. Secular society was organized on a system of *fealty,* or allegiance, so that powerful nobles controlled less powerful nobility, who swore allegiance to them in return for receiving the promise of protection. Common people owed allegiance to the nobility who controlled the lands on which they lived and worked.

During the Dark Ages, feudal lords often fought over land. Over time, by conquest and marriage, large areas came under the control of individual lords, and all of the people living in those lands owed allegiance to those lords. These consolidated holdings formed the basis for the kingdoms of Medieval Europe, and eventually formed the European nations such as Spain and France. The history of Europe in the Middle Ages includes numerous efforts by kings to achieve power over their subjects. It resulted in what are called *absolute monarchies,* in which kings became total despots with absolute authority over all the citizens in their realms. As a result, political power was redefined from meaning that citizens owed allegiance to their local nobility to meaning that everyone, nobility included, owed allegiance to "the Crown."

Italy, like Germany, was an exception to the general rule of power consolidation under a monarch that characterized most of Europe in the Middle Ages (Gramckow, 1995). Italy entered the nineteenth century as a disorganized collection of independent principalities and city states. During the nineteenth century a unified Italian state was formed that encompassed these independent governments and centralized national power in a constitutional monarchy. As Davis (1988) concludes, a central theme throughout this period was a crisis of law and order. Successive coalitions and government leaders relied on the law-and-order crisis to expand state power and to use the police and law to control not only the general public but also their political adversaries.

Because of its central location in the Mediterranean, Italy became a commercial center early in the Middle Ages. Italian city-states became trading centers and world economic powers. They also were among the first European areas to experience the urban blights of poverty and crime. It became unsafe to travel the city streets, especially at night. Increasing levels of crime and violence, not to mention general disorder, served to dampen commercial activity. As Johnson (1988:66–68) relates, commercial growth depends upon the maintenance of peace.

To protect social order, city leaders formed patrols headed by noblemen to "police" the cities. These patrols served largely to control the lower classes and protect the interests of the city nobility and the growing middle classes of manufacturers and businessmen who emerged in the commercial cities. For the most part, the police and the magistrates focused on controlling and punishing the lower classes, allowing those of higher status more freedom of behavior.

The Black Plague was added to this experience with urban crime during the fourteenth century. In several places, citywide curfews were imposed that prevented the citizens from leaving their houses. The goal of the curfews was to prevent contagion and the spread of disease, but enforcement required the establishment of a form of police, whose job it was to ensure that the curfew was not violated. In Italian urban areas, at least, there was a tradition of public policing prior to the Italian unification.

Another contribution to Italian policing came from the French, who had conquered a large part of northern Italy during the reign of Napoleon. The French instituted a central government, complete with the Napoleonic Code of laws and a French-style administrative bureaucracy that included police. In the early part of the nineteenth century, French rule created peace and prosperity in the Piedmont area of Italy. After the collapse of the French empire, the Piedmontese retained much of this French system, and it became the model for Italian government after the unification.

Davis (1988:139–140) states that the police of the Piedmontese underwent a transition in the early 1850s, becoming the "ever-wakeful eyes of the state." This change was a movement away from its role as a mediator and regulator of social disputes to a preventive and repressive function. The police increasingly became instruments of central government power to control public life. Over time, Davis (1988:186) writes, the "makers of the new Italy revealed only too often an indecent haste to . . . accept even the flimsiest of pretexts to justify the suspension of both civil liberties and civil laws."

In the long run, the use of a repressive central police failed in Italy, Davis suggests, because repression and the disregard for constitutional civil liberties created a "mass politics" that eventually overcame the powerful elites and led to a more democratic government. Interestingly, however, this more democratic government was replaced by the regime of Benito Mussolini and a "police state" within 30 years.

It may be that the traditional local, personal, and kinship patterns of social organization in Italy were not amenable to central police powers. Nonetheless, the development of central policing in nineteenth-century Italy illustrates how a formal policing organization is created and employed to support the interests of social and political elites. Further, the broad preventive and repressive powers of the unification-era Italian police further illustrate how the police function is shaped by perceptions of the nature of the crime-and-disorder problem.

Policing in France

Philip Stead (1983) extensively describes the development of policing in France. He notes that the French police have wide-ranging powers that surprise American and British observers. "Even more impressive, however, is the fact that so much police power is at the disposal of the central government, the political government of the day" (1983:9). That is, police powers in France are nationally organized and directed, and these powers serve the interests of that national government.

Stead attributes this centralization of police power to the pattern of violent political upheaval that has been apparent in French history since 1789. In a little over 200 years, France has been governed by three separate monarchies, two emperors, and five republics. Most of these government changes were the result of violent revolution or conquest. In this context, Stead (1983:11) asks, "With this pattern of constitutional reversals dominating France's political history since 1789, such reversals so often being the consequence of the government's loss of control of the streets, is there any wonder that French statesmen have become intensely conscious of the need to be well informed and to have immediately at hand the means of maintaining order?"

The history of France in the Middle Ages is a story of the struggle for power between the Crown and local nobility. In Medieval France the king achieved autocratic power over the nation, moving from a position as "first among equals" to that of undisputed leader. In making this transition, French kings changed the organization of the country from feudalism to a centralized political state.

Over the years, the kings introduced royal judges to exert control over feudal lords. They established a standing army, loyal to the king, that could be used to uphold royal laws and decrees. They also chartered independent cities across the land. These cities were not under the control of feudal nobility, and they were responsible solely to the king. The cities were empowered to maintain order within their walls. In rural areas the king's peace was upheld by the army.

A standing army required its own "police" for the maintenance of order within the ranks. Accordingly, within the French army, the office of **provost** developed, with a provost corps serving as the military police. Soldiers in the provost were called the **marechaussée,** or marshals. By the end of the Middle Ages, they had assumed civilian policing duties in the rural areas. As the capital city of the country and home of the king, Paris had long had a royal officer, the provost of the city, who was a combined governor, judge, and police chief for the city. Originally known as **gens d'armes** (men at arms), the best-known French police agency today is the **Gendarmerie Nationale,** administered by the Ministry of Defense.

With the revolution in 1789, the First Republic came to power, and with it, the idea of a civil police. A variety of organizations were attempted before Napoleon took power in 1804. Under his direction, the revolutionary Ministry of Police was expanded and institutionalized. With the fall of Napoleon, the civilian police were placed under the authority of the minister of the interior as the **Police Nationale.** Thus, France had established two separate, though related, police organizations that continue today.

As an agency of the ministry of the interior, the Police Nationale perform a variety of functions, including the enforcement of law and the making and enforcement of regulations. French citizens must apply to the Police Nationale for passports, identity cards, driver's licenses, and other documents. The police are also responsible for a variety of social services, including health inspections, public welfare, and firefighting. Thus, the centralized police of France exert a tremendous influence over the lives of French citizens.

During the political upheavals that have characterized French political history since the late eighteenth century, one of the primary functions of the police has been intelligence gathering. As Stead notes, there is little wonder that this is so. What is different about the police in France as compared with their English counterparts is the French degree of control and surveillance over citizens and their clearly national-level organization. For example, over 2,500 French police work in the Directorate of General Intelligence and Gambling (Stead, 1983:7). Their task is the collection of intelligence on political, social, and economic matters necessary for the information of the government. These tasks include infiltration of political, labor, social, and other groups by undercover police.

The tradition of France, especially the requirements of identity documentation and the realization of strong centralized power under the government, support the broad powers of the police. The focus of French police on order maintenance, with a relative deemphasis on individual liberty, reflects this tradition and the French history of political turbulence. The use of the French police by successive French governments to gather information about political adversaries and general threats of disorder indicates the role of the police in supporting the dominant elites. The police of France, then, reflect the cultural and political development of that country and the interests of its elites over time.

Policing in Meiji Japan

The history of police development in Japan presents a stark contrast to the creation of policing organizations in European societies. Like the police of Imperial Rome, France, and nineteenth-century Italy, the Japanese police have pervasive powers over the citizens and are responsible for a large number of public welfare functions in addition to order maintenance and crime control. Like the police in those other nations, the Japanese police developed to serve the interests of the elites by suppressing political and economic opposition. Similarly, the Japanese police developed from a strong feudal form of social organization and were part of a centralized national government. Unlike their European counterparts, the Japanese police did not evolve from earlier functions. The Japanese instituted a centralized, modern police organization within a span of about 10 years. In many ways, the Japanese "skipped" a period of transitional policing.

For centuries, Japan existed in relative isolation even from its Asian neighbors. It was largely a nonindustrial, noncommercial society organized along feudal lines. Japan was split into several *domains,* or principalities, under the direction of the **shogun,** or overlord. In theory, at least, the leader of each domain owed allegiance to the shogun, but in practice, Japanese political organization was local in nature. Despite having a large population, and being geographically dispersed, Japanese society was very much mechanically organized. Strong traditions of collective responsibility and class distinctions were in place. That the Japanese emperor held both political and spiritual leadership positions illustrates the relative lack of role differentiation in the society.

Essentially, as in Medieval Europe, Japanese society was organized into a few social classes based on status. The nobility, or **samurai,** were the upper class. Lower classes were composed of peasants, artisans, and merchants (Parker, 1987). Each domain was essentially self-policed, and a few samurai were assigned "police duties" as assistants to the shogun's magistrates. Most police functions were incorporated into other, general administrative offices. The shogun did create the office of **ometsuke,** consisting of five high-ranking samurai and their followers. The ometsuke were responsible for the surveillance and control of domain leaders and upper-level shogunate officials (Westney, 1987).

In the middle 1800s, the shogun was persuaded, in large part by threats, to open trade with Western nations. Beginning in 1859, six trading ports in Japan were opened to the West. The unfavorable conditions of these trading treaties, including the fact that Westerners would not be subject to

Japanese law, spurred opposition to the shogun. By 1867 the shogunate was overthrown, and a coalition of southern domains assumed power in what has been called the *Meiji Restoration*. One goal of the Meiji was to reestablish the dominance of the emperor. Accomplishing this goal required, among other things, the creation of a stronger, centralized national government.

Added to the desire for a stronger government was resentment at the arrogance of Westerners. Thus, the Meiji wished to bring Japan to a par with the Western nations and to make foreigners comply with Japanese law. To accomplish these goals, Meiji leaders set about the task of restructuring Japanese government services. They consciously studied Western organizations and practices in a variety of fields, including military, postal service, education, business, and policing. Based on these studies, they adopted Western organizational practices for Japanese agencies.

The first modern police force in Japan was instituted in the open-port city of Yokohama. Originally this was a composite force of English, French, and Japanese nationals under an English commander. Naturally, the force adopted the British style of patrols, ranks, and functions. In 1868 the Japanese government assumed control of this police force and continued to follow the English practices but with an all-Japanese police agency (Westney, 1987:38). Over the next few years, that force was reorganized, but it remained basically the same and provided a model for other cities.

In 1871 the central government abolished the traditional domains. Until then, the capital city of Tokyo had been patrolled by samurai recruited from the various domains. Now the Japanese government sought a new model for policing in Tokyo. A study of Western practices was commissioned, and in 1872 the Japanese selected the French model for the development of a new Tokyo police, and later a national police agency. As Westney observes, the French model appeared best suited to Japan:

> The perceived urgency of creating a standardized and effective police force throughout Japan in order to end extraterritoriality (Western immunity from Japanese law) . . . caused the government leaders to shy away from the Anglo-Saxon model's reliance on local initiative and local control. Moreover, the multifunctional nature of the continental model was very attractive to leaders in a society that was only beginning to construct a modern administrative system and where a standardized local government structure had not yet been formed, even on paper. Finally, the role of the continental police in political surveillance and control made it far more appealing to the Japanese, in an era of widespread antigovernment activism, than did the more circumscribed Anglo-Saxon model. (1987:41–42)

The Japanese developed a police system based on the French national model. The police were representatives of the national government and were

A nineteenth-century police officer in Nagasaki, Japan.
(The Bettmann Archive)

primarily responsible for the protection of that government. They had wide-ranging powers and duties, including annual (or more frequent) surveys of the population, health inspection, public health and morals regulation, public education, fire protection, and a variety of tasks not related to law enforcement.

The Japanese police adapted the French model to Japanese society. The design of Japanese cities made travel difficult, and the relatively large areas over which the cities spread led to the adoption of fixed posts for police officers. Police were originally recruited from the samurai class, and officers were assigned to posts away from their home neighborhoods. The separation of the police from their traditional neighborhoods and the strict regulation of officers, both on and off duty, built a sense of loyalty to the organization and to the government within the police. Between 1870 and 1900, the Home Ministry of the central government gained responsibility for the creation and administration of a national police force throughout all of Japan. The Japanese police reflect the cultural traditions and values of Japanese society (Alarid and Wang, 1997).

POLICING THE CENTRAL STATE

The brief sketches of police development in four non-English societies presented here are necessarily limited. They give an overview of the forces at work in each society that shaped the police structure and function, but they gloss over important variations. While similar in many respects, the police in Imperial Rome, nineteenth-century Italy, France, and Meiji Japan were also different from one another in many specifics. Space does not allow a detailed description of how the police in each society reflected the cultural traditions of their societies in functions and structure. But, remember, our goal is to identify common themes in police development, not idiosyncracies.

In all four cases, the police are a product of the consolidation of power in a national (or imperial) government. Members of the police organizations were selected, trained, and paid by the central government. Police agents owed loyalty not to kinship groups or local leaders, but to the central government. In each case, the police were created to protect and to demonstrate the power of the state. That is, the national police were charged with protecting the social order and, in doing so, established the power of the state to control the population. As Stead (1983:9) notes in reference to the French, these police served "the political government of the day." While governments might change over time, the police organizations endured, as did their central function of social control.

The influence of government structure and cultural traditions on the organization and function of the police is clearly visible in comparisons of policing in different countries (Das, 1997). The contemporary movement to community policing is worldwide, yet community policing will not be easy to develop in some countries (Gramckow, 1995). Changing philosophies of policing have some "crossover" influence between different policing models, but each philosophical change is adapted to the specific circumstances of the police organization (Feltes, 1994; Das, 1995).

Returning to the developmental model, the similarities of these police agencies can be related to the similar circumstances in which they developed. In general terms, these police emerged as part of a broader social change that saw the creation of relatively strong central governments and a realignment of the social class structure. As the patterns of social organization shifted from kinship and feudal loyalties toward a national identity, a range of government organizations developed, of which the police were one. Further, the earlier patterns of social organization did not emphasize individual liberty, but rather a structured class system where the powerful few (the *pater familias,* or feudal lords) held positions of relative equality and the majority of the population were in subservient positions. The basic change was thus one of consolidating power (social, political, and economic) into one role from several. For most citizens, their relationship to those in power did not change. That is, they were still subservient.

The importance of consolidating power into a central government for the ultimate structure and function of the police can also be seen in the colonial police used by the British. Although, as we shall see, the British police were tightly controlled by legal restrictions to protect the rights of citizens, police in the English colonies had a much different structure. British police in England were unarmed, but colonial police were issued firearms. The police force in Northern Ireland, for example, was markedly different in size, authority, and lack of civilian oversight from the police force in England (Weitzer, 1996). Colonial police often show the influence of elite interests, whether that elite be political—British imperialism—or economic. The way in which colonial policing differs from policing in the homeland contributes to a unique style of policing in ex-colonies after they have obtained independence (Deflem, 1994).

In colonial societies, the police assumed pervasive powers over the citizens in the name of a central government that had previously been wielded by a local or regional leader. As agents of the newly created central government authority, the police were generally given broad powers of arrest, regulation, and investigation, and were further instructed to focus attention on various "antigovernment" activities. In Italy (Davis, 1988:217–241), France (Stead, 1983:45–51), and Japan (Westney, 1987:40–45), the national police devoted considerable attention to the control and investigation of the government's political enemies. This attention included intelligence gathering, arrest and regulation of "subversives," and, occasionally, censorship and propaganda.

The police, in addition to this political use, were charged with a variety of public welfare and regulation tasks, plus crime control. The word *police* is French, but it originated in ancient Greece and Rome. It refers to the internal governance of the community, or **polity.** In the societies examined here, the police were central to the governance of the polity. Typically located within a broader government entity such as a ministry of the interior, the police came to be the repository of all sorts of regulatory functions, from street lighting and repair to health services. The police registered births and deaths, issued identity papers, conducted censuses, provided firefighting services, and generally administered government services at the local level. Consequently, they were an integral component of government and interacted frequently with all citizens.

The police developed in these societies as the pattern of social organization shifted from mechanical to organic solidarity. As a correlate of this change in social organization, political power was consolidated in the hands of an identifiable elite (the emperor, king, or government generally). Police were created to protect the interests of the social elites (leaders of government), and their focus was on those problems of crime and disorder that the elites associated with threats to the existing order. Growing from a tradition in which order concerns dominated interests in freedom, the police were granted pervasive powers over the citizenry and charged with a range of government activities.

Within this social and historical context, strong, centralized, national police organizations represented cultural products of the societies in which they were developed. When compared with the types of police that developed in English settings, these "continental" systems are clearly distinct. These police are more centrally structured and possess greater power over more aspects of daily life than do their English counterparts (Ma, 1999). The reasons behind the differences can be understood best by comparing the development of continental police systems with those of the English systems. As Stead (1983:163) observes, "Nations carve their police systems in their own likeness. The police style of London would not be effective in New York, any more than the style of New York would be effective in London. Neither would be effective in Paris." We must now turn to a description of English police development as the basis for our comparison.

CHAPTER CHECKUP

1. Describe Lundman's model explaining the development of different types of police.
2. What is the difference between mechanical and organic solidarity in societies?
3. As regards elite interests, how do Newman and Fried distinguish between the purposes of law and the state, respectively?
4. Identify four models of policing.
5. In what ways were the police of Imperial Rome, unified Italy, France, and Meiji Japan similar?
6. What were the roles of quaestores and vigiles in ancient Rome?
7. How did the British colonial police compare with the police of England?
8. Regardless of where policing develops, what is policing said to be?

REFERENCES

Alarid, L. F. and H. M. Wang (1997) "Japanese management and policing in the context of Japanese culture," *Policing: An International Journal of Police Strategies and Management* 20(4):600–608.

Applebaum, R. P. (1970) *Theories of social change.* (Chicago: Markham).

Bayley, D. (1996) "Policing: The world stage," *Journal of Criminal Justice Education* 7(2):241–251.

Brace, C. L. (1872) "The dangerous classes of New York," in L. W. Dorsett (ed.) (1968) *The challenge of the city 1860–1910.* (Lexington, MA: D. C. Heath):19–22.

Das, D. (1995) "Police challenges and strategies: The executive summary of International Police Executive Symposium (Geneva, Switzerland)," *Police Studies* 18(2):55–74.

Das, D. (1997) "Challenges of policing democracies: A world perspective. Executive summary of the second annual International Police Executive Symposium, Onati, Spain, 1995," *Policing: An International Journal of Police Strategies and Management* 20(4):609–630.

Davis, J. A. (1988) *Conflict and control: Law and order in nineteenth-century Italy.* (London: Macmillan Education).

Deflem, M. (1994) "Law enforcement in British colonial Africa: A comparative analysis of imperial policing in Nyasaland, the Gold Coast, and Kenya," *Police Studies* 17(1):45–68.

Durkheim, E. (1933) *The division of labor in society.* (New York: Free Press).

Feltes, T. (1994) "New philosophies in policing," *Police Studies* 17(1):29–48.

Fosdick, R. B. (1915) *European police systems.* (New York: The Century Co.).

Fried, M. (1967) *The evolution of political society: An essay in political anthropology.* (New York: Random House).

Germann, A. C., F. D. Day, and R. R. J. Gallati (1988) *Introduction to law enforcement and criminal justice.* (Springfield, IL: Charles C. Thomas).

Gramckow, H. (1995) "The influence of history and the rule of law on the development of community policing in Germany," *Police Studies* 18(2):17–32.

Johnson, H. (1988) *History of criminal justice.* (Cincinnati, OH: Pilgrimage).

Kappeler, V. (1996) "Making police history in light of modernity: A sign of the times?" *Police Forum* 6(3):1–6.

Klockars, C. B. (1985) *The idea of police.* (Beverly Hills, CA: Sage).

Klockars, C. B. (ed.) (1983) *Thinking about police: Contemporary readings.* (New York: McGraw-Hill).

Lundman, R. (1980) *Police and policing: An introduction.* (New York: Holt, Rinehart & Winston).

Ma, Y. (1999) "Comparative analysis of exclusionary rules in the United States, England, France, Germany, and Italy," *Policing: An International Journal of Police Strategies and Management* 22(3):280–303.

Newman, K. S. (1983) *Law and economic organization: A comparative study of preindustrial societies.* (Cambridge, England: Cambridge University Press).

Nisbet, R. (1964) "Kinship and political power in first century Rome," in D. Black and M. Mileski (eds.) (1973) *The social organization of law.* (New York: Seminar Press):262–77.

Olivero, J. M. (1990) "Research note: A new look at the evolution of police structure," *Journal of Criminal Justice* 18(2):171–175.

Parker, L. C. (1987) *The Japanese police system today: An American perspective.* (New York: Kodansha International).

Rawlings, P. (1999) *Crime and power: A history of criminal justice 1688–1998.* (London: Longman).

Robinson, C. and R. Scaglion (1987) "The origin and evolution of the police function in society: Notes toward a theory," *Law and Society Review* 21:109–153.

Shelley, L. (1994) "The sources of Soviet policing," *Police Studies* 17(2):49–66.

Silver, A. (1965) "The demand for order in civil society," in D. J. Bordua (ed.) *The police: Six sociological essays.* (New York: John Wiley).

Stead, P. (1983) *The police of France.* (New York: Macmillan).

Walker, S. (1983) *The police in America: An introduction.* (New York: McGraw-Hill).

Weitzer, R. (1996) "Police reform in Northern Ireland," *Police Studies* 19(2):27–44.

Westney, D. E. (1987) *Imitation and innovation: The transfer of Western organizational patterns to Meiji Japan.* (Cambridge, MA: Harvard University Press).

Wormser, R. A. (1962) *The story of the law.* (New York: Simon & Schuster).

chapter **3**

THE ENGLISH ROOTS OF AMERICAN POLICING

In contrast to the policing that developed in non-English societies as an extension of centralized authority, the tradition in England and the United States is one of local responsibility and control of policing. Unlike continental police systems, a continuing conflict between freedom and order is central to understanding the creation of police forces in England and America. This chapter examines the early evolution of policing in England and its effect on the American experience.

In earlier chapters we saw that the police are the state's agency of social control, empowered to use coercive force when necessary to enforce the laws and maintain order. The development and practice of policing correlate with community, organizational, and individual characteristics. In the previous chapter we saw that the consolidation of power by a single person or relatively small group of people correlates with a centralized, pervasive form of policing. Policing in the English tradition is decentralized and more responsive to the community. This chapter reviews the development of English and U.S. policing in order to identify the factors correlated with this more decentralized police structure.

Prior to the Norman conquest in 1066, England was essentially an agrarian nation comprised of a number of local communities. There was no strong central government in England and no standing army. Instead, the rural population was organized into multifamily units called **tythings.** *About 10 families in an area were grouped together into a government unit as a tything.*

Each tything was responsible for the maintenance of the king's peace, and each member of the tything (males over the age of 12) was obliged to ensure the peace. In practice, this meant that if a member of the tything committed a crime, the tything either had to turn over the offender to the king's court, or be held accountable as a group for fines and compensation. This form of communal policing and shared responsibility was called **kith-and-kin policing** *by Charles Reith (1938).*

Tythings, in turn, were organized into units of 100 families (roughly 10 tythings), headed by a hundred-man, *or royal* reeve. *These hundreds were also called* **shires.** *The* **shire reeve,** *under the king, had a general responsibility for preserving the peace of the shire. In time the shire reeve came to be known as the* sheriff.

In an agrarian society organized along family lines and composed of relatively small groups of people who shared similar beliefs and experiences, mutual responsibility was a workable system of "policing." However, by charging each tything with the responsibility of maintaining the peace, the English system created a precedent for local control of the police function.

Further, the lack of a strong central government, and relative isolation from the influence of Roman civilization and Roman law, allowed a different form of legal system, the **common law,** *to develop in England. Unlike Roman law, the common law not only placed greater emphasis on individual liberty, it did so by placing enforceable restraints on government officials. As Tobias (1975:99–100) reports, English peace officers could arrest or employ force, but only if they adhered strictly to all forms of the law, and only if the use of force was reasonable and necessary. Procedural errors and unnecessary force could be the grounds for a civil suit against the officer.*

Another important distinction between the Roman law on which continental systems were based and the English common law was the procedure for trial. The common law relied upon an adversary process where the state and the defense "competed" in court. Unlike the inquisitorial process on the Continent, English peace officers were subjected to rigorous and often hostile cross-examination by the defense. The legal and social roots of policing in England were quite different from those in non-English cultures. The English traditions stressed local autonomy and the rights of the individual.

The Evolution of English Policing

In 1066 William of Normandy successfully defeated the English forces at the Battle of Hastings and declared himself ruler of all England. To manage the occupation of England, and to reward his followers, William instituted a rigid feudal system in England. He granted control over large land areas to his barons in return for their promises of loyalty and support. In turn, the barons granted control over smaller areas to lower-ranking nobility in return for similar promises of loyalty and support.

Under the feudal system, each level of nobility promised to pay taxes, provide troops, and grant obedience to a higher level, the highest being the king. This was the basic system in place on the Continent, and it was imposed on the English by the conquering Normans. The lowest level of the feudal system was the manor, and the lord of the manor was the local representative of the king.

Building on the existing structure of local government in England, manorial lords tended to support the "shire." The sheriff, however, was less likely to serve as a representative of the 100 families than as an agent of the king. In addition to preserving the peace, the sheriff was primarily responsible for the collection of fines and taxes from the citizens and the preservation of Norman control.

As Critchley (1972:3) observes:

> In the early period after the Norman Conquest the juxtaposition of two alien cultures, with a master race holding its defeated enemies in subjection and attempting to impose a new language and foreign manners, strained the primitive means of law enforcement beyond their capacity. The Normans tightened up observance of the old system and required the sheriffs, who were royal officers, to supervise its workings by holding a special hundred court, which sat twice a year, to ensure that all who ought to be enrolled in a tything, and thus pledged for their good behaviour to one another, were in fact enrolled. This court of scrutiny, which came to be known as the "view of the frankpledge and sheriff's tourn," also served the ends of savage repression, for the Norman sheriffs were men of great power and little scruple. They extorted the payment of fines at the least opportunity or none.

The Normans added a new law-enforcement title—that of **constable**—to this traditional Saxon organization. The title came from the traditional Norman/French position of *comes stabuli,* or master of the horse. In Normandy the position was both civil and military, derived its status and power from the Crown, and was considered a high honor. As Critchley (1972:5) observes, the title constable was first recognized in a statute of 1252, but "had no doubt been in general use for many years earlier." The constable was equivalent to the Saxon position of the "hundred-man," an administrative and law-enforcement official of the community. Critchley (1972) states that for many years the titles *tythingman, borsholder, headborough,* and *constable* were used interchangeably, but that the title of constable, because of its Norman derivation, was considered superior.

In an examination of the office of the constable in England, Joan Kent (1986) argues that over time, the constable's office was the vehicle through which the police authority in England was centralized under the Crown. She notes that originally the office was nondifferentiated, as the constable was a representative of the people who served a variety of civil and military functions (1986:14–15). The constable was something of a cross between the mayor and the local chief. Over the years, however, the constable was increasingly given responsibility to central government, so that in 1331, "Edwardian legislation recognized the constable as well as the township as having obligations to the state for keeping the peace. Such officials thus became police officers of the crown" (Kent, 1986:17).

By the thirteenth century, England was undergoing substantial social change. The beginnings of the industrial revolution, the closing of the monasteries, the movement to less labor-intensive farming methods, and the growth of trade created a shift toward urbanism. England began to develop larger and less personal cities, and increasingly came to be a society of organic solidarity as the family ties and shared experiences that underlay the tythings were replaced by urban living.

The tradition of community responsibility for order maintenance and policing was not very effective in the increasingly impersonal towns and cities. In 1285 the Statute of Winchester was enacted by Parliament. This statute reaffirmed the position of constable and empowered "urban" constables to draft citizens as watchmen. These unpaid watchmen were required to keep watch over the city through the night. The statute further reaffirmed the **hue and cry,** requiring citizens to come to the aid of a constable or a watchman calling for help. All males between the ages of 15 and 60 were to maintain arms for the purpose of subduing offenders. Those who did not respond to the hue and cry were considered accomplices to the offender and punished as criminals (Critchley, 1972:6–7).

The requirements of watch service, response to the hue and cry, and the maintenance of arms were not well received by the citizenry. That the constable was required, and empowered, to enforce these provisions of the statute helped create a rift between the constable and the community. When the constable's role was broadened to include that of tax collector during the tax increases of the middle 1600s, the constable often became an unpopular figure in the community.

In time, the burdens of the office of constable became so great that anyone who could afford it paid a fee to be relieved from service. Constables were responsible for the presentation of charges against neighbors at court, the custody of arrested offenders, the arrest of suspected criminals, and the collection of taxes.

In addition, the constable was required to oversee almost all of the municipal needs of the community, including the operations of inns and public houses; the condition of highways, bridges, and buildings; the execution of court-imposed corporal punishments, such as whippings and brandings; and the general maintenance of the peace. In keeping with tradition, although the constable was allowed to retain fees and some fine payments, the job was unpaid. The demands and unsavory duties of this essentially voluntary position, along with the often hostile attitude of community residents, convinced most people that the job simply was not worth it. Still, Joan Kent suggests that for quite some time the constables did a commendable job. "The constableship under Elizabeth and the early Stuarts was a relatively effective embodiment of the principle of 'local self-government at the King's command,' and constables of that period seem to warrant a much more favourable press than they have usually received" (Kent, 1986:311).

Yet, Kent suggests that, as the primary official of local government, the office of constable was overwhelmed by the demands of governing growing villages and cities: "It seems likely that by the later seventeenth and eighteenth centuries the weight of local government could no longer be borne by such part-time officials" (1986:310). Thus, the duties of constable were increasingly entrusted to *deputies* hired by the person elected to office. These deputies were often unkempt, illiterate, and as "criminal" as the people they arrested. In short order, the office of the constable became corrupted and inefficient (Critchley, 1972:18–19), and the search for an alternative began. This search would result, during the reign of Queen Victoria, in what came to be known as the *new police.*

Despite Stewart-Brown's (1936) well-documented and reasoned argument that a form of professional police predated the Norman Conquest, most observers contend that the "new police" were the first professional police in England. The creation of the *Metropolitan Police* in 1829 represented a major reform in policing in England and served as a model for the later development of modern police forces in the United States.

THE NEED FOR A NEW POLICE

Passage of the Statute of Winchester indicated a breakdown in the traditional system of law enforcement and peacekeeping in England. The shift from an agrarian and rural society to a more commercial, industrial, and urban one, coupled with the continued decline of the office of constable, created a crisis of order in Britain (Rawlings, 1999). Increasingly, the highways and city streets of the nation came to be dangerous and crime ridden. As Tobias (1979:24) observes, "Agricultural and industrial progress was changing

The First Professional Police? Serjeants of the Peace

Most histories of the development of policing in England either state or imply that there was no formal police office prior to the creation of the Metropolitan Police in 1829. However, there is evidence that a forerunner to these more modern "professional police" existed in the form of the "serjeants of the peace." Ronald Stewart-Brown wrote (1936:93), "The inference is strong that, in order to enforce the Anglo-Saxon peace system in its various aspects, a machinery of public security officers must have been required and have existed."

His review of English history revealed that the "frankpledge" system of communal responsibility for keeping the peace was imposed by the Normans on only about two-thirds of England. In other areas, chiefly in the Northwest around Wales and Cheshire, a different system prevailed. Here the local lord was responsible for maintaining the peace, and citizens were accountable by personal surety, not communal pledge, to obey the king.

To preserve the peace, the lord had in his employ armed officers whose job was to enforce the law. These officers were known as "serjeants of the peace." Stewart-Brown (1936:103–04) argues that the existence of the serjeants of the peace, and the fact that communal policing was not imposed in areas where they operated, means that they were the first "professional" police. Thus, he suggests that in some areas of Britain, professional police existed at the time of the Norman Conquest.

the way of life of the people, and the forces of law and order were not adequate for the new tasks being thrust upon them."

This observation was perhaps no truer than in London and its immediate surroundings. Throughout the eighteenth century, the population of England began to concentrate in towns and cities. The city of London held 10 percent of the nation's people by early in the eighteenth century, and crime was rampant. Highwaymen (mounted robbers) and *footpads* (unmounted robbers) struck citizens traveling to and from the city on suburban roads. Pickpockets, burglars, and thieves operated with impunity in certain parts of the city.

The discovery and sale of gin during the middle part of the century added to the problems of crime and disorder by making drunkeness commonplace. Crime and violence were at home in the city. The traditional practices of the constable, aided by the hue and cry and the watch system, were incapable of combating the rising tide of criminality and disorder.

Social Organization and English Policing

The English tradition was one of local responsibility for order maintenance. English "peace officers" were selected from the local community and were not paid by the state. Thus, these officials reflected community values and had little incentive to represent the interests of the central government over those of the municipality. As the need for local government services rose, the duties of constables and similar officials were expanded. There was no clear identification of a police agent, but rather a general representative of the king. Interestingly, this same official often served to represent the concerns of local residents to the king. In this way, the British peace officer found himself in a precarious position somewhere between the government and the citizenry.

Unlike the people of Rome, France, Italy, and Japan, the British never saw the emergence of an all-powerful king. When dissatisfaction with King John reached the breaking point, the English nobility required that he sign the Magna Carta in 1215. In the latter part of the fifteenth century, the Wars of the Roses were fought to settle the question of which "house" (York or Lancaster) would control the throne. In the middle seventeenth century, the English Civil War was settled when Oliver Cromwell's Parliamentarians defeated the forces of King Charles I. These events illustrate the decentralized nature of elite power in Britain. The inability of any one person or group to consolidate power in England is an important contrast to the evolution of policing in continental systems.

Forerunners of the New Police

In some of the wealthier neighborhoods, citizens agreed to pay *levies* (taxes) for the support of paid watchmen (Tobias, 1979:43–45). In these areas the streets were patrolled, day and night, by uniformed, armed watchmen paid by public funds. However, there was no central police authority, and not all neighborhoods could, or would, pay their watchmen.

The office of *justice of the peace* was renamed the **magistrate.** Magistrates had general responsibility for judicial and police administration duties. They administered the watch systems in their jurisdictions. One of the most forward-looking magistrates was Thomas de Veil, who became magistrate of Bow Street Court. De Veil was succeeded by Henry Fielding in 1748.

Fielding organized a force of voluntary former parish constables to work under his command as thief takers. **Thief takers** acted like contemporary detectives, investigating crimes and bringing offenders to prosecution. Successful prosecution of an offender earned the thief taker a reward, and further, he could confiscate the possessions of the criminal. Additionally, the crime victim would often pay a reward to the thief taker for the return of stolen property.

This form of bounty system gave thief takers a bad reputation among the citizens. Most were freelance, and many, like the infamous Jonathan

Wild, were suspected of encouraging crimes for the purpose of solving them, or retrieving stolen property in order to gain the reward money.

"Fielding's people," as this group of ex-constables came to be known, were under public control, working on orders from Fielding. While they, too, relied on reward money for their livelihood, they were purposely selected by Fielding and centrally commanded by him. By 1753 Fielding was able to pay them a small stipend from discretionary funds. In time, these thief takers came to be known as the **Bow Street Runners.** On his death in 1754, Henry Fielding's office as Bow Street magistrate was filled by his half-brother, John.

John Fielding continued the Bow Street Runners as detectives. He, like Henry, advertised in the newspaper to encourage crime victims to provide information to the runners. He established the ***Police Gazette,*** a bulletin that presented details about known offenders and descriptions of stolen property. In addition, John Fielding believed in the usefulness of preventive patrol and periodically paid his runners extra to have them patrol the streets in times when crime seemed particularly common.

During his tenure as chief magistrate of London, John Fielding continued to support preventive patrols from time to time. At the direction of Parliament, in 1763 he organized the *mounted patrol,* which provided protection and crime prevention on the major turnpikes and highways around London. In 1792 a *foot patrol* was established that had jurisdiction over the streets of the city and was commanded by the Bow Street magistrate. In 1805 the horse patrol was resurrected to patrol the highways around the city. Reorganization in 1821 restricted the foot patrol to the paved streets, and a *dismounted horse patrol* was organized to serve the outlying districts between the jurisdiction of the foot patrol and that of the horse patrol.

All three of these preventive patrols were commanded by the Bow Street magistrate. The horse patrol and dismounted horse patrol were uniformed and were organized into one administrative unit. Members of all three units were paid, full-time police officers. In total, however, the number of patrol officers never exceeded 300 to 400. In 1821 the final addition to the Bow Street magistrate's policing complement was established as the *day patrol,* consisting of 27 officers (Tobias, 1979:52).

There was also a *Thames River police,* composed of some 70 officers who patroled the wharfs on foot and the river in row boats. Their job was to prevent theft from ships' cargoes, and to control smuggling along the river. This police force was independent of the Bow Street magistrate, and with the various Bow Street patrols, constituted the scope of "professional" police in the London metropolitan area before 1829.

As Tobias concludes, the various Bow Street and Thames River patrols were adaptations to the volunteer and neighborhood-paid watches of the past. He writes (1979:53–54), "When the level of crime proved to be a problem, the authorities of the day very sensibly turned first to a series of attempts to improve the policing system with which the metropolis was already equipped. Eventually, they were forced to the conclusion that these

The Bow Street Police Office in 1816, the home of the Bow Street Runners.
(The Bettmann Archive)

were not enough, and had to abandon virtually the whole system . . . and
supplant it by the new Metropolitan Police."

Thus, for about 75 years government officials and police reformers in
England attempted to respond to rising crime and disorder in London by
improving the traditional systems of policing. The watchmen and constables became salaried offices. Policing came under the control of judicial officers. The numbers of officers were increased and their duties more closely
defined. Special-purpose policing organizations were created as the need for
them arose. Still, in the end, the four patrols of the Bow Street magistrate,
the Thames River police, and the various parish and vestry watch and constabulary systems were not sufficient to maintain order and prevent crime.

One factor leading to the creation of the new police was the recognition that even the improvements on the traditional practices were insufficient to check the rise in crime. The existence of paid, full-time, preventive
police agencies, of course, provided a precedent for the development of a
central police force. However, in addition to a recognition that the problem
of crime and disorder required a more permanent solution, thinking and at-

titudes about police had to change. Philip Rawlings (1999:74) has commented that "The main argument against a shift from parochial (local) policing lay in the old belief that it would give too much power to central government and so threaten liberty, which was assumed to depend on power remaining in the hands of local authorities. . . ."

As Critchley (1972:35) observes, throughout this 75-year period, there was no groundswell of support for changing the traditional police practices. In fact, he states, "For many years the English people had no desire for a police institution; indeed, with few exceptions, they regarded its nonexistence as one of their major blessings." Whatever the level of crime and disorder, the people were apprehensive about the powers of a police institution and its implications for personal liberty. What was needed was a definition and description of a police agency that would be accountable to the citizenry and would support "the rights of Englishmen."

The New Science of Policing

Despite the efforts of the Fielding brothers, along with periodic legislation to organize special patrols and the development of paid watches in certain areas, most Englishmen opposed the notion of a police force. Opposition to police was grounded on two distinct, but related English values. The first was the importance placed on individual liberty. The second was the tradition of local self-government. Before a formal policing organization could be created it was necessary to reconcile these values with the functions of the police. This reconciliation was accomplished through the development of a science of policing. The primary architect of the science of policing was a Scottish magistrate, Patrick Colquhoun.

Colquhoun, a merchant from Glasgow, had been named lord provost of that city at the age of 37. A reform of the office of the justice of the peace around London, the Middlesex Justices Act of 1792, led to his appointment as a Metropolitan magistrate. His long-standing interest in policing led to the publication in 1797 of his *Treatise on the Police of the Metropolis*. In this book, Colquhoun argued that a well-regulated and -administered police was necessary.

He suggested that such a police force would serve not only to detect and apprehend offenders (a task the constables and other patrols were doing) but also would prevent crimes by their presence in public. Further, by example, enforcement, and distribution of literature, these police could help mold and correct public morals and manners. Importantly, Colquhoun believed the police should not be linked to the magistrates and justices as were the constables and patrols but should instead form a separate unit of government service.

Colquhoun was greatly impressed by the French police, whose organization and efficiency he admired. The problem with the French police was not found in their organization, but rather in their application as agents of government repression. Colquhoun argued that the preservation of order and protection of citizens through a police agency was consistent with the British constitution.

Critchley (1972:38–47) explains that Colquhoun's "science of policing" was a logical outgrowth of utilitarian philosophy. It was the utilitarians who managed to reconcile the idea of a police force with the protection of individual liberty. In doing so, they cleared a major philosophical obstacle to formal policing in England. By casting the police in the role of defenders of liberty, rather than as a threat to liberty, they enabled later reformers to counter arguments against the police idea based on fears of repression.

Briefly, **utilitarian philosophy** supports the contention that the best government action is that which provides the greatest benefit to the most people. If, for example, the creation of the police could prevent more harm (in the form of people becoming crime victims) than it caused (perhaps through restrictions on liberty), then policing is justified. Similarly, on an individual level, the idea is that people will do those things that benefit them most and avoid those things that cause them harm.

Utilitarian philosophy was the primary justification for laws in England. This philosophy supports the concept of *deterrence,* for example, by suggesting that if the penalty for crime outweighs the gains from crime, the individual will avoid criminal behavior. For many years there was a split among utilitarians. Some, like Colquhoun, desired police reform so that the patroling officers could prevent crimes from occurring by their presence in the community. Others who were jurists, like George Romilly, thought that revisions of the criminal laws to ensure that just penalties would be imposed would prevent crimes by deterring individuals through the fear of punishment.

In the early going, the jurist position was dominant, and Parliament revised the criminal laws. Many crimes that had carried capital penalties were rewritten to involve less severe sanctions. The rationale was that overly harsh punishments discouraged juries and judges from convicting offenders or caused citizens to question the justice of the law. For example, if a starving person stole a loaf of bread, no one would want to see the individual hanged. The combination of nonenforcement and lack of citizen support, the utilitarian jurists believed, undercut the deterrent effect of the law. If the laws could be "fixed" so that their deterrent effects might be realized, there would be no need for a preventive police. Since police cost money for salaries and equipment, and legal reform was essentially free, the utilitarians first tried to improve the criminal law.

Changes in the criminal laws and penalty structures during the late eighteenth and early nineteenth centuries did not result in great decreases in the levels of crime and disorder. Legal reform alone was insufficient. Now utilitarians who supported the creation of police gained credibility for their ideas. Cesare Beccaria (1763) had written that it was better to prevent crimes than to punish them in his *Essay on Crime and Punishment.* This concept had impressed the leading utilitarian thinker of the age, Jeremy

Bentham. Bentham agreed with the notion that prevention was far better than detection and therefore supported crime prevention as the goal of criminal law. With his disciple, Colquhoun, he supported the idea of a preventive police. If the police would protect the property and safety of English citizens by preventing crimes, the benefits of police would outweigh their costs.

The term *police,* as mentioned earlier, was of French origin, and that in itself hindered wide acceptance of policing. During the latter eighteenth and early nineteenth centuries, England and France were frequently at war, and French ideas were not particularly well received in most of England. Further, the role of the French police during the "terror" that followed the French Revolution convinced many British leaders that policing was contrary to the English heritage and British values.

The utilitarians, however, had managed to anticipate these objections. The idea of separating the police from the judicial authority, espoused by both Bentham and Colquhoun, created a structure in which the judicial officers could serve as a check or control on the exercise of police powers. The English utilitarians had proposed a "separation of powers" and "checks and balances" model of policing that differed from the French system. Not only was policing consistent with the British constitution because it protected the safety and property of citizens, it was also limited by structure so that it could not be a totally repressive force. Finally, the organizational structure of the new police included, at the end, two equal commissioners who directed the force under the supervision of the home secretary. This model prevented any one individual from commanding the police, and ensured Parliamentary oversight through the home secretary.

With these philosophical and practical provisions to overcome fears of repression, the last remaining obstacle to the new police was the tradition of local self-government. Unlike the nations of continental Europe, which had a history of centralized government from the Roman tradition, English government functions were local and decentralized. The organization and structure of the new police had to be sensitive to this tradition.

Sir Robert Peel, the "father of modern policing," resolved this problem. Although the practical need for improved policing had long been recognized, and others had supplied the arguments and rationale for the new police, it was Peel who succeeded in gaining the approval of Parliament. Peel's political skills were critical to the passage of the **Metropolitan Police Act of 1829.**

In April 1829 Peel, as home secretary, introduced his "Bill for Improving the Police in and near the Metropolis" in Parliament. Since taking office as home secretary in 1822, Peel had been developing a plan for improved policing. Indeed, his father had been involved in the first attempt to create a new police agency in 1785 (Walker, 1983:3). Over those years the son had become a masterful politician.

The initial bill to create a centralized police force was introduced in 1785, but it was withdrawn after sharp criticism in the press and from the

Sir Robert Peel (1788–1850)

Born in Lancashire, Robert Peel was the son of a manufacturer and a member of the middle class. Peel's father (also named Robert) was a member of the British Parliament and received a baronet from the Crown. This honor brought the title Sir, which the younger Peel inherited on his father's death in 1830. Peel was educated at Harrow and entered Oxford University in 1805, graduating with a Bachelor of Arts in 1808. In 1809, he was elected to the House of Commons. He began a steady and successful political career. Peel was named Home Secretary in 1822 and succeeded in passing the Metropolitan Police Act in 1829, creating the first modern police force. Peel continued his political career, serving as Prime Minister twice. He died in a riding accident in the summer of 1850. (Brown Brothers)

City of London. The City of London had its own government and covered only about one square mile of the London metropolitan area. The city, however, had powerful friends in Parliament, and through these friends, city leaders could influence the outcome of legislative efforts.

It took over 40 years for another bill to create a police force to be introduced, and again, a Robert Peel was its drafter. Prior to introducing the bill, Peel had created and chaired a committee to study the state of policing in the metropolis. The committee reported that existing policing practices were inadequate and almost universally condemned. Further, based on testimony from Edwin Chadwick (a disciple of Bentham and friend of Colquhoun), the report suggested the creation of an Office of Police. Peel used this committee report as the springboard for his police-reform bill.

Throughout his term as home secretary, Peel had repeatedly addressed Parliament about the problems of policing but had always noted the need to ensure the rights of Englishmen. He would frequently remark that he saw no way to reconcile the need for better policing with the maintenance of individual liberty. Through various committees and other activities as home secretary, Peel studied policing and honed his political skills. Critchley (1972:49) notes, "There is little doubt that Peel saw as his ultimate objective the creation of a police system throughout the entire country." But he proceeded cautiously and deliberately.

The 1829 reform bill expressly excluded the city of London proper from the jurisdiction of the metropolitan police. The scope of police reform was limited to the metropolitan area immediately surrounding the city; it was not national. Local justices and magistrates were stripped of their police administration duties but were not centralized. The costs for the new police were shared between the local parish and central government. The exact size, duties, structure, and equipment of the new police force were not specified. Finally, Peel proposed to treat the entire developmental process as an experiment, beginning slowly in a few districts close to London, and gradually expanding based on experience. In short, Peel compromised with possible opponents, protected local interests, and provided few specifics on which opponents could base arguments. By doing so, he ensured passage of the act. "It is one of the remarkable facts about the history of police in England that, after three-quarters of a century of wrangling, suspicion, and hostility towards the whole idea of professional police, the Metropolitan Police Act, 1829, was passed without opposition and with scarcely any debate" (Critchley, 1972:50). In the language of political science, Peel had successfully co-opted his opposition.

The Police of the Metropolis

The Metropolitan Police Act established a structure for the new police in which finances were administered by a receiver, who was appointed by the Crown. The home secretary appointed two "fit persons" as justices who would jointly control the Police Office and recruit "a sufficient number of fit and able men." These men would be sworn in as constables by one of the justices and would have the powers and privileges of constables at common law.

Peel selected Charles Rowan, a retired cavalry colonel, and Richard Mayne, a young attorney, as his first justices. In time the title "justice" was changed to that of "commissioner." The commissioners found offices in a building that opened onto a narrow lane in the rear known as **Scotland Yard.** In time, this address became synonymous with police headquarters so that the English referred to the police department as "Scotland Yard." In short order, the commissioners and Peel decided on the size, function, and equipment of the force (Tobias, 1979:78–86).

They decided that the force would operate 24 hours per day, a break with the tradition of separate night watches and occasional day patrols. They also decided the force would be uniformed in order to achieve the greatest crime-prevention benefit from patrol and to allow easy identification and supervision of officers on the streets. The uniform would be blue to distinguish it from the scarlet of military uniforms. The officers would be armed only with a truncheon and would carry no lethal weapons.

The metropolitan district was divided into 17 divisions served by 165 men each, for a total force of nearly 3,000 officers (Critchley, 1972:51). Each division was commanded by a superintendent who oversaw the activities of four inspectors and sixteen sergeants. The sergeants each supervised nine officers. The organizational scheme, with its military structure and use of some military titles, reflected the organizational and command experience of Colonel Rowan. This paramilitary organization and bureaucratic structure is characteristic of most police agencies in England and the United States today.

By careful selection of candidates and establishment of a pay rate at slightly less than what was earned by skilled artisans, Peel endeavored to fill the ranks of constables with working-class men like the majority of people whom they would patrol. Special efforts were made to exclude "gentlemen" such as retired military officers and patronage appointees from the force. In addition, Peel and the commissioners demanded impeccable behavior from constables, both on and off duty, and specifically instructed their constables to be civil and respectful to the public.

Not surprisingly, when the new police took to the streets in September 1829, they were seen as a threat by many citizens and despised by most. Turnover in constables was high in the early years, largely because of dismissals for improper conduct such as being drunk on duty. Numerous complaints against the officers were lodged by citizens, and Rowan and Mayne painstakingly investigated each complaint. In August 1830 the first constable was killed in the line of duty. In 1833 the police broke up an illegal political rally during which three officers were stabbed, one dying immediately. A coroner's inquest brought a verdict of "justifiable homicide" in the case, which was later quashed by a higher court. Obviously, the public and the new police were at odds.

Over the first 10 to 20 years of their existence, the new police gradually won the respect and acceptance, if not affection and admiration, of the citizenry. Their restraint in the use of force, their persistence against public ridicule and attack, and their constant civility in dealings with the public established the new police as an institution of order and liberty in England. In these years, the "**peelers**" who were feared and hated by the citizenry became the "**bobbies,**" who were respected and appreciated.

The Evolution of the British Police

In 1835 the Municipal Corporations Act required new towns and boroughs to create police forces, and the County Police Act of 1839 allowed the creation of police forces in the 56 remaining counties (Walker and Richards, 1995). By the middle of the nineteenth century, there were over 180 separate local police forces in Great Britain. Parliamentary acts over the next 100 years created a system where "central and local government became partners in providing police service" (Walker and Richards, 1995:42).

Near the end of the nineteenth century, social reformers increasingly pushed for the police to regulate morals. The police were called upon to deal with habitual offenders, drinking, gambling, and prostitution (Petrow, 1994). The police themselves sought to avoid much of this burden, feeling that increased intervention and control over individual liberty would result in a public backlash against the police. Rather, the Metropolitan Police concentrated on regulating immorality by focusing on the fraud and crime that surrounded such practices and limiting the opportunity for good citizens to come into contact with vice offenders. Coupled with the scrutiny of the judiciary, police reluctance to become the enforcers of morality managed to establish a workable balance between the demands for civility (proper or "moral" deportment) and liberty (freedom from state intervention).

By the middle 1990s, the number of separate police forces had dwindled to 43, with the largest being the Metropolitan Police. A new law, the Police and Magistrate's Courts Act of 1994, changed the structure of the local police authorities in ways that increased the potential influence of the central government and that made it much easier to consolidate local police forces (Walker and Richards, 1996). It remains to be seen whether and how these legislative changes will alter policing in England. The British police continue to operate as a cooperative venture between the central and local governments.

1066	Norman Conquest
1215	Magna Carta signed
1252	Statutory recognition of Constable
1285	Statute of Winchester
1331	Constable recognized as Peace Officer of the Crown
1748	Henry Fielding becomes Bow Street Magistrate
1754	Henry Fielding succeeded by John Fielding
1763	John Fielding creates Mounted Patrol in London
1792	London Foot Patrol established
1797	Colquhoun publishes *Treatise on the Police of the Metropolis*
1805	Horse Patrol re-instituted on highways around London
1821	Day Patrol established in London
1822	Robert Peel named Home Secretary
1829	Metropolitan Police Act passed by Parliament

FIGURE 3.1 The evolution of British police

The Metropolitan Police of London. (Culver Pictures)

THE EARLY AMERICAN EXPERIENCE

The English colonists in America brought with them traditional offices and practices of social control. The colonies typically had sheriffs and constables, among whose tasks were order maintenance and law enforcement. In the earliest years of the colonies, the agrarian nature of colonial society supported informal policing (Johnson, 1988:105–106). As cities grew, and as the population in the colonies increased, however, Americans began to face the same problems of order as did the British.

Policing in the United States reflects the same social forces at work as those that influenced the development of the new police in London. Unlike in England, however, police development in the United States resulted from three distinct law-enforcement traditions. Though eventually resulting in similar patterns of policing, law enforcement developed along different lines in the New England and Middle colonies, the southern colonies, and the frontier.

Policing in the Northeast

As a result of climate and geography, the colonies in the Northeast (from New England to Virginia) developed into commercial and industrial centers, while the southern colonies developed a plantation system of agriculture. Most farming in the northern colonies was of a subsistence nature, and fishing, timber, shipbuilding, and shipping industries developed. For these reasons, the colonists in the Northeast more quickly developed towns and cities than did their neighbors to the South.

As with the English, these colonists experienced a breakdown in the maintenance of order with the growth of urbanism. After independence, waves of immigrants, beginning with the Irish in the 1830s, swelled the population of cities. Crime, disorder, and riots became common in the cities, and the search was on for new ways of controlling the citizenry. The developing cities in these colonies followed the pattern first seen in London.

Initially, the sheriff was the most important colonial "police" officer, appointed by the governor of the colony and responsible for a range of duties from law enforcement to road maintenance. The sheriff, however, did not have a preventive or patrol mission. In the villages and towns, another official, the **marshal,** was also available. The marshal resembled the English constable in terms of duties and powers. Both the sheriff and the marshal were daytime jobs without patrol responsibility. Paid through a schedule of fees for specific duties such as serving subpoenas, housing prisoners, and the like, these offices were generally occupied by entrepreneurs more than public servants.

As cities grew, the ability of the marshal to control crime and disorder was revealed to be severely limited. Throughout the late 1600s and the 1700s, colonial cities such as Boston, New York, and Philadelphia experimented with voluntary and paid watches. For the most part, watches operated at night and were charged with a patrol and preventive mission. The nonpaid or very low-paying nature of the job, however, led most citizens to avoid watch duty and created a situation in which watchmen were drawn from the lowest classes of society. The lack of organization made the watch incapable of controlling mobs and riots, and as the cities grew in the early 1800s, civic leaders became alarmed at the levels of crime, riot, and general disorder in the cities.

As Walker (1983) notes, the American cities had few models from which to choose in reacting to the perceived breakdown in social order. The most attractive was that of the London Metropolitan Police. Still, there was substantial opposition to the creation of police forces in the United States. As had occurred in England, critics of the new police raised issues of cost and, more importantly, individual liberty. Again, as in England, the notion of crime prevention served to allay some of the fears about an oppressive police force, and the lack of alternatives led city leaders to grudgingly accept the costs.

Unlike the Metropolitan Police Act of 1829, however, the U.S. cities seeking to design a new police force now had the London model from which

A Watchman on patrol in New York during the colonial era. (Culver Pictures)

to begin. The London model, however, was not simply adopted. In American cities, the model of the London Metropolitan Police was adapted to reflect particularly American values and concerns. Each city independently created its own police force, leading to decentralization. Further, in an effort to control the possibility of a repressive police, there was generally no strong central administration. Finally, to ensure democratic values, the U.S. police were tied directly to the political processes in the cities.

The first modern police department in the United States was established in New York City in 1845. Its organization clearly illustrates the Americanization of the London model. The New York City police department was administered by a Board of Police Commissioners, but officers were selected from the political wards in which they would patrol. Ward aldermen nominated officers, who were appointed by the elected mayor. The 800 initial police officers were not uniformed and were not supposed to be armed.

New York created a local, municipal police organization with a weak central administration. Officers were selected through political processes and were specifically required to reside in the neighborhoods in which they would work. Unarmed and not uniformed, the first American police were to be of the people and for the people. Monkonnen (1981) states that this organizational structure reflects an American conception of representation as *actual*—the police actually came from the neighborhoods they policed. In England, representation was *virtual*—through Parliament, the interests of everyone were protected. The American police were designed to reflect the competing interests of specific groups, while the British police were organized to represent the common interests of all citizens.

With some variations, similar reforms of policing occurred in other northeastern cities. Boston, for example, had created a police organization as early as 1838 that was composed of six police officers. However, the city retained the constables and night watch. In time, a separate night police was added, and the day-police complement was increased. Similarly, Philadelphia experimented with various combinations of police, constables, and watch systems. This city also developed a citywide police force with patrol obligations in the middle 1800s.

Policing in the South

Colonists in the South of the United States faced a different set of social problems than those in the North (Hindus, 1980). In large part because of the existence of slavery, "that peculiar institution," policing in the South developed in a peculiar fashion as compared with the Northeast or with England. With a few exceptions, like Charleston, South Carolina, and other trading centers, the southern colonies were essentially rural, consisting of large plantations that used slave labor to produce cash crops like rice, tobacco, and cotton. Social-control problems in the southern colonies stemmed not so much from the concentration of a laboring class in cities as the need to maintain control over a captive work force on dispersed and independent plantations. The solution was not found in a police force, but in the creation of slave patrols (discussed shortly).

Whereas the northeastern colonies developed towns and cities relatively early, and local government became focused at a community level, the southern colonies relied more heavily on the county. The primary law-enforcement official in the South was the county sheriff, a tradition that continues in an altered form today. Contemporary southern sheriffs are much more important to local law enforcement than are their northern counterparts.

To guard against slave uprisings—a constant fear of the white slaveholders—and to provide for social control in the plantation areas, property owners banded together to form the **slave patrols.** These patrols rode circuits between plantations, ever watchful for escaped slaves. They also conducted periodic checks on plantations to inquire about the safety of the owners and to ensure that the slaveholders were maintaining proper discipline and control of their slaves. In this regard, the slave patrols operated to enforce the ban on educating slaves and the requirement that slaves away from their home plantations carry passes.

This peculiar (to the northeastern colonist) form of policing reflected the social concerns and order-maintenance needs of the South. The slave patrols did not have a general crime-prevention function and were not official government policing agencies in a contemporary sense. Formal policing was provided by the sheriffs and the police of the few cities that developed.

As Williams and Murphy (1990) suggest, however, in many ways the first modern police in the United States may have been the slave patrols of the southern colonies. They note (1990:3–4) that by the middle 1700s every southern colony had enacted legislation to create and require slave patrols. These patrols were focused largely on control of blacks and not on general crime prevention within the community. That is, the patrols were specifically designed for the control of slaves, not the enforcement of laws and maintenance of order within the white community.

As Johnson (1988:184) observes, the creation of slave patrols, consisting of armed, uniformed officers empowered to regulate slaves to the point of replacing discipline from the white master in the control of slaves, was accepted in the South. The conflict between liberty (in this case, the right of the master to control his or her slaves) and order (the prevention of slave uprisings) did not materialize. Rather, "public safety superseded any fear that southern leaders might have had of a military coup."

In Charleston, South Carolina, a city guard of armed officers to control the slave majority of the population was created. In 1846 the organization's name was changed to the city police. The slave patrol in both the urban and rural South was a clear forerunner of later police organizations. Members of the patrol were given great powers over citizens and established a tradition of repression and tension between police and policed that continues to the present.

Policing on the Frontier

For approximately two centuries, the American frontier offered opportunity to any brave enough to live there. From colonial times there was a gradual but constant stream of pioneers to the frontier. These settlers brought with them the policing institutions and traditions of their former communities. Where northerners settled and formed towns, they created marshals and police forces. Where southerners settled, they brought the concept of a sheriff aided by the community in the form of a **posse.** Additionally, the frontier also encouraged the refinement and development of alternative methods of law enforcement, including vigilantism and entreprenuerial policing.

Given that settlement preceded civilization on the frontier, in many places there simply was no formal law-enforcement machinery. In those areas, committees of vigilance might form to combat particular instances of crime or disorder (Brown, 1983). These groups of **vigilantes,** as they became known, were generally well organized and episodic. They would form, for example, to combat a rash of robberies or thefts. Members of the group, headed by leading and respected citizens, would seek out suspected offenders, provide some sort of trial, and impose punishments. At the conclusion of this focused effort to control an identified threat, the vigilantes would disband.

However, some committees of vigilance would become self-sustaining and exist beyond the resolution of the immediate problem. These vigilantes became self-proclaimed police and took on a preventive and interventive role of law and order. In some cases, rather than filling a void in law enforcement, vigilante movements would develop parallel to existing police systems. The vigilante movements provided cheaper law enforcement to the community and firmly established the normative expectations of residents about what behaviors would not be tolerated in a community.

Brown (1983) argues that the vigilante tradition of the United States is indigenous. Although vigilance groups existed in other nations, at other times, the nature of American vigilantes was peculiar to the United States. Further, Brown suggests that the vigilante tradition illustrates the U.S. ambivalence toward the law. Vigilantes typically break the law by taking it into their own hands for the purpose of enforcing their own law. Further, vigilantism indicates a willingness to take extralegal action if the official machinery of justice is perceived as ineffective or inappropriate. Finally, Brown observes (1983:71), "Perhaps the most important result of vigilantism has not been its social-stabilizing effect but the subtle way in which it persistently undermined our respect for law by its repeated insistence that there are times when we may choose to obey the law or not."

THE DEVELOPMENT OF MODERN POLICING IN AMERICA

These three traditions of policing in the United States—from the Northeast, the South, and the American frontier—continue to influence the organization of contemporary policing and U.S. attitudes about policing to this day. Johnson (1988:188–189) notes that policing developed from different traditions in the North, South, and West of the United States. In each case, the type of police reflected "the society that gave it birth." Whatever the form—voluntary, bureaucratic, vigilante—every society will have police. The police are a product of the society and heavily dependent on it.

Despite the diversity of origins, in the period between the Civil War and the turn of the century the similarity among police organizations in American cities was striking. Given a local orientation, and the existence of three distinct traditions, the fact that the police organizations in city after city were so similar requires explanation. It was to this issue that Eric Monkkonen (1981) turned his attention in *Police in Urban America 1860–1920*.

Monkonnen's basic thesis is that the spread and development of policing in the United States after the creation of modern police forces in the major cities of the East followed a diffusion-of-innovation model. That is, to a greater or lesser extent, the initial development of policing occurred in large cities over a relatively long period of time. As these new police agencies "worked out the bugs," smaller cities copied the innovation and applied it to their own circumstances. Opposition to the establishment of a police agency, and the time required for implementation, decreased with each succeeding wave of adoption.

This model is similar to an explanation of the spread of fashions: A designer may break with the trend by introducing a new look (longer or shorter skirts, wider or narrower ties), and at first only a few courageous souls are willing to make a bold fashion statement. In time, though, the idea catches on, and eventually this radical break with the existing fashion trend becomes the new trend.

Of course, not every new fashion will catch on or prove to be enduring. The difference with policing was that the problems of order and crime that first plagued large cities, spurring the creation of police, eventually troubled smaller cities. Seeking solutions, cities that developed police systems later had the advantage of learning from the experience of the earlier reformers; they had working models to adopt.

The first modern U.S. police departments borrowed heavily from the London Metropolitan Police. Later generations of American police, however, were more likely to follow American developments rather than British. "Up until the modeling of the early U.S. uniformed police on English precedents, one must maintain an Anglo-American perspective on policing, but after the establishment of the first few American departments, the paths of the police of the two countries diverged, those in American cities looking at each other rather than to London" (Monkonnen, 1981:40).

In the latter part of the nineteenth century, the development of policing in the United States became an increasingly American phenomenon. The influence of English practices and traditions continued, but their effects were less direct as a new American police profession and tradition developed.

Correlates of Police Evolution

The evolutionary model of police development outlined in Chapter 2 suggests that the form and structure of police agencies will reflect differences in social organization, elite interests, and perceptions of crime. The British and American history of police indicates that the creation of a formal police role and organization is associated with general role specialization in society and the breakdown of informal social controls. In both London and New York, for example, the creation of a police department represented a reluctant response to the failure of traditional social control to contain crime and disorder within acceptable limits. Police arose in cities where personal relationships among people were weak and where social, economic, and political conflict among groups was common.

The need for social control was not in itself, however, sufficient to lead to formal policing. The structure and organization of police forces also had to reflect community values. The emphasis on individual liberty and the fear of a repressive police correlates with weak command structures (two or more commissioners) and a narrower definition of the police task. In both England and the United States, the police are separate from the judiciary. The absence of a consolidated elite power correlates with a local, citizen-based policing under the control of elected public officials.

Despite their similarities, important differences developed between the police of England and those in the United States. The British police were essentially national and accountable to Parliament, reflecting a conception of virtual representation. The American police were accountable to local political officials and tied directly to neighborhoods, reflecting a conception of actual representation. As Monkonnen (1981) notes, within a few years after the first American police departments were started, the continued development of policing in the United States became increasingly independent of the English influence. In the next chapter, we turn our attention to the evolution of policing in the United States.

CHAPTER CHECKUP

1. How was "policing" accomplished in pre-Norman England?
2. What is the Statute of Winchester?
3. Who were the Fielding brothers, and what was their contribution to policing?
4. What was Patrick Colquhoun's contribution to policing?
5. How was Robert Peel able to obtain passage of the Metropolitan Police Act of 1829, and what were the provisions of the act?
6. When and where was the first modern American police force created?
7. What three traditions influenced American police development?
8. What were the duties of slave patrols?
9. What is *episodic policing*?
10. How did the idea of policing spread in the United States?

REFERENCES

Beccaria, C. (1763) *On crimes and punishment,* translated by H. Paolucci (1977). (New York: Bobbs-Merrill).

Brown, R. M. (1983) "Vigilante policing," in K. Klockars (ed.) *Thinking about police.* (New York: McGraw-Hill):57–71.

Critchley, T. A. (1972) *A history of police in England and Wales,* 2nd ed. (Montclair, NJ: Patterson Smith).

Hindus, M. S. (1980) *Prison and plantation: Crime, justice and authority in Massachusetts and South Carolina.* (Chapel Hill, NC: University of North Carolina Press).

Johnson, H. A. (1988) *History of criminal justice.* (Cincinnati, OH: Anderson).

Kent, J. R. (1986) *The English village constable 1580–1642: A social and administrative history.* (Oxford: Clarendon Press).

Monkonnen, E. H. (1981) *Police in urban America 1860–1920.* (Cambridge, England: Cambridge University Press).

Petrow, S. (1994) *Policing morals: The Metropolitan Police and the Home Office, 1870–1914.* (New York: Oxford University Press).

Rawlings, P. (1999) *Crime and power: A history of criminal justice 1588–1998.* (London: Longman).

Reith, C. (1938) *The police idea: Its history and evolution in England in the eighteenth century and beyond.* (London: Oliver).

Stewart-Brown, R. (1936) *The serjeants of the peace in medieval England and Wales.* (Manchester, England: Manchester University Press).

Tobias, J. J. (1975) "Police and public in the United Kingdom," in G. L. Mosse (ed.) *Police forces in history.* (Beverly Hills, CA: Sage):95–113.

Tobias, J. J. (1979) *Crime and police in England 1700–1900.* (New York: St. Martin's Press).

Walker, D. and M. Richards (1995) "British policing," in W. Bailey (ed.), *The encyclopedia of police science,* 2nd ed. (New York: Garland):41–48.

Walker, D. and M. Richards (1996) "A service under change: Current issues in policing in England and Wales," *Police Studies* 19(1):53–74.

Walker, S. (1983) *The police in America: An introduction.* (New York: McGraw-Hill).

Williams, H. and P. V. Murphy (1990) *The evolving strategy of police: A minority view.* (Washington, DC: U.S. Department of Justice).

chapter 4

THE EVOLUTION OF POLICING IN AMERICA

CHAPTER OUTLINE

With the creation of the New York City Police Department in 1845, the path was cleared for the spread of formal, preventive policing throughout America. During the last half of the nineteenth century, every major city in the nation created a uniformed police. Eric Monkonnen (1981) suggests that the spread of policing from city to city was such that the idea of uniformed, formal police organizations must have been contagious.

The first cities to develop police departments, such as New York, Boston, and Philadelphia, relied on the model of the Metropolitan Police in London. Smaller cities then more or less quickly imitated the larger ones,

*based on a general growth of urban services and the exchange of informa-
tion between cities. Monkonnen concludes that the relatively rapid spread
of policing throughout America was the product of a general movement toward increased government services. He writes (1981:55), "The growth of
uniformed urban police forces should be seen simply as a part of the growth
of urban service bureaucracies." He supports this contention by noting
(pp. 56–57) that while the police in some cities were created in response to
riots or fear of crime, police agencies spread to even more cities where there
were no riots or increases in crime.*

*The police were one of many "rationalized" services that evolved in ur-
ban government during the latter half of the nineteenth century. In other
places, such as France, Italy, and Japan, these services were grafted onto existing national police offices. In the United States, with its lack of national
services and emphasis on local autonomy, the police developed along with
city fire departments, sanitation departments, water departments, and similar services. The history of the American municipal police is, then, not about
the origin and spread of the idea of police to prevent crime but rather about
the evolution of the functions and control of institutionalized public-service
bureaucracies.*

*George Kelling and Mark Moore (1988) break the history of American
policing into the political era, reform era, and community problem-solving
era. They suggest that these periods are distinct, based on the dominance of
a particular strategy of policing, yet they recognize that the boundaries between the eras are not clear. They suggest that the political era stretched from
the 1840s through the early 1900s. This era was followed by the reform era,
which lasted until the late 1970s. The latest era, that of community problem
solving, began at the end of the reform era and continues today.*

THE POLITICAL ERA

At their start, American police agencies were political entities. As we saw
with the New York police, officers were selected from the wards or neighborhoods they would serve and were appointed by elected political officials.
The lack of strong central administration, the influence of political actors,
and the neighborhood ties between the officers and the people they policed
ensured a partisan process of policing. Kelling and Moore (1988:3) observe
this early structure of policing meant that the most important policy decisions were made at the precinct and street levels.

Decentralization, including both the absence of a strong chief administrator and the use of precincts or neighborhood-based police stations,
served to fragment police services within cities. Thus, local political leaders
could exert pressure on precinct commanders, and citizens could approach
patrol officers directly for police services. As a result, both police administrators and police officers were sensitive to public concerns. As Samuel

Walker (1977:8) notes, "American police departments reflected the general style of local government."

This style had a distinctly neighborhood-based and service-provision flavor. As Conley (1995:560) puts it, "Political disorders were local in nature, and the threat was to local institutions." Thus, local government, of which the police were a part, was expected to meet the needs of citizens in the neighborhoods in which they lived. Recruiting officers from the wards in which they would work, and locating police administrative units (precincts, districts, and stations) in those neighborhoods, meant that the police would reflect the interests of neighborhood residents more than citywide needs that might cross neighborhood lines. As Lane (1975:119) points out, the police maintained order by catering to the needs of neighborhoods, often in the form of providing social services to the poor.

Conley (1995) suggests that these social services helped to ameliorate the problems of urban residents and thereby mediated conflicts. During the political era, the provision of services to citizens was more important than the control of crime. A number of researchers have noted that the police in the last half of the nineteenth century were closely tied to politicians (Richardson, 1970; Lane, 1975; Monkonnen, 1981). The requirement of constituent service, giving voters what they want, led politicians to press the police for social services. Kelling and Moore (1988:3) write, "Partly because of their close connection to politicians, police during the political era provided a wide array of services to citizens." One effect of service provision, of course, was the maintenance of social control by meeting the needs of citizens in an orderly and predictable manner through police intervention.

Nineteenth-century police operated soup kitchens and shelters for the homeless, assisted citizens in finding employment and securing medical care, and generally helped the poor or unfortunate. The recruitment of neighborhood residents as police officers and the local administration of police precincts ensured that the police would reflect community values in their law-enforcement and order-maintenance decisions. For example, despite the existence of laws against drinking, if residents tolerated the behavior, the patrol officer was unlikely to enforce the law (Fogelson, 1977). Thus, the police not only served to mediate ethnic and class conflicts over material resources, but also to enforce the law in a way that reflected community standards of morality.

Conley (1995) states that the local and political nature of American policing is critical to our understanding of how the police operate and what are the limits of reform. He writes (1995:560), "Police authority emanates from the political majority of the citizens, not from abstract notions of law. Authority rested on a local and partisan base, within limited legal and symbolic standards, and was legitimized by informal expectations."

In their early development, the police in America were a general-purpose service agency that reflected the desires and interests of the citizens. The ability of the police to secure citizen compliance rested on their ability to provide services to the people. From the outset, the police in the

United States depended upon public approval of their work, not public acceptance of the law, for their authority. In the decentralized and local organization of urban police, this meant that the police in each neighborhood attended to the concerns of the dominant political faction in that neighborhood. In poorer areas, the police provided services; in commercial areas, the police provided security; in middle-class neighborhoods, the police maintained order. As a whole, the police department attempted to balance the demands of the various factions throughout the city (Conley, 1995:562).

Factionalization of the police was both their strength and weakness in this era. Sensitivity to the values and interests of different groups within the city gained support for the police and allowed them to serve the pressing needs of citizens. On the other hand, responding to problems as defined by others denied the possibility of self-direction to the police. Conflicts among the various factions within a city also meant that the police were consistently embroiled in political controversy. In time, the weaknesses of factionalization came to be seen as outweighing its strengths. The movement to reform the police that began in the late 1800s was, in the main, an attempt to defactionalize city policing in America (Monkonnen, 1981:59).

The American Police Circa 1900

Toward the start of the twentieth century, the police in American cities were well established but faced growing criticism. Craig Uchida (1989:19–20) describes the police officer of this period as a political operative. The officers were selected on the basis of their political connections and service. Officers were responsible to political leaders in the wards and precincts rather than to the city government. Further, unlike their British counterparts, American police officers were given greater individual discretion and had to establish their own authority with the citizens.

The London "bobby" represented government authority and the power of Parliament. The American "copper" depended on personal authority. Too often, this personal authority was achieved through violence as the police officer established his authority based on his ability (and willingness) to beat anyone who challenged him (Miller, 1977:84–86). The relative independence and isolation of the American police officer led to a tradition of brutality that haunts the police to this day (Johnson, 1981).

In addition to physical force, officers also had wide personal authority over the use of legal force through making arrests. Miller (1977) observes that the U.S. police made far greater use of arrest on suspicion and arrest for disorderly conduct than did their London counterparts. He argues that this difference in use of arrests in situations defined by the officer (as opposed, for example, to burglary, which is defined by the law) indicates the greater discretionary personal power of the American police.

A counterpoint to the ability to use arrest to achieve personal ends, as in arresting a disrespectful citizen on a charge of disorderly conduct, is the ability not to arrest when the law would seem to require it. Thus, the neighborhood patrol officer was able to select the laws he would enforce so that they supported community expectations. In fact, since the patrol officer relied on personal support from citizens, he was not likely to enforce laws that were unpopular within his patrol area (Uchida, 1989:20).

The freedom of the individual officer from responsibility to either legal constraints or the direction of a central administration created a context ripe for corruption. The factionalization of the police, along with their central role in the partisan politics of American cities, virtually required political corruption in the police. Police officer positions were distributed as part of the patronage system of machine politics in American cities. Officers

The link between local politics and policing was often criticized in American newspapers. (The Bettmann Archive)

owed their allegiance and their jobs to the political party that appointed them, so they often promoted the interests of political leaders. For example, police protected the illegal operations of political leaders and those who supported the dominant politicians. At election time, officers also prevented supporters from opposing parties from casting their ballots by intimidation, arrest, and sometimes even physical force. The police of the political era were deeply enmeshed in corruption from political interference (Fogelson, 1977; Lane, 1975).

Corruption of a more mundane sort was also rampant. The police of the latter nineteenth and early twentieth centuries frequently accepted bribes and graft on their own. So not only did police do the bidding of political leaders, they also took advantage of the numerous opportunities available to supplement their city salaries. Mark Haller (1976) observes that the Chicago police, and very likely those in other cities, were part of a large-scale and well-organized system of rackets in the city. Corruption was rampant not only within police departments but throughout all of city government. In many departments, promotions and even beat assignments were auctioned to the highest bidder within the department. Similarly, illegal operations such as bordellos, gambling houses, and saloons contributed to monthly "pads" that were divided among all the officers in the precinct, including the precinct commanders.

The brutality, political manipulation, and corruption of city police in America during this era gave rise to calls for reform. Recognizing the deficiencies of the police as they then existed, a number of reformers from both outside and within police agencies sought to improve the police. During the next era of American police history, several forces worked to professionalize the police.

THE REFORM ERA

By virtue of their ties to and direct role in city politics, the U.S. police have always been at the center of political controversy (Johnson, 1981). The reform movement of the early twentieth century was itself another act in the politics of policing. The limitations and defects of the police, including corruption, brutality, and political manipulation, were attributed to the influence of partisan politics. The ultimate goal of police reformers was the removal of politics from the police. As Samuel Walker (1977: 31) notes, however,

> Despite its claims to nonpartisanship, then, the idea of professionalization was itself a movement that served partisan ends. The attempt to remove the influence of politics was essentially an effort to supplant one political element with another. To achieve this end, however, the reformers undertook both a wholesale restructuring of police departments and a redefinition of the police role.

The Politics of Police Reform

The primary proponents of police reform were found among the elites of urban society and within the ranks of a new breed of police administrators. Often in conflict with each other, these two groups shared a similar vision of the police for the future. The elites of America's cities wanted to break the power of the political machines and, especially, to control the morals and behavior of the immigrant poor in the cities. The police administrators sought independence from city hall so that they could develop policing into an autonomous, efficient, and respected field of practice. What these groups shared was the abstract notion of an independent police.

Independence for the police was expected to mean freedom from partisan political interference and the corruption such influence bred. The elites wanted a police agency that enforced all of the laws evenly at all times. They were opposed to the practice of allowing illegal operations run by political favorites to be immune from regulation. They were particularly interested in using the police to curb what they saw as dangerous behavior among the working classes, including gambling, drinking, and other vice offenses. For police administrators, independence meant the freedom to control police officers and priorities without regard to political affiliation.

The redefinition of the police role from one of public service and order maintenance to that of law enforcement and crime control helped clarify the police function. Rather than meeting a broad, social-service need, the police would concentrate on crime-related matters. As a result, police involvement in providing temporary shelter to the homeless, locating missing children, and other service activities declined later in the nineteenth century (Monkonnen, 1981).

Pressure for reform coming from outside the field of policing was not sufficient (Kelling and Moore, 1988:4). It required the joint efforts of urban elites external to police agencies and the police administrators themselves to create police reform. Initially, reform efforts by middle-class citizens were simply attempts to take control of the police from the lower classes and thereby weaken political machines. While similar, the goals of elite reformers and those of police administrators differed in one important respect: Police administrators desired autonomy to shape and control police practice, whereas elites wanted police who would serve their interests rather than those of their political opponents.

Reform Goals of Urban Elites.   The late nineteenth and early twentieth centuries were periods of dramatic change for the United States. In this era, the nation became increasingly urbanized and industrialized. One effect of these social changes was the development of an urban middle class composed of those who managed the factories, stores, and transportation systems. These middle-class Americans grew dissatisfied with urban government in general and with policing in particular.

The nature of urban politics, being decentralized constituent service, meant that these business leaders of American cities had little control over city government. Political machines based on the bloc voting of ethnic neighborhoods did not need to concern themselves with the interests of the middle class. As taxpayers, property owners, and managers, however, the growing and increasingly socially conscious middle class began to work for government reform. This reform effort would culminate near the turn of the century in what has been called the Progressive Era.

David Johnson maintains that the Progressive reform agenda for the American police was simply an extension of the general effort to improve cities and city government. He writes (1981:64),

> Decentralized power within the cities encouraged duplication of effort, waste, and, of course, outright corruption. In these circumstances housing shortages, crime, disease, and population congestion appeared to be intensifying rather than diminishing. Law enforcement, which was

Civil service selection and specialized training and education for police characterized the reform era. (The Bettmann Archive)

only one aspect of a far broader set of problems would inevitably be affected by the wider debate over urban problems.

The outcome of that debate was to identify the political boss system as the major cause of urban troubles. Naturally, then, the solution would be to eliminate the political machines.

Essentially, the core of the issue became a conflict between universalistic and particularistic styles of government and policing. The political machines were based on a **particularistic style** where decisions were made case by case. Thus, appointment as a police officer, the decision to enforce a law, and the like, were based on whom one knew or who one was. The Progressives sought **universalistic,** rational decisions based on legal standards or other stated criteria. This meant police officers should be appointed on the basis of their qualifications not their connections. It also meant laws should be enforced as they were written, not as they were interpreted by the officer.

In contrast to the fragmented, particularistic style of policing common during the political era, reformers proposed centralized, universalistic procedures. Original fears of concentrating too much power in the hands of a police administrator were overcome by the experience with bosses. If decentralized administration led to corruption and inefficiency, creating a strong central police administrator could do no worse. The task of police reform within the Progressive movement was to centralize police administration, improve the quality of police personnel, and destroy the power of the political bosses.

The Progressives believed it was the obligation of government to improve the living conditions of the people and to promote moral standards. A major stumbling block, in their eyes, to fair and orderly life in the city was the political corruption of the police. They were particularly concerned with the failure of police departments to enforce controls on drinking. As Walker (1981:9) states, "The failure of the police to enforce the laws was the mainspring of police 'reform' in the nineteenth century." The influence of the largely immigrant working class over police-enforcement practices was evidence of their political power.

The goal of urban elites was to seize control of the police from the political machines and the working class. In their earliest efforts, they were unsuccessful in this endeavor largely because of the effectiveness of the machines. To put it simply, within the confines of the cities, the middle classes simply did not have enough political power. Thus, for several decades reformers in a number of cities attempted to control the local police through state government. Indeed, Johnson (1981) argues that the Massachusetts State Police were created by Boston elites for the purpose of forcing the Boston Police Department to enforce liquor laws. In general, these efforts at control had little effect on the practice and quality of policing in the cities.

Reflecting their broader interest in social reform and in molding the immigrant working class into their image of moral Americans, the Progressive reformers wanted to use the police to control immigrant behavior. As a result, "officers were often required to enforce unpopular laws foisted on immigrant ethnic neighborhoods by crusading reformers" (Kelling and Moore, 1988:4). This effort to reshape the police into an instrument of class control was insensitive to the ethnic and neighborhood composition of police agencies. It also failed to recognize the structural inability of police administrators to control the decisions and actions of patrol officers on the streets. The mere changing of commanders or governing bodies for police agencies was not enough to change the police. Early American police were part and parcel of the communities they served. The link between the police and the community had to be broken first, so that police behavior could be made more independent of community values, not just of political influence.

Reform Goals of Police Administrators.   Police administrators during this era also began to seek reforms in policing. They sought autonomy and status, and they viewed a general improvement in the practice of policing as the means to achieve both increased independence and increased respect. In the earliest statements of the need for police reform, police chiefs sought more efficiency, but that efficiency depended on protecting the police agency from partisan political influence and improving police personnel qualifications. Attaining autonomy and better personnel would enable the police to better enforce laws and maintain order.

The typical chief of an American police department in the latter part of the nineteenth century was at best a figurehead. The officers owed their primary allegiance to the local political leaders, not to the police agency or the profession. Police command, in practice, was decentralized to the precinct level, if not the ward. In any conflict between the desires of the chief and the political bosses, the chief was going to lose.

As James Richardson (1995:557) observes, "At top levels, such as among board members and commissioners, political winds could blow harshly . . . [state] legislatures stepped in and replaced individuals holding senior administrative positions." Police administrators were often caught in the middle of conflicts between what different groups felt was the role of the police or were involved in political battles between city and state governments.

These conflicts over the ends of policing might have been tolerable had they been infrequent and minimal. The problem, from the point of view of the police chief, was that conflict was the core of the police role. In any city there were those who wanted the police to enforce the laws and those who wanted the police to provide social services. There were also those who wanted the police to enforce only certain laws or to enforce laws in only certain neighborhoods. As the titular head of the department, the chief was a lightning rod for conflicts.

A common organization of the police was to have the department headed by a chief, who in turn was accountable to some elected official. At some times, in some cities, boards or commissions of elected officials oversaw the police. In other instances, the mayor or city council had oversight authority. Thus, the chief was a go-between for the policy-making authority (mayor, commission, council, etc.) and the precinct captains. The chief had little power over the police officers and precinct captains, yet was responsible for their actions. When membership on boards and commissions or when mayors changed, chiefs were often expendable (Maniha, 1970). Although chiefs had little actual authority over the actions of their subordinates, they were held accountable by their superiors (Fogelson, 1977:65).

The Lexow Committee investigation of the New York City police in 1894 is a classic example of the plight of the police administrator. The Democratic machine of Tammany Hall controlled New York City, and the sensitivity of city government to local concerns allowed drinking, gambling, and other offenses to flourish despite state laws against them. A local minister, Charles Parkhurst, began a crusade against vice from his pulpit. Parkhurst's efforts caught the attention of the Republican party, which sought to embarrass the Democrats. When the Republicans gained control of state government in 1893, they set the Lexow investigation in motion. As a result of the widespread corruption uncovered in that investigation, the state legislature in 1895 passed a law creating a bipartisan commission to administer the New York City police. Theodore Roosevelt was a member of that commission. He resigned his position two years later (Walker, 1977:46). By that time, Roosevelt was convinced that a bipartisan, multimember commission was not capable of bringing about change in the police.

As Walker (1977:26) observes, "The struggle among different political factions resulted in a continuing series of changes in the administrative structure of police departments in the nineteenth century. The sequence of events in New York City found its parallel in virtually every other city." The net result of these political maneuvers was that police administrations, regardless of structure, remained under the control of some partisan group. Further, actual control of the police still rested at the neighborhood level so that changes in administrative structure had little impact on police practices. "Reform-minded administrators," as Theodore Roosevelt soon discovered, "could not translate their ideas into policy." They lacked both an ideological rationale that could counter the idea of the police as a partisan instrument and a means of asserting uniform control over the rank and file (Walker, 1977:28). The notion of police professionalism could overcome the deficiencies of reformers such as Roosevelt.

The Rise of Police Professionalism

The goals of police administrators and urban elites were compatible in regard to removing the influence of partisan politics from policing. The idea of a police profession, growing from the emerging science of policing and rapidly developing technological changes in society, provided a vehicle for police reform. Reformers felt that policing was far too important to be left to the whims of politicians and amateurs. Professional policing was the answer to reformers' prayers.

Among other things, the concept of a *profession* includes the expectation that its practice requires special training, skills, or abilities, that its members are committed to the vocation, and that they are entitled to some degree of autonomy in their activities. These are precisely the expectations of police officers that would serve to insulate them from political influence. They were expected to be qualified for the job of policing and committed to that occupation, not to a particular political party. Equally as important, they would themselves decide on enforcement strategies and priorities. Police professionalization became possible in the early twentieth century as a result of a growing "science of policing," the spread of civil service, and the adoption of bureaucratic structures.

One of the primary tasks in developing police professionalism was to define the police role. Here a number of forces converged to make law enforcement and crime control the principal functions of the police. As Monkonnen (1981) notes, by the latter part of the eighteenth century, police agencies were divesting themselves of many social service duties as alternatives such as welfare agencies emerged. Urban elites, of course, had been concerned primarily with the lack of enforcement of the law by the police, and thus they supported the definition of the police role as law enforcement. Police administrators, too, accepted crime control as their function, in part because it was the one thing for which the police alone had responsibility. As a result, the legitimacy and purpose of the police during the reform era became that of crime control. Kelling and Moore (1988:6) succinctly describe this change: "Police agencies became *law enforcement agencies*. Their goal was to control crime."

The Science of Policing.   One hallmark of a profession is the development of a specialized body of knowledge. For the police, this body of knowledge was relatively slow to develop, beginning with the publication of a number of memoirs and departmental histories in the 1880s (Walker, 1977). Police administrators formed a professional association and exchanged information beginning in 1893 at the first meeting of the National Chiefs of Police Union. This organization, which became the International Association of Chiefs of Police (IACP), conducted annual meetings at which police administrators shared observations, information, and experience.

In 1871 Chief James McDonough of the St. Louis, Missouri, police department convened a meeting of police chiefs. Although those in attendance had agreed to meet the following year in Washington, D.C., that meeting was never held. In 1893 William Seavey, chief of police in Omaha, Nebraska, and Robert McLoughrey, superintendent of the Chicago police, convened a meeting of police chiefs in Chicago, where the participants organized the Chiefs of Police Union (Dilworth, 1976).

The Chief's Union met annually, and attendees listened to discussions of police practices and problems from around the country. These meetings, and resolutions passed by this body, helped form a professional identity among police administrators. The presentations were recorded in volumes of proceedings that added to the growing police literature.

Initiated and led by police administrators, police science was primarily concerned with administrative and organizational matters. During this same time period the science of **criminalistics** was developing in Europe. The investigation and prosecution of crime were universal police problems. By virtue of their national organization, police systems on the European continent were administered by highly educated executives who were familiar with and accepting of scientific procedures. Alphonse Bertillon became the records clerk of the Paris Prefecture of Police, in which capacity he was responsible for recording descriptions of all arrested offenders. In short order Bertillon devised a system of descriptions composed of four components: meticulous measurements of the offender's body (foot length, cranial size, arm length, etc.); specific notation of such features as scars and tattoos; photographs; and fingerprints. This **"Bertillon system"** of identification allowed police to determine the identity of criminals regardless of the names they gave at arrest. In the late nineteenth century, the Bertillon system was the standard form of identification records used by police throughout the world (Bailey, 1995).

Another European, Hans Gross, improved investigation practices. An Austrian prosecutor, Gross studied and wrote about evidence collection and handling. He urged that police study criminal psychology and employ any technical and scientific procedures that would assist in crime-scene analysis. Gross wrote an important text on criminal investigation that supported the use of scientific experts in criminal investigations and stressed the importance of physical evidence. Later, as a professor at a number of universities throughout Europe, Gross trained a generation of criminalistics experts.

A third important innovation was the development of **fingerprint** technology for criminal identification (as adopted by Bertillon) and for investigation. Although several amateur scientists independently studied and perfected fingerprinting, Sir Francis Galton is generally regarded as the founder of criminal fingerprinting. Galton proved that fingerprints did not change over a person's lifetime and that the prints were unique to each individual. These facts meant that people could be positively identified by fingerprints and that fingerprints could be used to identify persons who had been present at a crime scene. Therefore, if the investigating officers collected fingerprints from a crime scene, they could use them to match against prints on record from earlier crimes to seek out the offender.

The respect and appreciation shown by earlier police leaders for technological advances seems quaint by modern standards. It is difficult to understand how telegraph service, patrol wagons, cameras, or automobiles revolutionized policing, but they did. Speaking in 1903, Francis O'Neill, chief of police in Chicago, recounted the scientific and technological strides taken by American police. He concluded his speech on "What Science Has Done for the Police," with these words: "The forward stride from the lanterned night watch, with staff, announcing the passing hours, to the uniformed and disciplined police officer of the present, equipped with the telegraph, the telephone, the police signal service, and the Bertillon system of identification, is indeed an interesting one to contemplate (in Dilworth, 1976:11)."

At the start of the twentieth century, policing was developing into a distinct, specialized field of practice. The creation of an association of police administrators helped develop an occupational identity among the police and provide national leadership, at least in the form of notifying police leaders about the latest developments. In regard to criminal investigation, at least, police practice was at the cutting edge of science and technology. Police-reform advocates could now claim that effective policing required specialized knowledge and experience.

Civil Service. A second step on the path to police professionalization was accomplished with the institution of civil service procedures for police selection and promotion. In theory, at least, civil service testing of police applicants, and the appointment of officers based on their civil service rankings, meant that the police would now be composed of officers selected for what they knew, rather than for whom they knew. This important distinction meant that police officers would be qualified for their positions and would be committed to the occupation rather than to a particular political power.

By the end of World War I, civil service appointments of police officers were the norm, particularly in the largest cities (Johnson, 1981:68). Civil service had long been a goal of the Progressives, who wanted a disinterested and efficient government administration (Richardson 1974:62). Removal of partisan political interference from appointment and discipline of officers, they believed, would create this professional police service. Though not without its problems and loopholes, civil service procedures for police weakened the role of politics while strengthening that of the police chief in the control of patrol officers. It also created an occupational identity among police officers for the first time. As Fogelson (1977:116) observes, it was after the application of civil service procedures that police officers came to think of themselves as a distinct group: "Not until the early years of the century did the big-city police perceive themselves as policemen, as opposed to laborers, clerks, and railway conductors temporarily employed as policemen."

Civil service testing was only a beginning step in the effort to upgrade police personnel (Johnson, 1981:68–69). In addition to using tests to select only qualified applicants, many reformers and police administrators called for improved training and education for police officers. Some, such as August Vollmer (discussed later), wanted to add requirements for appointment that included psychological testing, residency requirements, and background investigations. The police were to be not only minimally qualified for their positions, but an elite organization composed of the "best and the brightest."

By 1920 the American police were poised on the brink of professionalization. The efforts of Progressive reformers coupled with those of police administrators had laid the groundwork for a police profession. A technology, if not a science, of policing existed that fit the revised definition of the police role as that of crime control. Police organizations were gaining independence from outside interference, which also meant that entry into the occupation was not restricted. Finally, police administrators had developed a professional identity and had been granted the authority necessary to direct and control their bureaucracies.

Police historian David Johnson (1981:105) calls the period between 1920 and 1965 the "Triumph of Reform" because during this era police professionalism reached its height. Achieving professionalism was a goal of the early reformers, but it was achieved by a later generation of police leaders who were able to continue the reform movement and capitalize on changes in American society.

The Professional Police.   The ideal of police professionalism probably owed as much to the institution of bureaucratic police organization as it did to the efforts of various reformers. The bureaucratic organization of the police stemmed from the centralization of administrative power and the natural development of police agencies. Monkonnen (1981:58–64) suggests that the creation of uniformed police led inexorably to the development of bureaucracy. Further (Monkonnen, 1981:59), "the defactionalization of the police was simply an aspect of the growth of police professionalism, which was itself a built-in consequence of the move to rationalize and uniform the police." One product of the attempt to centralize administrative power, the creation of specialized bureaus within police departments, similarly supported both bureaucratic and professional development.

Monkonnen's argument is that the creation of uniformed police represented the creation of a specialized institution for social control in the cities. The creation of a social institution sets the stage for its evolution as an organization. That evolution includes the narrowing of focus, the development of roles and norms, and the development of a formal organizational structure. Given that the police were created for a purpose—social or crime control—it follows that they would develop a functional structure. Bureaucracy is nothing if not a functional structure, and thus police bureaucracies are a natural, if unintended, outgrowth of the creation of formal police organizations.

A second support for the evolution of police bureaucracies was the product of intentional changes in police organizational structures. The movement to create strong police executives sought to decrease the powers of precinct captains (and, by extension, ward leaders). One means of strengthening the position of the police chief was to place responsibility for certain activities under his direct supervision. This was particularly true for those tasks that were the most troublesome to reformers and the most important to political bosses, such as vice control.

Consequently, during the early twentieth century, many police departments started special **vice squads.** In most cities, these vice squads were soon complemented by other special squads for alcohol, gambling, detectives, personnel, and a host of other problems. As Fogelson (1977:78–79) describes the proliferation of such squads, they offered something for nearly everyone and drew little opposition. Citizens felt the squads allowed the police to use special skills and knowledge and that they removed temptations to corruption from the patrol officer. The officers liked the squads because they saw plainclothes work as an improvement over uniformed patrol. Police administrators approved of squads because they allowed greater control over police activity and enabled the chief to "do something" in the face of public pressure about certain issues. For example, if a clamor arose over traffic safety, the chief could create a "traffic squad."

The emergence of squads both reflected and supported functional specialization in police agencies. Rather than retaining the neighborhood focus that had led to the concentration of authority at the precinct level, functional specialization supported a structure sensitive to problems instead of geography. The bureaucratic police agency had a citywide burglary squad and did not focus specifically on the problem of prowlers in a specific neigh-

Policewomen in 1924 train for self-defense. (UPI/Bettmann)

borhood. Further, the special squads "were an important part of the image which the police were trying to cultivate as professionals who understood how to fight crime" (Johnson, 1981:71).

By the 1920s policing in the United States was on the brink of professionalism. Civil service appointments and the National Police Chiefs Union had created an occupational identity and restricted entry into the vocation. The application of science to the problem of crime detection and control, coupled with the redefinition of the police as primarily responsible for law enforcement, identified the professional obligation, or "occupational mandate" (Manning and Van Maanen, 1978), of the police. Stronger central administration and bureaucratic organization of police agencies insulated the police from political interference and rendered police activity more rational and universalistic. What remained was for the police to consider themselves professionals, and to achieve professional recognition from the citizenry.

A second generation of police leaders led the movement toward professional recognition. One of the most influential of these police administrators was August Vollmer. In 1905 Vollmer was elected town marshal of Berkeley, California. He served as marshal for four years, and when the city of Berkeley created a police department in 1909, Vollmer was named its first chief. He retired from the chief's office in 1932, after 27 years leading the Berkeley police.

Vollmer was an innovator. He believed that the first obligation of the police was crime control and public protection. To that end, he supported the use of all applicable technology and the provision of scientific training to police officers. In 1908 he began a police school for deputy marshals in Berkeley by enlisting professors from the University of California at Berkeley to instruct classes.

In 1929, with several state and local commissions investigating policing and criminal justice administration, President Herbert Hoover reacted to a national concern about crime and disorder, creating the **National Commission on Law Observance and Enforcement.** He appointed the Attorney General, George Wickersham, to head the commission, which came to be known as the **Wickersham Commission.** The commission conducted a broad study of crime and criminal justice. August Vollmer was given responsibility for reporting findings about policing when the commission concluded its work in 1931. Not surprisingly, the Wickersham Commission proposed sweeping changes in American policing that included improvements in police selection, training, organization, and management. The recommendations of the Wickersham commission, although not adopted everywhere, provided a "blueprint" for creating a professional police.

Alfred Parker (1972:39) writes that by the time he retired, Vollmer was heralded as the "father of modern police administration" because of his innovations. He was the first police administrator to employ or develop "a signal system; a workable *modus operandi;* the use of bicycles, motorcycles, and automobiles equipped with radios; the lie detector, a scientific crime investigation laboratory, selection of college men for his police force; a police

August Vollmer (1876–1955)

Orphaned in childhood, Vollmer was born in New Orleans. After service in the Spanish-American War, he was elected marshal of Berkeley, California, in 1905. Vollmer devoted his career to improving the police in America and was active as a consultant, administrator, and college professor for 50 years. He was primarily responsible for drafting the *Report on Police* of the Wickersham Commission and has been called the Father of Modern Professional Policing. He pioneered the use of the automobile in police patrol, advanced education for police recruits, and instituted a variety of other progressive practices. He committed suicide in 1955. (Courtesy of the Bancroft Library)

school in his department, attracting many scientists to teach his officers and encouraging many colleges to offer courses for future policemen." Vollmer was the first to urge the development of a record bureau in Washington, D.C., a bureau that eventually became the FBI.

Vollmer consulted with other police agencies and wrote extensively about the police. He was president of the International Association of Chiefs of Police in the 1920s and from that position spread the message of police professionalism. In a variety of ways and from a number of positions, Vollmer pressed the American police toward professionalism. As a police chief and professor of police administration at Berkeley, Vollmer recruited and trained the next generation of police professionals, including such luminaries as O. W. Wilson, who himself significantly advanced the cause of police professionalism (Bopp, 1977).

The Great Depression of the 1930s had a tremendous impact on the police by making the position of police officer attractive to a class of people who otherwise would not have considered it (Johnson, 1981:117). The job security offered by police employment, coupled with the lack of alternatives at that time, encouraged better-educated, middle-class men to apply for police officer positions. This new pool of applicants also enabled police administrators to upgrade the entry-level educational requirements for police jobs.

After World War II, O. W. Wilson founded the first college-level school of criminology at Berkeley, and was a national spokesman for police professionalism. His position at Berkeley lent credence to the idea of a police profession by virtue of the fact that colleges and universities now offered

Orlando Winfield Wilson (1900–1972)

Born in Veblen, South Dakota, Wilson was the son of an attorney. After moving to California, he graduated from the University of California at Berkeley in 1924. While at Berkeley, Wilson worked as a police officer under August Vollmer. With the support and encouragement of Vollmer, Wilson served as police chief in Fullerton, California, and Wichita, Kansas. In 1939, he accepted a position on the faculty at the University of California at Berkeley, again with Vollmer's active support. He served as Dean of the School of Criminology between 1950 and 1960. In 1960 he accepted the position of Chief of Police in Chicago, where he remained until retiring in 1967, Wilson was a leading proponent of modern management in policing and his text, *Police Administration,* is a classic in the field.

specialized courses in police science and administration. The returning veterans of the armed forces joined police departments in large numbers and, taking advantage of G.I. Bill benefits, many enrolled in college courses.

By 1960, then, policing in the United States had achieved something of a professional status. Rather than a job, policing was considered to be a career by both its members and the general public. The task of the police was primarily crime control, and the police were generally recognized as the crime-control experts. The long fight for independence from political interference had ended with a strong proscription against political and even public control over police activities. The adoption of technology, including the telephone and automobile, had changed the relationship between the officer and the public. Now an anonymous citizen could telephone the police, and an equally anonymous and impersonal officer would be dispatched. The neighborhood cop had been replaced by the motor patrol officer.

The American Police Circa 1960

By the early 1960s American police were largely autonomous and isolated from the communities they served. They had achieved a level of professional respect, if grudgingly granted, from the public. The "Keystone Cops" of earlier days had been replaced in the entertainment media with *Dragnet, The FBI Story,* and *Adam-12,* all of which depicted competent, professional police officers. This very professional status, and its independence, were the roots of a new movement to reform the police. As Walker (1981) notes, one effect of professionalism was to isolate the police from the public.

The 1960s were turbulent years in the United States. During this period the Civil Rights Movement gained momentum and frequently flared

A professional police officer of the 1960s.

into riots. Opposition to the war in Vietnam spurred large-scale demonstrations, campus unrest, and further riots. The professional police were caught in the middle of these changes. Further, one underlying theme of the 1960s was a general questioning of the social status quo.

As a social-control organization, the purpose of the police is to protect the status quo. As movements for social change swept the country, the police were consistently placed in the position of repressing protestors and demonstrators. The courts, especially the U.S. Supreme Court under Chief Justice Earl Warren, redefined *civil liberties* in what has been called a *due-process revolution*. Most of the early decisions in this revolution involved court oversight of the professional police. In case after case, the courts decided that the police had behaved inappropriately. Police powers, like the military, were too important to leave in the hands of professionals. The courts would serve to control police excesses.

Another serious threat to the professional police during the 1960s was the annual increase in the crime rate. The professional crime fighters, it appeared, were not able to fight crime very well. Year after year the number of serious crimes occurring in the United States increased. It was becoming clear that something more than professionalism was needed to improve the police. Increasing rates of crime and fear of crime, repeated confrontations between the police and the public, and the imposition of restrictions on police by the courts fed a new movement for police reform. This new reform effort sought to re-establish the connection between the police and the community (Fogelson, 1977).

THE COMMUNITY PROBLEM-SOLVING ERA

Kelling and Moore (1988) identify the contemporary period as the community problem-solving era of American policing. Beginning in the late 1960s and through the 1970s, police officials, researchers and political leaders recognized the problems and limitations of professional police. Events of the 1960s, including the riots and due-process revolution, caused a change in thinking about the police. People began to see the need for a closer link between the community and the police.

A number of forces merged to identify the isolation of the police from the citizens as a problem. The President's Commission on Law Enforcement and Administration of Justice (1967) supported the basic concept of police professionalism but urged greater police–community cooperation. Police agencies across the country instituted community-relations programs and experimented with alternative organizational structures such as team policing and neighborhood storefront station houses. An effort was made to involve the public in the "war on crime."

Experiments in police effectiveness during the 1970s indicated that traditional, professional police strategies for crime control were not effective. Routine motor patrol and rapid response time to calls for police service were not related to either levels of reported crime or citizen perceptions of safety. Traditional police responses to crime did not appear to appreciably reduce rates of crime or materially add to the ability of the police to solve those crimes that did occur.

A related development was the recognition that fear of crime was at least as important to citizens and the quality of their lives as was the actual level of crime. The police could perhaps best serve the public by making them feel safe. Further, feelings of safety were related to disorder more than to levels of crime. The order-maintenance function of the police that had been overshadowed by crime control in the professional reform again gained importance.

Police agencies adapted to the new situation by expanding the definition of the police role. As Kelling and Moore (1988:1) explain, "Accepting the quality of urban life as an outcome of good police service emphasizes a wider definition of the police function and the desired effects of police work." Crime control is still an important, if not dominant, function of the police. But in the era of problem solving, the emphasis is on crime prevention through general police efforts to improve the conditions of urban life. The problem-solving police, according to this view, should engage in advocacy to secure needed services for residents, such as employment, sanitation, and health care. The police should be as interested in maintaining order (and perhaps the appearance of order) as in solving crimes. The effect of these efforts will be to make people feel safer so that they are more active in their communities. People being more active in the community will mean that more law-abiding citizens will be on the streets, and consequently, being on the streets will become safer.

The community problem-solving era is evolving now, and thus it is too soon to tell whether it will endure, much less what its long-term effects might be. On the surface, it would appear that this new era is no more than a return to the policing style of the political era, with the important difference that the police now consult with the community rather than taking orders from politicians. Proponents of the problem-solving police believe that having learned the lessons of the past, including those of the political era, today's police will not revert to those types of corruption and partisanship typical of the past. Some, however, see the current expansion of the police role beyond crime control and law enforcement as troublesome. Michael Buerger, Anthony Petrosino, and Carolyn Petrosino (1999) catalogue potential negative consequences of police problem solving. They write (1999:142), "Aside from the ever-present danger of corruption, there are several potential dangers to expanding the problem-solving role." Among these dangers are officer burnout, personalization, overidentification, and overcommitment. These are all described as problems that stem from officers being too closely involved with the people and communities they police. If the current era represents at least a partial return to the levels of police involvement with communities that characterized the political era, we may see a future call for returning to more impersonal, professional policing.

CORRELATES OF POLICE REFORM IN AMERICA

The evolution of policing in the United States from its creation in the middle nineteenth century through the present reflects the influence of the same factors we found to be correlated with its initial development. The original political police reflected the lack of a dominant elite, the existence of a fragmented and conflicting social order within cities, and the presence of a personal relationship between the police officer and the people being policed. Originally created in response to perceived intolerable levels of crime and disorder in the largest cities, police departments served as a model for smaller cities as well.

The spread of policing and the sharing of information among police leaders led to the development of a "professional" identity among police administrators. The initiation of a formal association of police leaders created a political elite that sought to influence the evolution of American policing. Within the broader society, the growing middle class also developed its own political power. These two elite interest groups defined police reform as an important issue and collaborated in efforts to change the police.

1845	New York City creates large, modern police department
1883	*Pendleton Civil Service Act* creates federal civil service model
1893	First meeting of National Chiefs of Police Union
1894	Lexow Committee investigates New York Police
1905	August Vollmer elected Marshal in Berkeley, California
1919	Boston police strike
1931	Wickersham Commission releases *Report on Police*
1960	O.W. Wilson named Chief of Police in Chicago
1966	U.S. Supreme Court decides *Miranda v. Arizona*
1967	President's Commission on Law Enforcement and Administration of Justice
1969	Kerner Commission on the Causes and Prevention of Violence
1973	National Advisory Commission on Criminal Justice Standards and Goals
1982	"Broken Windows" published
1994	Violent Crime Control and Law Enforcement Act passed by Congress

FIGURE 4.1 Evolution of American policing

Thus, it appears that changes in American policing correlate with the emergence of new elites, both nationally and within cities. The sharing of information between cities and among police administrators correlates with changes in the definition of the principal purpose of police and the identification of the best methods of achieving that purpose. On a broad level, at least, police structure and functions correlate with social, economic, and political changes in American society. However, the specific nature of these correlations is revealed at a community, or municipal, level. Thus, although in this chapter we have traced general trends in American policing, we must recall that our analysis thus far has focused on the *idea* of police, not on the specific practice of policing in America.

HISTORY AS PROLOGUE TO THE PRESENT

This chapter reviewed the evolution of the police in the United States. From this history, we are able to gain an understanding of the present position of the police. Throughout its development, policing in America has adapted to changing social, political, and economic forces. The structure of American communities, we have seen, is directly related to the structure and practices of police agencies. In this regard, the police will reflect the times and communities in which they operate.

In addition to realizing that the police are inexorably linked to the social setting in which they occur, we have also seen that several ideas and developments influenced the evolution of the American police. The idea of police independence is powerful and persists today. The movement to involve the community in policing is not to be confused with an effort to allow nonpolice to dictate the police function. It is generally agreed that the community should provide information and guidance, but that decisions about the allocation of police resources should remain the province of police administrators. So, too, the idea that policing should be done by professional, career police officers is deeply ingrained. Neighborhood watch programs may assist the police, but vigilantism is not condoned. Finally, police agencies have a life of their own. Bureaucracy is the core of police organization, and thus those who would study or change the police must contend with the realities of bureaucracy. It would be very difficult to return to the particularistic and decentralized styles of policing typical of the middle 1800s even if we wished to do so.

A final observation that can be made as a result of this review of U.S. police history is that the only constant is change. In less than two centuries, the American police have experienced three major reform efforts. In any period, there are those who push for a reformation of the police or the police task. In this regard, history shows that the American police are similar to the weather: Nearly everyone has an opinion about them, but no one seems able to control them.

The next section of the book examines the contemporary status and organization of policing in the United States. In particular, we will examine the various types of police and law-enforcement agencies currently operating in the United States. This examination includes a study of the organization, structure, and personnel of both public and private law enforcement and policing in America.

Chapter Checkup

1. What three eras of American policing do Kelling and Moore (1988) identify?
2. Where did the authority of American police emanate, according to Conley (1995)?
3. What was the ultimate goal of police reformers?
4. What were the goals of urban elites and police administrators for police reform?
5. What does Monkonnen (1981) mean by a movement to defactionalize the American police?
6. According to Fogelson (1977), how did civil service selection of police officers affect their commitment to their jobs?

7. What separates the community problem-solving era from earlier eras in American policing?

8. How has the role definition of the police been changed and expanded in this most recent era?

REFERENCES

Bailey, W. G. (ed). (1995) *The encyclopedia of police science,* 2nd ed. (New York: Garland).

Bopp, W. J. (1977) *O. W.: O. W. Wilson and the search for a police profession.* (St. Louis: Kennikat Press).

Buerger, M. E., A. J. Petrosino, and C. Petrosino (1999) "Extending the police role: Implications of police mediation as a problem-solving tool," *Police Quarterly* 2(2):125–149.

Conley, J. A. (1995) "The police in urban America, 1860–1920," in W. G. Bailey (ed.) *The encyclopedia of police science,* 2nd ed. (New York: Garland):558-564.

Dilworth, D. C. (ed.) (1976) *The blue and the brass: American policing 1890–1910.* (Gaithersburg, MD: International Association of Chiefs of Police).

Fogelson, R. M. (1977) *Big-city police.* (Cambridge, MA: Harvard University Press).

Haller, M. (1976) "Historical roots of police behavior: Chicago, 1890–1925." *Law and Society Review* 10(Winter):303–324.

Johnson, D. R. (1981) *American law enforcement: A history.* (St. Louis: Forum Press).

Kelling, G. L. and M. H. Moore (1988) *The evolving strategy of policing.* (Washington, DC: U.S. Department of Justice).

Lane, R. (1975) *Policing the city: Boston, 1822–1885.* (New York: Atheneum Press).

Maniha, J. K. (1970) *The mobility of elites in a bureaucratizing organization: The St. Louis Police Department, 1861–1961.* Unpublished Ph.D. dissertation. (Ann Arbor, MI: University of Michigan).

Manning, P. K. and J. P. Van Maanen (eds.) (1978) *Police and policing: A view from the street.* (Santa Monica, CA: Goodyear).

Miller, W. R. (1977) *Cops and Bobbies: Police authority in New York and London, 1830–1870.* (Chicago: University of Chicago Press).

Monkonnen, E. H. (1981) *Police in urban America 1860–1920.* (Cambridge, MA: Cambridge University Press).

Parker, A. E. (1972) *The Berkeley police story.* (Springfield, IL: Charles C. Thomas).

President's Commission on Crime and Administration of Justice (1967) *The challenge of crime in a free society.* (Washington, DC: U.S. Government Printing Office).

Richardson, J. F. (1970) *The New York police: Colonial times to 1901.* (New York: Oxford University Press).

Richardson, J. F. (1974) *Urban police in the United States.* (Port Washington, NY: Kennikat Press).

Richardson, J. F. (1995) "Early American policing (1600–1860)," in W. G. Bailey (ed.) *The encyclopedia of police science,* 2nd ed. (New York: Garland):553–558.

Uchida, C. D. (1989) "The development of American police: An historical overview," in R. G. Dunham and G. P. Alpert (eds.) *Critical issues in policing: Contemporary readings.* (Prospect Heights, IL: Waveland):14–30.

Walker, S. (1977) *A critical history of police reform: The emergence of police professionalism.* (Lexington, MA: Lexington Books).

Walker, S. (1981) *Police in America: An introduction.* (New York: McGraw-Hill).

PART TWO

THE LAW-ENFORCEMENT INDUSTRY IN AMERICA

Consider the following hypothetical case: A man armed with a machine gun enters the post office on campus and robs the clerk, taking cash, postage stamps, and mail. He then steals a car parked outside and drives away. He gets on the interstate highway and flees across state lines.

Which American law-enforcement and police agencies would have an interest in this case? The answer, of course, would depend on the specific structure of police agencies in the area, but with the information available to us, we can identify several likely agencies, at all levels of government— federal, state, and local.

FEDERAL AGENCIES

There are a number of federal law enforcement and police agencies in America. The involvement of these agencies in criminal matters is governed by a definition of federal interest that helps demarcate the jurisdiction of each agency. In our case, the suspect was armed with an automatic weapon and so would attract the attention of the U.S. Bureau of Alcohol, Tobacco and Firearms (ATF) of the Treasury Department. Since mail was stolen and a U.S. Post Office robbed, the Postal Inspection Service of the U.S. Postal Service would become involved. Since a stolen car was transported across state lines, the FBI of the U.S. Department of Justice would also be interested in the case.

STATE AGENCIES

State police and law-enforcement agencies have a variety of organizational structures and specialized functions. If the campus involved was a state college or university, and if this school had a campus police, those police would be involved in the case. All states except Hawaii provide traffic patrol on interstate highways. Thus, our escaping robber would attract the attention of at least two state highway patrol/police departments.

LOCAL AGENCIES

Local agencies include all police and law-enforcement organizations operated by counties, cities, villages, and townships. At a minimum, our post office robber could attract the interest of two county sheriffs (crossing the state line means he has been in at least two counties—one for each state). This case would also be the responsibility of the local police for any other municipal government through which the fleeing offender passed. At least one local agency, the police of the town or city in which the college is located, could become involved.

Based on a conservative estimate and the sketchy facts provided in our example, we can identify at least three federal, two state, and three local law-enforcement agencies that could respond to this single criminal event. Depending on the circumstances, many more agencies could be added to this list of eight. For example, a private college might not operate a campus police department but might instead contract with a private security firm, adding another category of agency—private security—to our list. Many public schools also have contracts with private security firms.

As this illustration indicates, law enforcement and policing in America is a complex industry consisting of a large number of distinct organizations in both the private and public sectors and at all levels of government. These organizations frequently have overlapping jurisdictions and similar interests. Many times they have competing interests as well.

In Part II we present an overview and description of the law-enforcement and police industry in America. Federal and state police agencies are examined in Chapter 5. In Chapter 6 we review special-purpose and private police agencies. The final chapter of this Part returns to an examination of local police, with an emphasis on general-purpose, municipal police agencies.

A final note is in order before we begin to study today's police industry. If our hypothetical robber were to elude all of the law-enforcement agents and agencies in pursuit, he could attract the attention of yet another federal agency. Failing to disclose the proceeds of this robbery on his tax return could interest the Criminal Investigation Division of the U.S. Internal Revenue Service!

FEDERAL AND STATE POLICE

CHAPTER OUTLINE

In addition to the thousands of local and municipal police departments that were our primary focus in Part I, police agencies are also organized at the federal and state levels of government. These agencies have become larger and more important in the American police industry since the beginning of the twentieth century.

Prior to 1900 the federal government played a very limited role in law enforcement. Crime and crime control (as well as general order maintenance) were considered to be local issues, beyond the scope of the federal government. Thus, federal law-enforcement efforts were latecomers to policing in America. The tradition of local autonomy removed most crimes from federal interest.

Federal police developed as the role and functions of federal government expanded. The history of the federal police is a story of crisis management.

As new federal responsibilities were recognized or assumed by Congress and the President of the United States, little attention was paid to enforcement problems. Over time, as various crime problems were recognized, such as mail fraud in the postal service, the federal government would react by creating a "police" agency. This practice of reacting to problems as they arose rather than anticipating and planning for them led to the creation of numerous federal police. Donald Torres (1995a:287) observes, "Since the earliest days of the Republic, Congress has chosen to spread responsibility for enforcement of federal laws and regulations among many agencies rather than to concentrate them in one, all-encompassing police or investigative force. The result is a diversity of agencies with police or quasi-police powers."

*In a similar fashion, state governments came to identify problems and circumstances that seemed to require the creation of state-level policing agencies. As on the federal level, most states developed a number of distinct policing organizations with statewide **jurisdiction,** or responsibility. The development of federal and state policing represents, in many ways, attempts to "fill gaps" in policing that resulted from the local nature of American police organization. This chapter examines the history, distribution, and organization of federal and state policing in America.*

FEDERAL POLICING

There are over 70 distinct federal agencies that exercise police powers or serve police functions. Some of these are familiar to us, such as the Federal Bureau of Investigation and the Drug Enforcement Administration (DEA). Others, such as the National Zoological Park Police, The National Gallery of Art Protection Staff, the U.S. Mint Police, and the U.S. Supreme Court Police, are virtually unknown to most Americans. In a like vein, few of us consider the fact that the Metropolitan Police Department of the District of Columbia is a federal agency under home rule.

The relatively precise, or specific, titles of these agencies give us a clue about the structure and organization of federal policing. As the need for policing services emerges around a federal issue, the traditional response of Congress has been to create a distinct policing agency to deal with that need. Similarly, each federal agency is typically charged with responsibility for "policing" itself. Thus, the office of an **inspector general** exists within such organizations as the Environmental Protection Agency, National Aeronautics and Space Administration, Office of Personnel Management, and Veterans Administration. Agents of these offices are charged with investigating allegations of fraud or illegal activities against the agency, or improper and illegal practices within the agency. In general, federal police and law-enforcement agencies have narrowly defined responsibilities that focus on detection and investigation. Relatively few federal agencies have the general-purpose mandate of local police to maintain order, provide services, and enforce the law. Thus, many federal officials are not uniformed, and most federal agencies do not have a patrol responsibility. In 1998, only about 20 percent of federal law enforcement officers were reported to have primary responsibility for patrol (Reaves and Hart, 2001:2). The growth of federal police and law enforcement has followed the general growth of the federal government throughout American history. As the agencies of the federal government have become more important in society, so too have the roles of federal police.

The first Congress, meeting in 1789, passed several laws creating policing needs for the federal government. **The Judiciary Act of 1789** authorized the position of U.S. Marshal. **The Tariff Act** of that same year created the position of customs collector. The appointment of watchmen for the Capitol (1801) and for the president's residence (1802) laid the groundwork for the later development of the Capitol Police and the U.S. Park Police. Yet, in the earliest days, the federal government left most crime-control and policing obligations to the states. As we saw in Part I, the states in turn left the problem of policing to local government. The U.S. Marshal was the general-purpose law enforcement officer for the federal government for many years.

The U.S. Marshals Service.   Under the provisions of the 1789 Act, the President was authorized to appoint a marshal for each federal judicial district. George Washington originally appointed one marshal for each of the 13 judicial districts. Stanley Morris (1995:796) reports that each marshal "was empowered to execute 'all lawful precepts issued under the authority of the United States.'" Marshals were authorized to hire as many deputy marshals as needed, and to deputize citizens in special circumstances, such as a posse. This broad authority meant that the marshals were able to respond to almost any law-enforcement or policing problem facing the federal government. Morris (1995:797) recounts that "as the federal government's principal law-enforcement agency, the marshals of the first half of the nineteenth century enforced the laws applicable to counterfeiting, neutrality violations, the African slave trade, robbing the mails, and murder on the high seas on federal property. When new territories were established, marshals were appointed to provide a federal presence. In the Indian Territory (present-day Oklahoma) and in the territory of Alaska, the marshals were essentially the only law enforcement authority."

U.S. Marshals were responsible for court administration, leasing courtroom space, paying court expenses including witness fees, salaries of federal attorneys, court clerks, bailiffs, and other such charges. Modern marshals still pay the fees for witnesses and jurors. They also have primary responsibility for security at U.S. federal courts and protection of court personnel. The U.S. Marshals Service also operates the Federal Witness Protection Program, provides custody for federal prisoners from arrest until delivery to prison, escorts federal prisoners, and apprehends federal fugitives. Beginning in the middle of the nineteenth century, U.S. Marshals increasingly came under the control of the U.S. Attorney General. The historically decentralized offices of the U.S. Marshals were finally consolidated in 1969 into the U.S. Marshals Service. Still, reflecting their earlier tradition, the Marshals retain much independent authority over their respective offices. As Morris (1995:797) notes, "As federal law-enforcement officers, marshals retain the broadest authority and jurisdiction." They are still authorized to execute "all lawful precepts of the United States."

Historian David Johnson (1981) describes the early development of federal police in America as being "haphazard." There was little public notice of, or debate over, the emergence of federal police. Johnson attributes this lack of attention to the fact that (1) the executive branch was relatively small and unimportant to the daily lives of most Americans and (2) the policing problems of federal agencies emerged in a piecemeal fashion.

Federal police problems emerged separately in each department, and these problems were not necessarily shared by all departments. Further, the problems did not appear simultaneously, so a centralized federal police agency was never considered. Yet another important factor in the evolution of federal policing was the absence of an attempt at crime prevention. Federal crimes (such as counterfeiting or mail fraud) were not likely to be prevented by a federal police. Therefore, federal law-enforcement officials were organized to detect and investigate offenses rather than prevent them through practices such as patrol. Federal police agencies evolved as each federal department evolved and were therefore fragmented and often hidden from public view. As Johnson notes (1981:74), "In these circumstances, the national government's policing activities probably seemed unimportant to the general public."

Unlike the police in America's cities, federal officers were not organized into a separate organization and lacked a general mandate to uphold the law and maintain order. Each developed within a specific bureaucracy, where they investigated a rather narrow range of crimes specific to the function of that bureaucracy. As David Johnson observes,

> The Treasury Department, for instance, did not have a detective corps which was responsible for enforcing all the laws related to revenue. Instead, it had special agents assigned to two of its branches, the Customs Service and the Bureau of Internal Revenue, and the department also controlled the Secret Service. Each of these law enforcement details worked at separate tasks instead of combining their efforts. The federal law enforcement effort was therefore fragmented from its inception (Johnson, 1981:75).

This fragmented nature of federal policing not only served to limit public awareness and concern over national policing, but also created the circumstances under which a continual expansion of federal policing would occur. As each federal agency encountered policing problems, it would develop its own investigatory unit. Similarly, as the size and functions of the federal government grew over the course of American history, the number of law-enforcement problems encountered by federal agencies would also grow. The net result has been a continuous expansion of federal police and policing.

Federal Police Expansion

One of the first federal agencies to experience policing problems was the U.S. Post Office (Johnson, 1981). No other federal agency touched the lives of so many citizens in the early years of the nation. Further, individuals and businesses consistently used the mails for conducting business, usually in cash. The large sums of money transported by mail made post offices and mail carriers attractive targets for burglars and highwaymen. The large amount of cash cycling through the mails also proved too great a temptation for many postal employees.

After the Civil War, the mails experienced two additional policing problems. Mail fraud became more and more common as swindlers offered a variety of "services" to consumers via the mail. Citizens could invest in stocks, commodities, and real estate; participate in lotteries; purchase merchandise; and receive racing "tip sheets" by mail. Unfortunately, all too often the customer sent money and received no service. Finally, Congress passed legislation to control "obscene mail" in 1865, leaving its enforcement to the U.S. Post Office.

Faced with the problems of burglary, robbery, employee theft, mail fraud, and control of obscenity, the Postal Inspection Service grew in size and function. Local police were not (and could not be expected to be) concerned with many of these issues. Over the years, the postal inspectors developed a reputation for effectiveness. They also created an informal system of cooperation with local police agencies in which they shared information.

The second federal law-enforcement agency to develop was the **Secret Service** (Johnson, 1981). The primary task of the Secret Service was the control of counterfeiting. Congress authorized paper currency in 1861, which was the heyday of counterfeiting. Not realizing the skill, ingenuity, or scope of counterfeiting in America, no provisions were made for the protection of federal currency. That this new paper money, unlike state banknotes, was legal tender throughout the nation attracted the immediate attention of counterfeiters. By 1865 a permanent force to deal with counterfeiters was created.

Prior to the organization of the Secret Service in 1865, the Treasury Department had used reward offers and contracts with private detectives

Members of the Secret Service accompany President Theodore Roosevelt in 1901.
(Courtesy of Library of Congress)

to fight counterfeiting. The Secret Service carried on this tradition when
the first agents were all hired from the ranks of private detectives. Lacking
clear policies and mandates, the Secret Service engaged in a variety of questionable investigative practices, including the use of informants, "deep
cover" investigations, and illegal searches. The relatively high success rate
of investigations, however, prevented much criticism.

Over the next 40 years, the Secret Service, by virtue of its broad mandate, came to be the general investigative and law enforcement arm of the
federal government. For example, when Congress created the Justice Department in 1870, it did not authorize any special agents. The Justice Department was forced to contract with private detectives, or "borrow" agents
of the Secret Service to conduct investigations.

The use of Secret Service agents to investigate land frauds for the Department of the Interior at the turn of the century finally attracted con-
gressional attention. The success of the Secret Service investigations led
many local officials in the affected states to pressure their congressmen for
protection. In 1908 Congress passed legislation that restricted the Secret
Service to dealing with counterfeiting and protecting the president.

In 1907 the Justice Department had asked for authorization to hire its
own investigators, but Congress refused. The attorney general, Charles J.
Bonaparte, then created a **Bureau of Investigation** on his own authority,
and President Theodore Roosevelt ordered the transfer of eight Secret Ser-

vice agents to this bureau. Bonaparte appointed William Burns, president of his own, well-known private detective agency, to head this bureau.

Historian David Johnson (1981) breaks the twentieth-century history of federal policing into three eras of expansion. The first occurred during the Progressive era, early in the century. The second corresponded with the Great Depression and World War II. The third era was a response to the unrest and pressures of the 1960s, including the Civil Rights Movement and the anti–Vietnam War activities.

In a thoughtful history of the Federal Bureau of Investigation, Tony Poveda (1990) identifies eras of development for that agency as well. His eras likewise correspond with the recognition of crises of national security and public order, including the Great Depression and turbulent era of the 1960s. The Bureau of Investigation was the forerunner of the modern Federal Bureau of Investigation (FBI). Its history indicates the social forces that shaped federal policing in the twentieth century.

The Federal Bureau of Investigation.   The Bureau of Investigation, founded in 1907, expanded during the years between 1910 and 1924 (Johnson, 1988). Passage of the Mann Act (1910) made it a federal offense to transport women across state lines for immoral purposes. The Bureau of Investigation was responsible for enforcement of the law, and grew to number over 300 agents. During World War I, fear of spies and sabatoeurs led to the Sedition Act, which involved all federal law-enforcement agencies but specifically the Bureau of Investigation in domestic intelligence operations. The Red Scare that followed the Russian Revolution of 1917 also supported investigations of suspected communists and anarchists. The bureau developed a General Intelligence Division in 1918, and a young attorney, J. Edgar Hoover, was named its director.

A series of widescale investigations and raids, including the "slacker raids" aimed at draft dodgers in World War I and the "Palmer Raids," resulting in the arrests of hundreds of suspected radicals, shook public confidence in the bureau. To this record of abuse were added charges and revelations of corruption and abuse of office by the bureau's director, Burns, and the attorney general. Burns resigned, and Hoover succeeded him as acting director. In December 1924, Hoover was named director of the bureau, which was renamed the Federal Bureau of Investigation (FBI) in 1935 (Torres, 1987).

Hoover spent the first decade of his directorship quietly improving the policies and procedures of the Bureau of Investigation and upgrading the quality of its agents. During the New Deal era, the role of federal government in the daily lives of citizens experienced another expansion similar to that during the Progressive era at the turn of the century. Hoover had by the 1930s created a measure of political independence for the bureau, gained control of all federal fingerprint files (1930), established a national crime laboratory (1932), and opened a national law-enforcement academy

John Edgar Hoover (1895–1972)

J. Edgar Hoover was the first director of the Federal Bureau of Investigation after serving during four years (1921–1924) as assistant director of the Bureau of Investigation under William Burns. He remained as director of the agency for 48 years. Hoover received a law degree from George Washington University in 1916 and insisted on high academic achievement among those hired as special agents of the FBI. Under his leadership, the FBI became an internationally respected model of professional, efficient law enforcement. In later years, Hoover and the FBI were criticized for violations of civil rights and abuse of power. (UPI/Bettmann)

(1935). A firm supporter of police professionalism, Hoover required college education and rigorous training of all new FBI agents.

During the 1930s, the Lindbergh kidnapping, a rash of bank robberies in the wake of federal bank deposit insurance, the Dyer Act (making interstate transport of stolen vehicles a federal offense), and other federal innovations supported an expanded role for the FBI. The FBI embarked on a vigorous enforcement and public relations campaign. As agents caught or killed those offenders placed on its "Most Wanted" list, or otherwise identified as "public enemies," the reputation of the FBI grew. It emerged from the Depression years as a model of professional—and effective—law enforcement, not only for the nation but for the world.

The focus on criminals, however, was short lived, as the outbreak of World War II and the repeal of prohibition helped reduce the importance of crime as an issue. During the war and immediately following it, the concerns of national security against the threats of spies or communists again involved the FBI in domestic intelligence. The bureau continued to expand as its efforts at ferreting out suspected communist sympathizers and spies grew in the early 1950s. This era was followed almost immediately by the emergence of the Civil Rights Movement and antiwar activities.

The FBI was pressed into service under the Kennedy and Johnson administrations to investigate discrimination and harassment of African Americans and to investigate and prosecute radicals during the Nixon administration. Hoover's death in 1972 resulted in revelations about FBI abuses of individual rights, especially in regard to domestic intelligence operations (Powers, 1986). Tony Poveda (1990:2–6) described the key points in the FBI scandal of the early 1970s. Between 1971 and 1976 past FBI abuses were made public, including the facts that Hoover's FBI conducted investigations of thousands of citizens, that Hoover himself maintained secret files on thousands of citizens, and that FBI agents routinely committed burglaries to further their investigations.

More recently, FBI involvement in high-profile cases such as the assault on the Branch Davidian compound in Waco, Texas (Kopel and Blackman, 1997), the shooting at Ruby Ridge, and the espionage trial of Dr. Wen Ho Lee have further damaged the reputation of the FBI. Added to these problems are recent revelations of improper operations in the FBI Crime Lab, the failure to disclose evidence in the Timothy McVeigh Oklahoma City bombing case, and the discovery that a high-ranking FBI counterintelligence officer was actually acting as a foreign agent (Klaidman, 2001).

FBI agents at the World Trade Center bombing in New York in 1993. (Tom Kelly)

Still, despite these revelations, the FBI maintains a solid, if tarnished, reputation as a premier law-enforcement agency.

Federal Law-Enforcement Functions

Most federal law-enforcement agencies are specialized in terms of both mission and method. With few exceptions, federal policing organizations do not provide general patrol and order-maintenance functions. Rather, federal agencies concentrate on investigation and crime control. Thus, most federal law enforcement employees have the title "agent" as opposed to "officer," and relatively few federal services are uniformed.

Federal policing activities are constrained by some definition of **federal interest.** Thus, local police are generally interested in robbery as a crime and respond to robberies ranging from street muggings to banks. Federal agents are limited to enforcing federal laws, so they react to robberies on federal property or robberies of federally protected persons or institutions such as federally insured banks.

The development of federal policing agencies, as we have seen, was a process of reacting to enforcement problems as they became recognized. The U.S. Postal Service, Secret Service, FBI, and other agencies were created and authorized to respond to particular problems. As the role of the federal government expanded throughout the twentieth century, the numbers and size of federal police agencies also grew. Still, these agencies remained decentralized. For example, in the U.S. Department of Justice alone, there are four distinct "policing" agencies; U.S. Marshals, Immigration and Naturalization Service, FBI, and Drug Enforcement Agency (Torres, 1995a). These four agencies employed over 36,000 agents and officers in 1998 (Reaves and Hart, 2001).

In general, federal enforcement agencies serve to enforce federal laws and regulations and protect federal property and institutions. The highly decentralized structure of federal agencies reflects the incremental development of federal policing as a response to specific problems. Police agencies operating at lower levels of government often have neither the interest nor the ability to effectively respond to national problems.

As shortcomings in municipal and state police protection came to light, Congress authorized the creation of federal solutions to national enforcement problems. For example, it would be unreasonable to expect municipal police to shoulder responsibility for enforcing customs duties or immigration regulations. Not only would these tasks be unevenly distributed among America's municipal police agencies, they would compete with local demands for police service. To ensure the protection of federal interests, it is necessary to develop federal police. After the terrorist attacks on the World Trade Center and Pentagon on September 11, 2001, concern about controlling terrorist threats has become a well-recognized federal interest.

Anti-terrorism.   Shortly after the attacks of September 11, 2001, President Bush created the Office of Homeland Security under the directorship of former Pennsylvania Governor Tom Ridge. This new office has the responsibility of coordinating the efforts of forty-six separate federal law enforcement and intelligence agencies in efforts to detect and prevent terrorist acts or to apprehend those who have committed such acts. In response to the 9-11 attacks and the subsequent anthrax attacks, federal resources have been mobilized to combat terrorism. The F.B.I., Bureau of Immigration and Naturalization, and other federal agencies have conducted broad investigations, arresting 1,200 men of middle-eastern descent for questioning (Thomas and Isikoff, 2001). Investigators from the U.S. Postal Inspection Service, along with other federal agencies, continue to investigate the anthrax mailings.

It is still too soon to tell what the long-term effects of the war on terror might be for American law enforcement, and federal law enforcement in particular. In keeping with the tradition of federal police development, the Office of Homeland Security is not empowered to direct federal agencies, only to coordinate efforts. Each federal agency is still autonomous and its interests in anti-terrorism reflect the specific mandate of the agency. It is possible that continued concern over terrorism could lead to an attempt to centralize federal law enforcement. At present, the focus on terrorism of federal policing has been sharpened, some federal agencies will expand like increasing the number of marshals flying on commercial aircraft, and some new federal police agencies may be created.

Federal agency roles are less complicated than those of the municipal police. Most federal enforcement agencies have a clearly identified role and mission, stated in the legislation creating the organization. By far, the primary role of these agencies is law enforcement through investigation. Relatively few federal police are charged with the ambiguous responsibilities of order maintenance or service delivery.

Another primary function of federal law-enforcement agencies is the general support of state and local police. The FBI, for example, provides training and fingerprint identification. It also serves as an information clearinghouse for U.S. police agencies. Federal law-enforcement agencies cooperate with local police on issues of mutual interest and generally share intelligence about criminal matters.

Federal Police and Law-Enforcement Agencies

Maguire and Pastore (1999:17) report that in 1992 the federal share of law enforcement and policing amounted to slightly less than 16 percent of the total spent on police functions in America. The federal government employs approximately 9.4 percent of all sworn law-enforcement and policing personnel in the country. The Bureau of Justice Statistics

TABLE **5.1** Federal Agencies Employing over 500 Law-Enforcement Officers

Immigration and Naturalization Service	17,654
Bureau of Prisons	13,557
Federal Bureau of Investigation	11,523
Customs Service	10,522
Drug Enforcement Administration	4,181
Secret Service	4,039
Administrative Office of U.S. Courts	3,599
Postal Inspection Service	3,412
U.S. Marshals	2,735
Internal Revenue Service	2,726
National Park Service	2,188
Alcohol, Tobacco and Firearms	1,967
U.S. Capitol Police	1,199
U.S. Fish and Wildlife Service	888
Federal Protective Service	803
Diplomatic Security Service	617
U.S. Forest Service	586

Source: B. Reaves and T. Hart (2001) *Federal Law Enforcement Officers, 2000.* Washington, DC: Bureau of Justice Statistics, p. 2.

(Reaves and Hart, 2001:1) reported that federal agencies employed more than 88,000 law enforcement agents and officers. The estimated federal budget for direct expenditures for police services in 1992 was set at over $8 billion (Maguire and Pastore, 1999:4). But it is exceedingly difficult to obtain an accurate picture of the federal law-enforcement and policing apparatus (Table 5.1).

Most cabinet-level departments and major administrative agencies have an **office of the inspector general** charged with the responsibility for overseeing the operations of the unit (Table 5.2). Inspector generals are authorized in the Departments of Agriculture, Commerce, Defense, Education, Health and Human Services, Housing and Urban Development, Interior, Labor, Transportation, and Treasury. Similar offices exist in the Environmental Protection Agency, General Services Administration, National Aeronautics and Space Administration, Office of Personnel Management (formerly the Civil Service Commission), Government Printing Office, and Veterans Administration. At least part of the work done by each of these offices involves investigations of alleged criminal acts by employees of the agencies or committed against those agencies.

Uniformed police and/or security forces are maintained by other federal agencies, including the military police or security police of the armed

TABLE 5.2 Offices of Inspector General Employing More than 50 Federal Officers

Dept. of Treasury, Tax Administration	352
Dept. of Defense	322
Dept. of Health & Human Services	303
Dept. of Housing & Urban Development	248
Social Security Administration	238
Dept. of Agriculture	217
Dept. of Labor	135
Dept. of Justice	119
Dept. of Transportation	91
Dept. of Veterans Affairs	77
Dept. of Education	59
General Services Administration	59
Environmental Protection Agency	52

Source: B. Reaves and T. Hart (2001) *Federal Law Enforcement Officers, 2000.* Washington, DC: Bureau of Justice Statistics, p. 6.

services, a uniformed division of the Secret Service, and the National Zoological Park, National Art Gallery, U.S. Capitol, U.S. Supreme Court, Federal Aviation Administration, U.S. Park, Mint, and Treasury Police. Each federal reserve bank and many individual federal office buildings also employ uniformed security officers.

In the Department of Transportation, the U.S. Coast Guard could be considered to be the nation's largest law-enforcement agency, with over 37,000 enlisted and officer personnel. Given the Coast Guard mandate to enforce laws and regulations relating to navigation, much of their responsibility involves the performance of "police" duties (Mueller and Adler, 1996). Other uniformed federal officers are employed by the U.S. Customs Service, U.S. Marshal's Service, U.S. Forest Service, Veterans Administration, and Federal Protective Service of the General Services Administration.

Nonuniformed investigators are employed by the FBI, Secret Service, Bureau of Alcohol, Tobacco & Firearms, Bureau of Immigration and Naturalization, Department of Defense, Internal Revenue Service, and various federal commissions. Commissions with regulatory and law-enforcement responsibilities typically employ their own investigators. Examples of these agencies include the Federal Trade Commission, Interstate Commerce Commission, and Federal Communications Commission.

A survey of federal agencies in 2000 revealed that the federal government employed about 88,000 personnel authorized to carry firearms and make arrests (Reaves and Hart, 2001). This survey did not include military police, the U.S. Coast Guard, and federal officers serving in foreign countries or U.S. territories. Of these personnel, approximately one-quarter

CHART 5.1
Primary duties of federal law
enforcement officers (percentages).
(*Source:* B. Reaves and T. Hart
(2001) *Federal Law Enforcement
Officers, 2000.* Washington, DC:
Bureau of Justice Statistics, p. 2.)

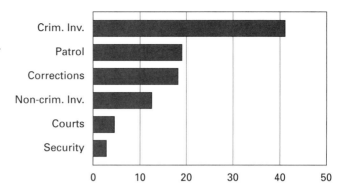

worked in corrections (Bureau of Prisons, Federal Probation and Parole Officers, etc.) or performed court operations (U.S. Marshals). About 20 percent performed duties related to police response and patrol. The most common assignments for federal personnel (41 percent) related to criminal investigation and enforcement, with another 13 percent assigned to non-criminal investigation and enforcement (Chart 5.1).

STATE POLICE

As with their federal counterparts, state police organizations emerged in reaction to perceived limitations of local policing. Throughout the nineteenth century, most law enforcement and order maintenance were left to local communities. With only a few exceptions, state governments did not operate police forces because there was no need for them. This situation began to change, however, near the end of the century.

David Johnson (1981) argues that state police forces developed in response to a broadening of problems of order. As industrialization and improvements in transportation progressed, a new gap in policing emerged. Although most cities had police forces, police service ended at the city limits. As American cities became more interdependent, it was apparent that problems of order outside any given city could seriously affect the lives of the city's residents. This was particularly true of strikes and other labor actions (Falcone, 1998). Factories and their workers were located in cities and were therefore policed. Raw materials, like coal and other ores, however, were mined in rural, unpoliced areas. In time a need to police the areas between cities was recognized and, further, was seen as a responsibility of state government.

State-level law enforcement and policing developed as a hybrid between the federal and local experiences. Most states created some form of general-purpose state policing agency, but they also created a variety of specific organizations in response to enforcement problems as they arose.

Thus, all states but Hawaii have a statewide police or highway patrol. These organizations differ in terms of the breadth of their mandate. **Highway patrols** typically are charged with the regulation of traffic and enforcement of traffic regulations on interstate and state highways. **State police** have a general-purpose policing mandate. Nevertheless, both highway patrols and state police are typically authorized to enforce all state laws and to assist local law-enforcement agencies on request (Germann, Day, and Gallati, 1988: 102).

As with federal law-enforcement and policing agencies, states also have created a variety of investigative units or agencies within larger state bureaucracies. These agents are typically concerned with more limited forms of offenses or with the investigation and prosecution of narrow ranges of crimes and violations of agency regulations. Examples of such limited-purpose organizations include state revenue and tax officers, state liquor control agents, and state fire marshals (Torres, 1995b:726).

Because of the variety of functions and agencies of state policing that have evolved in the 49 states having such organizations—recall that Hawaii has no state-level police (Torres, 1995b)—it is difficult to obtain a count of the number of these organizations in existence. For example, the Bureau of Justice Statistics (1980:7) reports nearly 1,000 state law-enforcement agencies operating in 1977. Of these, 670 separate agencies were listed as employing at least one sworn officer. Yet, Torres (1995b:728) identifies only 84 state agencies classified as either state police, state highway patrols, or state criminal investigative agencies. This more restrictive definition excludes liquor control agents and park rangers. More recent national surveys (Bureau of Justice Statistics, 1999) of state law-enforcement agencies have focused on the 49 primary state law enforcement agencies (state police or highway patrols). The majority of states, then, have both general-purpose policing agencies similar to local police and decentralized investigative, security, and police forces similar to those of the federal government.

The Development of State Police

The creation of the Pennsylvania State Constabulary in 1905 marked the start of modern, general-purpose, uniformed state police. Prior to the Pennsylvania Constabulary most states did not have policing agencies, although many had developed offices or positions with some law-enforcement responsibilities. Of these, perhaps the most famous were the Texas Rangers, started in 1835, before Texas was even a part of the United States.

As a Province of Mexico, Texas relied on the Mexican government for protection from criminals, attacks by Native Americans, or other threats to order and security. Mexico, however, was unable to guarantee protection to all of its northernmost provinces. In 1823 the Mexican government consented to allow the settlers in Texas to raise unofficial militia companies

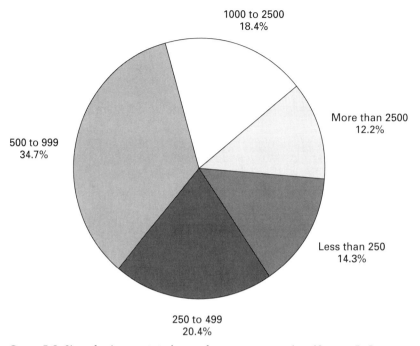

CHART 5.2 Size of primary state law-enforcement agencies. (*Source:* B. Reaves and A. Goldberg (1998) *Census of State and Local Law Enforcement Agencies, 1996.* Washington, DC: Bureau of Justice Statistics, p. 11.)

to repel raiders. These militia became the legendary Texas Rangers (Johnson, 1981:156).

The initial success of the Rangers in preventing and responding to raids by bandits and war parties from Native American tribes was warmly received by the settlers, who all benefited from this protection. After 1835, however, Texas declared its independence from Mexico, and the Rangers became a permanent force in the Republic of Texas in 1840. By the 1850s the Rangers had acquired a variety of general policing duties in addition to their original purpose of community defense. The Rangers patroled the Mexican border, provided law-enforcement services to isolated communities, and took responsibility for the control of runaway slaves and illegal Mexican immigrants, and for public safety on the roads, trails, and open spaces of Texas.

The Rangers were famous for their brand of "frontier justice," and many wanted criminals were killed "trying to escape," or resisting arrest. In particular, Johnson (1981:157) notes the discriminatory practices of the Rangers: "Negroes and Mexicans suffered the most, as Rangers shot and beat them with impunity." This type of action, however, was popular with the white Americans who dominated Texas at the time. After a period in

Bruce Smith (1892–1955)

The son of a banker, Bruce Smith was born in Brooklyn, New York. Smith earned both a master of science and a law degree from Columbia University in 1916 and began work for the New York Bureau of Municipal Research. He conducted a study of the Harrisburg, Pennsylvania, police department. Through the years (interrupted by Army service in the air corps during World War I), Smith became an expert and consultant on policing. He was instrumental in developing the *Uniform Crime Reports* and wrote *Police Systems in the United States,* published in 1940. He spent his career studying the police and trying to improve policing in America. (Courtesy of The Institute of Public Administration, New York)

which the "frontier justice" for which the Rangers were noted became increasingly unacceptable, in 1935 they were reorganized and placed under a department of public safety.

The second state police agency was the Massachusetts State Constables, created in 1865. The primary purpose of the state constables was the control of commercial vice, according to Bruce Smith (1969:23). In fact, these constables were organized in an effort to force the Boston police to control drunkenness and other vice crimes in that city. As Johnson (1981:158) notes, "A clash of cultural values lay at the heart of the controversy."

Native-born citizens wanted to control the large, Irish Catholic population of Boston. Boston politicians and, by extension, the Boston police, however, were required to attend to the desires of their constituents and were reluctant to enforce laws against drinking. A coalition of rural residents, middle-class Bostonians, and prohibitionists succeeded in passing state-level legislation to create the constables' positions. After some initial successes, in 1879 the state constables were reorganized into an essentially detective force called the Massachusetts District Police. In 1920 this organization was placed under a department of public safety. In 1903 Connecticut created a similar state constable organization, composed of a relatively small number of detectives and investigators.

Ethnic and cultural differences also underlay the development of the Pennsylvania State Police. A tradition of violence and disruption in the coal mining regions of Western Pennsylvania led the state legislature to authorize Iron and Coal Police in 1865. These "police" were employees of the mining companies, and were used extensively in antilabor activities. A major coal strike in 1902 attracted the attention of the federal government, and President Theodore Roosevelt appointed an investigating commission. The commission blamed much of the problem on the lack of effective law enforcement by the state.

In 1905 Governor Pennypacker's proposal for a state police was enacted into law. The governor based the model for the Pennsylvania State Police on the Philippines Constabulary—an essentially antiguerilla defense force in operation in the U.S.-controlled Philippines. Pennypacker selected John Groome, a National Guard officer and ex-member of the Philippines Constabulary, as the first commander of the police. Groome carefully selected 228 officers, most of them former soldiers. He admonished them that "one state policeman should be able to handle one hundred foreigners" (Johnson, 1981:160).

Over the next decade, Groome's officers brought a degree of order to the mining region. The force impartially controlled labor violence, allowing neither the miners nor the owners to use violent tactics. In other areas, however, the anti-immigrant stance of the police was seen, as they routinely conducted "sweeps" of immigrant communities for weapons and maintained a high-visibility presence at gatherings to intimidate participants into behaving in "appropriate" ways. After World War I, the Pennsylvania State Police were able to concentrate more on rural crime and rural policing, having brought order to the mining regions.

Ohio, on the other hand, presents a different story. Swart (1974) describes the development of state police activity in Ohio as a struggle. The state legislature virtually refused to take action on state police for over half a century. Instead, over the years the Ohio legislature recognized and authorized police powers for railway conductors and railroad police (employed by the private railroad companies) as well as cemetery guards. It also recognized certain vigilante associations in its "Antihorsethief" legislation (Swart, 1974:42–47). In 1883 the legislature approved a two-officer statehouse police charged with security at the capitol.

Lobbying pressure from insurance companies (including the agreement of these companies to a special tax that covered all costs) led to the creation of a **state fire marshal** in 1900, but this office was concerned with arson investigation and building code violations, not general policing. The imposition of taxes and licensure on the liquor trade in 1912 led to the creation of a liquor control board and inspectors in 1913. These agents, however, lacked general police powers. The passage of the Eighteenth Amendment to the U.S. Constitution mandated nationwide prohibition, and in 1921 Ohio created a Prohibition Commission. In time the Prohibition Department employed "police," but these officers did not have general police powers or responsibilities.

In the same year, Ohio started a Bureau of Criminal Identification and Investigation. This bureau was largely concerned with developing and maintaining records and providing other forensic aid to local Ohio police agencies. However, by 1927 some investigators of the bureau had engaged in active fieldwork. Still, there was no statewide police agency. Eventually the growing pressures of transportation problems in a state composed of several large cities scattered over an otherwise agricultural region were too strong. In 1928, almost as a last resort, the Ohio State Highway Patrol entered active service.

In the end, the Ohio State Legislature simply had to do something about traffic control and the intercity travel of wanted offenders. Proposals for a state-level policing agency with general powers continued to be rejected. Instead, the legislature opted for a more restricted highway patrol that had limited police jurisdiction of its own. In later years, of course, state-level policing in Ohio expanded, as it did in other states, including a much broader investigative role for the Bureau of Criminal Identification and Investigation.

David Falcone (1998) describes the development of the Illinois State Police as representative of the broader movement in the United States to create state police agencies in the early part of the twentieth century. Falcone (1998:61) argues, "The effects of the tumultuous history of the early twentieth century (i.e., World War I, prohibition, massive immigration and labor and evolutionary unrest) are identified as catalysts for policing at the state level." The Progressive movement's demand for professional policing supported the adoption of a military structure for state police agencies, and efforts to remove the state police from partisan political influence. Fear of labor unrest, and especially of immigrant populations supported the creation of centralized state police. These early police organizations were "predicated on a masculine ethos, driven by nativitistic attitudes and values." Falcone believes these characteristics, coupled with the heavy reliance on a military model, kept state police insulated from the public. Fear of centralized policing, especially at the state level, worked to limit the authority and jurisdiction of state police so that the mission or purpose of the state police is muddled. The mission of most general-purpose state police agencies often overlaps and conflicts with local, other state, and even federal law-enforcement agencies. Falcone (1998:79) concludes that state police agencies can not "lay claim to a well-defined mission that is exclusive and stable."

State Police Agencies and Functions

In his hallmark study of the state police, originally published in 1925, Bruce Smith limits his examination to those agencies that employed a large and permanent body of officers who regularly exercised general police powers throughout the state (1969:47–48). Applying this definition, Smith found eight such agencies operating in the United States. However, he notes that the term *state police* had no precise meaning and was applied to a variety of organizations (1969:47).

Torres (1995b) echoes this observation more than 60 years later. As with federal policing and law enforcement, states developed a variety of special-purpose policing organizations as the need for them arose and was recognized. The most familiar state police agencies are highway patrols, which have primary responsibility for traffic enforcement and criminal law enforcement along major state highways. Additionally, there are 23 state police agencies, which typically act as highway patrols but also have general police powers and responsibilities in the unincorporated areas of states. Thus, state police provide services, in the main, in those areas of states that do not have their own, local police agencies. These police services often overlap with the policing responsibilities of the county sheriff (Weisheit, Wells, and Falcone, 1995).

The International Association of Chiefs of Police (1975) identifies a variety of functions served by state police agencies. In addition to direct police services of patrol, enforcement, and investigation, a common role for state police agencies is to support local police. This support function comes in the form of (1) direct assistance, as in cooperation in investigations, and (2) record management and forensic aid (crime labs), as well as the provision of basic and ongoing training to local police officers.

Beyond general-purpose police agencies, most states also have some specialist units. For example, states typically operate park police or ranger units in state parks and provide security for state buildings and facilities. States also have investigative and enforcement agencies with narrow ranges of authority, such as liquor control, gaming (or gambling), and athletic commissions. In this regard, state investigative agents are frequently authorized within specific state bureaucracies and charged with enforcing the laws and regulations of relevance to that unique agency.

Very little has been written about the variety and functions of state law-enforcement and policing agencies, especially about those not exercising general police powers. If you think about your own state, however, the range of agencies and functions may become clear. From what agency does one obtain a hunting or fishing license? What happens to someone who fails to pay state income taxes? What agents of state government will respond to consumer complaints about businesses in the state? Perhaps it is the very variety of agencies and functions, coupled with their lack of general polic-

Colorado State Patrol officers policing a demonstration at the Rocky Flats nuclear weapons plant. (UPI/Bettmann)

ing powers, that explains why so little is known about the numerous, independent security and enforcement agencies in the 50 states.

The Bureau of Justice Statistics reports that the state share of police-protection employment and expenditures in 1994 were nearly equal. States employed approximately 10.8 percent of all police personnel and carried about 11.6 percent of the total police-protection direct expenses for that year (Maguire and Pastore, 1999:4, 17). Based on data from the primary state police agencies in 49 states (excluding Hawaii), state police agencies averaged about 1,680 employees, of whom over 1,100 (65 percent) were sworn officers. These data do not include information about the multitude of other state policing and law-enforcement agencies (Reaves, 1996:1). In 1994 states spent about $5.3 billion on police protection (Chart 5.3).

PROS AND CONS OF FEDERAL AND STATE POLICING

Like any organization or social institution, federal and state police in America have both strengths and weaknesses. What constitutes a strength or weakness, of course, depends in part on the views held by those making

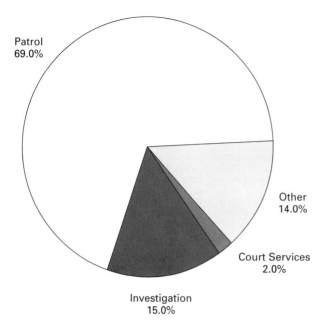

Patrol
69.0%

Other
14.0%

Court Services
2.0%

Investigation
15.0%

CHART 5.3
Duties of sworn officers in primary
state law-enforcement agencies
(*Source:* B. Reaves and A. Goldberg
(1998). *Census of State and Local Law
Enforcement Agencies, 1996.*
Washington, DC: Bureau of Justice
Statistics, p. 11.)

judgments, but in general they can be summed into two words—size and scope. State and federal policing, as compared with local policing, benefit from the size of their jurisdiction. These agencies typically draw personnel from a broader pool of national or statewide applicants, are funded by a larger (and usually richer) tax base, and can thus both afford and find better-qualified applicants (Chart 5.4). They can also provide their personnel with extensive training. For example, in 1997, 20 percent of large (employing 100 or more personnel) state police agencies required at least a two-year degree of new recruits, while fewer than 10 percent of local police departments had degree requirements. Similarly, the median large state police agency required new officers to receive 823 hours of training, while the median requirement for local police was 760 hours (Reaves and Goldberg, 1988).

The scope of many of these agencies is, however, limited. From the point of view of the agency, this is a strength, because the goals and purposes of the organization's activities are often narrowly defined. Rather than responsibility for an ambiguous task of maintaining order, federal and state police are frequently required only to enforce a specific set of laws and regulations. Instead of somehow preventing crime, these agencies are expected to investigate only offenses that have occurred.

This very narrowness of task, though, can be a weakness. Despite what expertise or resources an agency such as the FBI might bring to a problem, its authority to act is limited by definitions of federal or state "interest." The large area of jurisdiction may also mean that the agency is un-

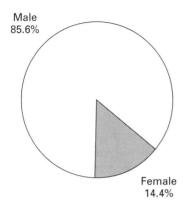

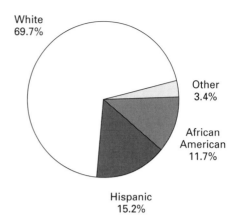

CHART 5.4 Gender and ethnic distribution of federal law-enforcement officers (*Source: B. Reaves and T. Hart (2001) Federal Law Enforcement Officers, 2000. Washington, DC: Bureau of Justice Statistics, p. 7.*)

able to assemble sufficient personnel for an action and must then rely on assistance from other, local agencies.

The jurisdictional and resource implications of state or federal authority, however, make these agencies particularly well suited to centralized functions and management of coordinated efforts. It is no accident that one of the main functions of state and federal police agencies is the provision of support services to local police (Falcone, 1998). A statewide agency achieves the requisite "economy of scale" to justify a state-of-the-art crime lab, while a small-town police department cannot. Similarly, by drawing classes from the statewide pool of police agencies, state organizations can operate training academies more efficiently than the vast majority of local police departments. On the other hand, these organizations are removed from direct public, or community, control and influence. They are sometimes seen as external forces and not considered by citizens to be a part of their everyday lives or concerns.

Thus, the size and scope of agency jurisdiction and mandate make federal and state policing well suited to centralized tasks, limit the range of activities in which they must or can become involved, and provide them with a rich resource base in terms of both personnel and money. These factors are reflected in the correlates of policing identified in Chapter 1 as they apply to federal and state police.

SOME CORRELATES OF FEDERAL AND STATE POLICING

Earlier we identified three sets of correlates that help explain policing: the people, the organization, and the community involved in the policing issue under study. There is no doubt that federal and state policing differ from local policing in important respects, and that these differences appear to correlate with differences in the people, organization, and community involved.

People

The people involved in policing occupy at least three relevant positions: police officials, clients of the police, and audience of the police. We have noted that federal and state police draw personnel from larger pools than do most local police departments. As with employees of national corporations, state and federal police officials owe a higher allegiance to "the company" than they do to their neighborhoods or places of residence. Transfers of personnel across the country or the state are relatively common in these agencies. In this regard, one of the greatest constants in the life of agents is not where they live but where they work. Therefore, federal and state police officials are more likely to be independent of local or hometown concerns and biases.

The clients of these agencies—both those who call on them for help and those who are affected by them as arrestees and suspects—are often removed from the agency itself. Unlike the local police, where one may know the officer or even the chief, there is not likely to be a personal relationship between the client and the official. Similarly, the narrower purpose of these agencies serves to structure the expectations of clients about what officials might do. Whereas we may hope to get away with a warning from a local officer for speeding, we expect the highway patrol officer to issue a citation.

A study of public perceptions of the state police in the state of Washington was conducted by Mark Correia, Michael Reisig, and Nicholas Lovrich (1996). They found that the characteristics associated with less positive perceptions of state police differed from those associated with perceptions of municipal police. Specifically, age and education were not important, but with state police, women were less likely to report positive perceptions than were men. Correia and his associates concluded that perceptions of state police may be more reflective of general perceptions of the legal and political system. That is, unlike local police, citizens associate state police with the larger society. If they feel disadvantaged in the larger society (women and minorities), they are less satisfied with the state police. In a later work, Reisig and Correia (1997) suggest that how citizens view the police differs according to level of government (state or local), based on citizen expectations of police service at those different levels.

Organization

Their narrower range of responsibilities translates into a clearer organizational mission for most federal and state policing agencies than exists in many local police departments. The operations of the organization are more predictable, and it is easier to establish rules, regulations, and procedures that can govern the on-the-job activities of employees. Administrators of federal and state police organizations can more easily control and direct the activities of their employees than is the case in most local police departments.

Of course, in some instances a state or federal police agency operates like a municipal police agency. While the police of the District of Columbia are a federal police agency, the functions and structure of that organization are more similar to those of other city police than to those of most federal law-enforcement agencies. Likewise, in rural settings where the state police are responsible for general policing functions, it is likely that each local post or troop will be somewhat independent from the central state agency. Mastrofski and Ritti (1996), for example, found this to be the case in a study of driving-under-the-influence (DUI) enforcement in Pennsylvania. They reported important differences between two separate troops of state police. Further, they suggested that those differences were reflections of organizational support for DUI enforcement based on different community characteristics.

Community

In many ways, most federal and state police agents and organizations are free from community influence. This is not to suggest that local issues do not affect operations in any way, but rather that the community served by these agencies is so large and so diverse that informal norms and expectations play a less significant role in shaping agency practice than they do for local agencies.

For the most part, federal and state policing agencies have a reputation for impartial, courteous behavior. This reputation is perhaps a result of the fact that these organizations are removed from the various communities. Officials have little opportunity to learn and be guided by commonly accepted notions of proper conduct, or cues to differential status among the persons with whom they interact. In response, the agents typically treat everyone in a similar fashion and intervene in similar circumstances regardless of geographic location.

CONCLUSION

Policing in America is largely a local responsibility and phenomenon. Over time, however, gaps in the ability of local policing to serve the needs of federal, state, and rural areas have led to the creation of federal and state policing agencies. These agencies typically have a narrower, crime-control, and investigative focus than do local police departments. The differences in functions and organization of federal and state police give these agencies certain advantages and disadvantages when compared with local police.

The development of federal and state policing was influenced by the same forces that shaped the development of local police, but with some important differences. Local police developed in keeping with an American tradition of local autonomy and a general distrust of government power. When policing problems surfaced in federal and state agencies, Congress and state legislatures reacted by creating specialized policing and investigative offices. Rather than a multipurpose social-control agency like the local police department, the state and federal governments tended to create a variety of decentralized, narrowly focused enforcement agencies. Where broader powers were granted, the common practice was to instruct these agencies to assist local police rather than to supplant them. Thus, federal and state policing in America reflect our apparent preference for local police.

CHAPTER CHECKUP

1. How can the development of federal police agencies be described as "haphazard?"
2. Why did federal police agencies expand after the Civil War, and especially in the twentieth century?
3. Identify four specific federal police agencies.
4. Briefly describe the history of the Federal Bureau of Investigation.
5. In what ways was the development of state police similar to that of federal police?
6. What forces led to the development of state police?
7. How do state police and state highway patrols differ?
8. Briefly describe the history and development of state police in two states.
9. What are two major strengths and two weaknesses of both federal and state police?
10. Briefly identify some correlates of federal and state police.

REFERENCES

Bureau of Justice Statistics (1980) *Justice agencies in the United States: Summary report.* (Washington, DC: U.S. Government Printing Office).

Bureau of Justice Statistics (1999) *Law enforcement management and administrative statistics, 1997: Executive summary.* (Washington, DC: U.S. Department of Justice.

Correia, M., M. Reisig, and N. Lovrich (1996) "Public perceptions of state police: An analysis of individual-level and contextual variables," *Journal of Criminal Justice* 24(1):17–28.

Falcone, D. N. (1998) "The Illinois State Police as an archetypal model," *Police Quarterly* 1(3):61–83.

Germann, A. C., Frank D. Day, and R. R. J. Gallati (1988) *Introduction to law enforcement and criminal justice.* (Springfield, IL: Charles C. Thomas).

International Association of Chiefs of Police, Division of State and Provincial Police (1975) *Comparative data report, 1974.* (Gaithersburg, MD: IACP).

Johnson, D. (1981) *American law enforcement: A history.* (St. Louis: Forum Press).

Johnson, H. A. (1988) *History of criminal justice.* (Cincinnati, OH: Anderson).

Klaidman, D. (2001) "A prosecutor's hardest case," *Newsweek* (July 16):25.

Kopel, D. and P. Blackman (1997) *No more Wacos: What's wrong with federal law enforcement and how to fix it.* (New York: Prometheus).

Maguire, K. and A. Pastore (1999) *Sourcebook of criminal justice statistics—1998.* (Washington, DC: U.S. Department of Justice).

Mastrofski, S. and R. Ritti (1996) "Police training and the effects of organization on drunk driving enforcement," *Justice Quarterly* 13(2):291–320.

Morris, S. E. (1995). "United States Marshals," in W.G. Bailey (ed.) *The encyclopedia of police science,* 2nd ed. (New York: Garland Publishing, 1995):796–798.

Mueller, G. O. W. and F. Adler (1996) "Hailing and boarding: The psychological impact of U.S. Coast Guard boardings," *Police Studies* 19(4):57–68.

Poveda, T. (1990) *Lawlessness and reform: The F.B.I. in transition.* (Pacific Grove, CA: Brooks/Cole).

Powers, R. G. (1986) *Secrecy and power: The life of J. Edgar Hoover.* (New York: Free Press).

Reaves, B. (1996) *Local police departments, 1993.* (Washington, DC: Bureau of Justice Statistics).

Reaves, B. and A. Goldberg (1998) *Census of state and local law enforcement agencies, 1996.* (Washington, CD: Bureau of Justice Statistics).

Reaves, B. and T. Hart (2001) *Federal law enforcement officers, 2000.* (Washington, DC: Bureau of Justice Statistics).

Reisig, M. and M. Correia (1997) "Public evaluations of police performance: An analysis across three levels of policing," *Policing* 20(2):311–325.

Smith, B. (1969) *The state police.* (Montclair, NJ: Patterson Smith).

Swart, S. L. (1974) *The development of state-level police activity in Ohio, 1802–1928.* (Ann Arbor, MI: University Microfilms). Doctoral dissertation, Northwestern University.

Thomas, E. and M. Isikoff (2001) "Justice kept in the dark," *Newsweek* (December 10, 2001):37–43.

Torres, D. A. (1987) *Handbook of federal police and investigative agencies.* (Westport, CN: Greenwood Press).

Torres, D. A. (1995a) "Federal police and investigative agencies," in W. G. Bailey (ed.) *The encyclopedia of police science,* 2nd ed. (New York: Garland Publishing):287–291.

Torres, D. A. (1995b) "State law enforcement agencies," in W. G. Bailey (ed.) *The encyclopedia of police science,* 2nd ed. (New York: Garland Publishing):726–730.

Weisheit, R. A., E. Wells, and D. Falcone (1995) *Crime and policing in rural and small-town America: An overview of the issues.* (Washington, DC: National Institute of Justice).

chapter **6**

PRIVATE AND SPECIAL-PURPOSE POLICE

CHAPTER OUTLINE

Private Police
  The Development of Private
    Police
  Private Police Functions
  Private Police Agencies
Special-Purpose Police
  The Development of Special-
    Purpose Police
  Special-Purpose Police
    Agencies and Functions

Pros and Cons of Private and
    Special-Purpose Police
Some Correlates of Private and
    Special-Purpose Police
  People
  Organization
  Community
Conclusion

Thus far we have concentrated on the police agencies and officers of common government units. This chapter broadens our perspective by examining two additional types of police. First, there are numerous private police and security agencies at work in America. These agencies have gained increasing importance in the maintenance of order, provision of services, and control of crime, although they typically lack the full powers and authority of a general-purpose police. Second, there are an unknown but considerable number of special-purpose, or special-district, police in America. These agencies typically have general police powers but are organized by a special government district or agency.

    As with federal and state police, special-purpose police were often developed in response to perceived limitations of a local agency. One obvious solu-

tion appeared to be the creation of special police. For example, many government services, such as transportation, housing, schools, and parks, do not typically fall under the routine supervision of the local police. What can be done if there is crime or disorder on a moving subway train, or in the hallways of a housing project, the school cafeteria, or the picnic area in a large park?

One response, of course, is to call the local police and await their arrival. These police, however, will typically not provide patrol services inside the buildings or on the train. The unwillingness or inability of the local public police to adequately meet the order-maintenance and crime-prevention needs of people using and administering these services poses a problem. In many places the solution to this particular problem was found in the creation of transit authority, housing authority, school district, or park district police.

Even though the local, general-purpose police agency has jurisdiction (and responsibility) to provide police service throughout a given area, it is often seen as not "suitable" for specific functions such as parks, schools, housing projects, and the like. General-purpose local police are geared to provide a range of services in support of the varied activities that the general public engages in. Special-district police, by contrast, are geared to providing a more limited range of police services to a relatively specific population engaged in a more limited range of activities. By virtue of the more focused environment (population and their activities defined by the purpose of the special district), these special-district police face a different mix of police problems than do the general-purpose police. For example, transit authority police may be more focused on managing the homeless, university police on providing security, and housing authority police on handling residential problems than on dealing with traffic or commercial offenses.

Private policing has also gained importance in recent years. Private police have a much longer history than do most special-purpose police. You will recall that the public police are an outgrowth of an earlier tradition of private responsibility for law enforcement and order maintenance. In many ways, the original watch-and-ward system of London and colonial American cities was a form of private police (Becker, 1995).

This chapter examines the development and spread of private and special-purpose police agencies in America. We describe the history, distribution, organization, and role of private and special-purpose police, and discuss the relationship between these specialized police and their counterparts in general government service.

PRIVATE POLICE

"In recent decades, private security has outstripped the growth of the public police; more people are now working in private security than in public policing" (Reiss, 1988:1). The current increase in private policing is not so much the development of a new form of policing as it is a resurgence of a past practice. Public policing grew from private roots, and in more recent times, private policing has emerged in response to the perceived limitations of public police.

Private police is an imprecise term that includes all sorts of privately operated and funded safety and security services. In an early study of private-sector services, the Rand Corporation (Kakalik and Wildhorn, 1972) used the term *private police* to refer to this component of the police industry. But, as Cunningham and Taylor (1985:101) observe, "Private security . . . involves very comprehensive programs of physical, personnel, and information security to protect the assets of an organization and to reduce losses." That is, the private security industry is quite diverse.

It is difficult to determine exactly how many private police agencies operate in America. Gallati (1983) estimates that over 70,000 such firms exist. In 1980 Cunningham and Taylor (1985) reported surveying managers of nearly 1,300 security firms. Surveys of local firms were limited to only 150 of the nation's metropolitan areas. Whatever the actual number, it is safe to say that the private police outnumber the public police, both in terms of agencies and personnel.

Let us consider the variety of organizations, structures, and personnel denoted as private police. It is common, for example, to have security guards stationed at construction sites and retail shopping malls. Anyone who has gone to a major amusement park or attended a concert or professional sporting event is also likely to have seen private police. Hotels often operate their own security office, as do industrial plants. Armored car services are typically operated by private security companies such as Brink's and Wells Fargo. Retail stores often use undercover security officers to control shoplifting and employee theft. Larger corporations frequently employ an investigative force to counteract industrial espionage, perform background checks on employees, and other such functions. Finally, we are probably all familiar with the "private eye," who is, essentially, a general-purpose detective for hire.

The Development of Private Police

As described earlier, private police existed before there were any public police, but the subject of this chapter is those more recent private police who engage in "policing" for a fee under a private-enterprise model. Thus, the "watch-and-ward," in which citizens were obliged to police their communities by the government does not truly fit this definition. Yet, even in that arrangement it became common practice for wealthier citizens to hire others to take their places on the watch. Many, like Daniel Defoe, paid fees to avoid service as constable.

More similar to the modern private police were the bodyguards (or footmen) and gameskeepers who worked for the wealthy before the advent of public police. These people were forerunners to today's private security officer. Others, such as the "thief takers" who became Fieldings' Bow Street Runners, supported themselves on the reward money granted for successful apprehension of criminals.

The first and largest private security firm in the United States was started by Allan Pinkerton in 1855 as the Northwest Police Agency (Morn, 1995). Pinkerton had gained considerable fame for his handling of several cases involving counterfeiters and for a time worked as a special deputy or agent of the Cook County (Chicago) Sheriff and U.S. Treasury Department. For nearly five years, between 1850 and 1855, he occupied a precarious position between private agent and government employee. In 1855 he decided to open his own business.

Perhaps Pinkerton's biggest customers were the railroad companies in Illinois. Because local police were lacking in rural areas, railroad facilities and equipment there were unprotected from vandalism and theft. Railroad employees, particularly conductors, handled large amounts of cash from

---

 Allan Pinkerton (1819–1884)

The son of a police officer, Pinkerton was born in Glasgow, Scotland, and emigrated to the United States in 1942. Settling in the Chicago area, Pinkerton followed his trade as a cooper, or barrel maker. While searching for wood for barrel staves, he discovered a camp of counterfeiters. When he returned with the sheriff to arrest them, he became an instant hero and thus began a career in law enforcement. Pinkerton investigated counterfeiters for the Treasury Department, kidnapping for the Cook County Sheriff, and mail theft for the Postal Department. He opened his own agency in 1855, and during the Civil War served as both an intelligence officer for the Union Army and a bodyguard to President Lincoln. Flamboyant and prolific, Pinkerton did much to excite public interest in criminal investigation and to help create the mystique of the detective. (The Bettmann Archive)

ticket sales and frequently stole from the company. Pinkerton and his employees set to work protecting railroad property and detecting employee theft, for a fee. Pinkerton and his agency became world famous, largely for their work with the railroads and other corporate entities.

In the earliest years of formal policing, most police departments did not have investigators, or detectives. The lack of a detective capability was not accidental. In the middle of the nineteenth century, only three models of detectives existed: informants, thief takers, and agents provocateur (Klockars, 1985). None of these were acceptable to a public that distrusted government power and the police. Informants are often disliked today—nobody likes a "tattletale." Thief takers, as we saw, were motivated by the reward, and did not serve the interests of the poor or the general public. In fact, they were often found to be in league with criminals. The agent provocateur was the most frightening of the three, however. These agents would encourage innocent people to commit crimes in order to make the arrest and gain the reward.

Several forces combined in the middle of the nineteenth century to encourage the development of private policing (Maahs and Hemmens, 1998). The lack of detectives and investigators among the new police, coupled with the growth of business and industry, meant that criminals who were not caught in the act by a patroling officer faced little risk of arrest. The "public" role of the police also meant that crimes committed inside factories, such as employee theft, were similarly riskless. The growth of the labor movement, of course, threatened the interests of elites, but the political power of the workers often made it difficult for local police to break strikes. Finally, transportation and communication improvements meant that travelers and shipments were without police protection as they moved through rural areas between cities.

Many of the nation's most well-known names in private security got their start in this period. We have already seen the importance of the railroad business to the growth of Pinkerton's agency. Wells Fargo began by providing security for its own stagecoaches. Often it was necessary to show that a shipper had security in order to obtain contracts to transport the mail. In addition, passengers were more willing to travel when they felt they would be protected by the transportation company. Brink's, Inc., which is most famous for its armored car service, began operations in 1859 to provide secure transport for valuable shipments.

In 1909 William Burns and William Sheridan started the Burns and Sheridan Detective Agency. They secured a contract with the American Bankers Association and provided protection to its 11,000 member banks. Within a year, Sheridan left the company, and it became the William Burns' National Detective Agency. Today Burns International is one of the largest private security companies, providing both uniformed and nonuniformed security personnel to a wide variety of private companies and associations. Recall that Burns was J. Edgar Hoover's predecessor as chief of the Bureau of Investigation. He had been employed as a federal agent from the time he started his agency in 1909 until he retired from federal service in 1925.

In more recent years, private policing has benefited from a growing perception that the public police cannot adequately protect citizens. Just as the inability (and, occasionally, the unwillingness) of public police to meet the needs of growing companies in the nineteenth century created a market for private policing, this perception of limited public police ability has spawned a number of private policing/security developments.

Becker (1995) reports that many observers believe the rising crime rates since the 1940s have supported increased reliance by both businesses and individuals on private security measures and private police. Hou and Sheu (1994) studied participation in private security systems in Taiwan. They found that one factor that explained citizen contracting with private security providers was a perception that the public police were ineffective. Private police are contracted to provide police services in a variety of settings, from residential neighborhoods to hospitals. As Germann, Day, and Gallati (1988) put it, "Private police may be employed by almost anyone or almost any organization." In some places, smaller municipalities have contracted with private police for the entire police responsibility of the jurisdiction (Germann, Day, and Gallati, 1988).

 William John Burns (1861–1932)

Founder of Burns International Detective Agency, William Burns was born in Baltimore, Maryland. His father became police commissioner in Columbus, Ohio, beginning his interest in criminal investigation. Burns joined the Secret Service in 1889, where he remained for 14 years until transferring to the Department of the Interior. In 1909, he founded his private detective agency with partner William Sheridan and took sole ownership when Sheridan left the business in 1910. Burns was named head of the Bureau of Investigation in the U.S. Department of Justice in 1921 and retired in 1925. (UPI/Bettmann Newsphotos)

Private Police Functions

As might be expected from the size and diversity of the private policing component of the U.S. police industry, private police serve a variety of functions. In brief, private police do what they are paid to do. Hence, the importance of different functions, such as order maintenance or crime prevention, is defined by the person or organization hiring the private police.

Regardless of the type of structure employed, private security and policing serve three basic functions (Cunningham and Taylor, 1984): physical security, information security, and personnel security. **Physical security** involves the protection of people and property and can be seen in the use of security officers at the entranceways to factories and construction sites or at ticket counters at attractions. **Information security** normally involves attempts to prevent industrial espionage activities and often is accomplished through preventive means, like computer access codes, limited access to facilities, and "counterespionage" investigations. **Personnel security** includes the protection of workers and executives from terrorists, demonstrators, kidnappers, or others who pose a threat. This function also includes background investigations, drug testing when appropriate, and the enforcement of safety regulations.

Within these three broad categories, private policing employs both uniformed patrol personnel and plainclothes detectives and investigators. Often these personnel serve as an adjunct to the public police, as when retail security officers apprehend a shoplifter in order to turn the offender over to the local public police. On other occasions, however, private police are independent of the public police, particularly where sanctions do not require the criminal law. For example, rowdy behavior at an amusement park (isn't that the point of an amusement park?) can be punished by expulsion or by arrest for disorderly conduct. If the practice is to expel unruly patrons, the private police of the park may have little interaction with the public police.

A growing trend, though, is in the direction of using private police as an adjunct to the public police. Walsh and Donovan (1989) report on one community in Brooklyn, New York, that operated its own general-purpose preventive patrol in private security. In the community of Starrett City, the realty company that owned the property on which this planned community existed operated a 59-officer security force within the city of New York. The security officers provided preventive patrol that included maintenance of order, delivery of services, and control of crime. The community was also served by one patrol beat of the New York City Police Department. Similar use of private security to augment existing public police has been seen in other locales (Shearing and Stenning, 1983).

A variation on this use of private police in a role that is typically associated with the public police occurs in contract policing. A number of municipalities across the nation enter into contracts with other jurisdictions for the provision of basic police services or to augment existing local, public police. Thus, a township might contract with a neighboring city or the county sheriff for police protection. Similarly, a community with its own small police agency might contract for additional police protection from another government unit.

Peter Colby (1995) reviewed the status of contract policing. As of 1994, Colby reports, only 74 percent of cities and counties surveyed provided police protection entirely on their own. Five percent used intergovernmental contracts, 3 percent used contracts with private companies, and 2 percent contracted with nonprofit organizations such as neighborhood associations. The remaining 16 percent apparently employed a combination of contracted and self-provided police protection.

Private Police Agencies

Private policing is provided in two basic forms. First, in **proprietary security** form, the principal, or person wishing to receive service, hires security personnel directly. This model is common in industrial settings, hospitals, and amusement parks, where the business hires, trains, and operates its own security force. Second, in **contract security** form, the principal rents services from an independent security company. This model is commonly used at places such as construction sites, where the general contractor for the construction project subcontracts with a private security company for site-protection services. Thus, if you wanted to protect your home, you could employ your own security force (proprietary) or contract with an agency like Burns or Pinkerton for security coverage.

Richter Moore (1987) reports that one effect of the growth of private policing has been an increased effort on the part of states to regulate this part of the policing industry in America. Thirty-five states required licensure of security personnel or agencies, but few states require any specific educational level for security personnel, and only 22 states had any mandatory training requirements. Ten states had no regulations for the private security field. Even in those states with requirements, it was common for agents of a proprietary security force to be exempt from state training and licensure requirements. More recently, Jeffrey Maahs and Craig Hemmens (1998) report that state statutes regulating private security personnel existed in 43 states, but that the content of statutes varies considerably across states. They note (1998:131), "there is no uniform pattern from state to state not only in what issues are regulated, but also in the specific requirements for each one." Only Hawaii and Michigan set a minimum educational level (eighth grade), and even the most common restrictions—fingerprinting and criminal records checks—are required in fewer than two-thirds of states.

No available national data allow a classification of the thousands of private policing organizations. Thus, it is impossible to determine how

Hospital security officers using closed-circuit television to police the grounds.
(Larry Mulvehill/The Image Works)

many are proprietary versus contract, or how many are private detective agencies versus uniformed patrol agencies. Nevertheless, given the nature of this part of the industry, and the large number of retail, hotel, industrial, and similar security operations, it is probable that the majority of organizations fall into the proprietary category. Yet the smaller number of contract agencies include very large organizations like Pinkerton, Burns, Wells Fargo, and others. It is quite likely that these contract agencies, while smaller in number, actually serve more clients than do proprietary agencies.

SPECIAL-PURPOSE POLICE

Yet another component of the police industry in America is composed of special-purpose public police agencies (Moore, 1995). We have already mentioned a few of these agencies earlier in our discussion of state police, but over 1,600 such special-purpose or special-district police exist in the United States (Reaves, 1996). A number of public police agencies have been created since the mid 20th century that are granted general police powers but given a relatively narrowly defined task. These agencies are charged with providing police services to government-run housing, transit, education, recreation, and similar services.

Some police agencies have special functions, such as the harbor or river patrol.
(Larry Kolvoord/TexaStock)

For example, several states that operate statewide police or highway patrol forces also provide **park police** for state parks or **campus police** for state colleges and universities. Similarly, in a number of cities and counties, the municipal governments operate special-purpose police agencies in addition to general-purpose city or county police. These agencies augment the local police and are important to determining the total level of police protection and services available in a community. Brian Reaves (1989:8) describes special-purpose police agencies this way: "They usually have full police powers but within limited jurisdictional boundaries. Airports, public housing, public parks, public schools, and transit systems are the most common examples of special police jurisdictions."

In this manner, special-purpose police are organized along a functional line that reflects the public service they are to protect. Housing authority police typically have full police powers and jurisdiction within the boundaries of the government housing authority's properties. Transit authority officers are given jurisdiction over the transportation facilities of the government transit authority. A city police officer might have jurisdiction covering the entire city, but the housing authority officer might be limited to public housing buildings and grounds.

On the other hand, whereas a city police officer is typically restricted to the area within the city's boundaries, special-police officers often have jurisdiction in areas that cross municipal or even state borders. The New York

TABLE **6.1** Types of Special-Purpose Police

TYPE OF JURISDICTION	NUMBER	PERCENT
College/University	699	53.1
Public School District	117	8.9
Airport	84	6.4
Natural Resources/Conservation	79	6.0
Criminal Investigation	72	5.5
Parks/Recreation	68	5.2
Medical School/Facility	42	3.2
Waterways/Harbors/Ports	38	2.9
Transportation	28	2.1
Government Buildings	24	1.8
Alcoholic Beverages	17	1.3
Public Housing	13	1.0
Other	35	2.6

Source: B. Reaves and A. Goldberg (1998) *Census of State and Local Law Enforcement Agencies, 1996.* Washington, DC: Bureau of Justice Statistics, p. 12.

City Transit Authority police, for example, had jurisdiction over the facilities of the Port of New York, which includes areas in two states (New York and New Jersey), and numerous municipalities.

Like their private police counterparts, special-purpose police developed to supplement general-purpose local police. Like federal and state police, special-purpose police were created as needs for additional police protection were recognized. Thus, the development of government services that crossed city lines, such as transit authorities and some parks, led to the creation of police organizations that could cross jurisdictional boundaries. As municipal government broadened the range of services it provided, such as housing, transportation, and public education, specialized police services were authorized.

The Development of Special-Purpose Police

Perhaps the first example of a special-purpose police agency in the Anglo-American experience was the Thames River police of London. As we saw in Chapter 3, concern about crime and disorder along the docks and wharves of London led to the development of a special police agency to patrol and protect the river and harbor area. Additional police were available within the metropolitan area, including the mounted patrol, dismounted horse patrol, and day patrol. These patrols, however, were not sufficient to control smuggling and theft on the waterfront, and so the Thames River police in the early 1800s became perhaps the first "harbor patrol."

In the United States, Yale University in 1894 organized a campus police consisting of two commissioned New Haven police officers. These officers policed the campus of the university and were separate from the New Haven Police Department. After the 1920s, similar campus guards, or watchmen, became more common (Petersen and Bordner, 1995), but full-scale campus policing was rare. In the 1950s college administrators increasingly created public safety or police agencies, typically under the command of retired police officers.

These new campus police officials tended to adopt the organizational model with which they were most familiar—that of the municipal police agency. Thus, throughout the middle of the twentieth century, campus police as a special-purpose agency became increasingly common. Public institutions, especially larger ones, tended to operate full-service police agencies on campus. Private institutions tended to operate either contracted private policing (smaller schools) or proprietary security and police agencies (larger schools). In all cases, however, despite their origins (public or private) and the range of powers (full police powers or private security designation), campus police focus on the maintenance of order, delivery of services, and enforcement of laws within the boundaries of the campus. Their clientele and purpose are different from those of municipal police. Campus police tend to emphasize service and order-maintenance activities over law enforcement (Petersen and Bordner, 1995).

The New York City Transit Police provide another illustration of the development of special-purpose police. Beginning in 1936, the mayor of New York created the post of special patrolman on the subway system (Murphy, 1995). In 1940, when the city took over all the private subway lines, more special officers were hired, and in 1947 these officers were granted peace officer status.

Over the next three decades the transit police grew. The department employed over 4,000 officers assigned to 12 zones by the early 1990s (Moore, 1995:722–723). The transit police patroled the subways and trains. They conducted sweeps, plainclothes antirobbery efforts, and emergency medical and rescue services. A detective unit in the department investigated major crimes occurring within their jurisdiction. In 1995 the transit police were merged into the New York City Police.

Similar transit authority police exist in other cities, such as Boston, Washington, D.C., and San Francisco. While not as large as was the New York Transit Police, these departments provide similar police patrol service on the transit systems in their respective cities. Housing authority police have developed along similar lines, with the largest such agency again being found in New York City.

Special-Purpose Police Agencies and Functions

Special-purpose police agencies represent a kind of catch-all category for those police organizations that do not fit neatly into the federal, state, private, or local police categories. We are concerned here primarily with police agencies that have a relatively limited mission typically reflecting a clearly defined geographic area, such as a park or transit system. However, some observers have included in this category state game wardens, liquor control officers, and other special police whose mission is limited more by the laws they routinely enforce than by geography.

The confusion over exactly which organizations should be included in special-purpose police makes it difficult to determine how many such agencies exist in the United States. The Bureau of Justice Statistics (1983) estimated that over 1,100 such agencies were in operation. In Ohio alone, however, the governor's office of Criminal Justice Services (1987) reported 100 special-purpose agencies. The Ohio report included a number of state law-enforcement agencies, counting all of those that required peace officer training of personnel except the state highway patrol, which was considered to be a state police agency. A 1996 survey by the Bureau of Justice Statistics (Reaves and Goldberg, 1998) reported more than 1,300 special police agencies employed over 43,000 sworn officers.

It is not possible to say with confidence how many and what types of special-purpose police agencies exist. We have seen that the police are an arm of government (exclusive of private police, by definition). Special police can therefore be considered police agencies operated by and for special, or specific, government functions. They are government agencies designed to provide police protection to fairly narrowly defined areas. The Bureau of Justice Statistics survey revealed that the largest categories of special police served public colleges and universities, conservation and natural resources, public schools, and transportation systems.

Whereas a general-purpose police agency will be responsible for policing a government jurisdiction (such as a city, township, or village) in general, the special police agency is responsible for generally policing a more specific area. Thus, transit authority police might protect buses, bus lines, trains, and subway lines, yet not be responsible for the surrounding neighborhoods. In the same fashion, housing authority police might be responsible for public housing buildings and grounds but not for the area across the street. Park police might enforce the laws, maintain order, and provide services within the boundaries of the park itself but not be authorized to intervene in situations outside the park gate.

The distinction between special-purpose police and the traditional municipal police is largely one of geography and variety. Special-purpose police are often police generalists who perform all of the functions of police—order maintenance, service provision, and law enforcement. They do not differ from the general-purpose police in terms of what they do. The difference between the two relates to where they do it. However, the nature of the special functions served by the government agencies and districts that employ special police tends to reduce the variety of tasks performed by them. Housing authority police, for example, are not likely to respond to many bank robberies or traffic accidents.

A city police agency has jurisdiction over the entire city and therefore polices everywhere within the city. A special-purpose police agency, on the other hand, has a more limited mandate and polices only those areas within the city that are assigned to it. These special-purpose police are, then, general-purpose police who work in special places with distinct populations. In the end, the combination of limited areas and populations results in a more limited version of "general policing."

One fairly common type of special-purpose police agency is the campus or university police department. Reaves and Goldberg (1996) report a survey of four-year colleges with 2,500 or more students enrolled. Of 682 campuses meeting these criteria, 680 reported having some type of police or security agency. Three-fourths of them employed officers who were granted general arrest powers by state or local government. Larger and public colleges were more likely to use sworn officers than smaller or private schools. Most campus security/police agencies used armed officers, but less than one-third of private campuses employed armed officers. Nearly one-fourth of all campuses contracted for security/police services, and most of these hired private security firms. Arrest powers were limited to campus for about two-thirds of private college law-enforcement officers and for nearly one-half of public college officers. In some cases officers were granted authority to arrest on campus and in an area around campus. However, public campus law-enforcement officers were usually granted arrest powers for the entire municipality, county, or state in which the campus was located.

PROS AND CONS OF PRIVATE AND SPECIAL-PURPOSE POLICE

Like their federal and state counterparts, private and special-purpose police have both strengths and weaknesses when compared with general-purpose local police organizations. These reflect the specialized functions and organization of private and special-purpose police agencies. By their very nature, these types of agencies receive a narrower and more clearly defined mandate than does the general-purpose local police agency.

Although a special police organization might perform the same three general tasks of the local police, it does so in a different context. Airport police, park police, housing authority police, campus police, transit authority police, and the like, are run by special government entities that themselves have a limited purpose. For example, a housing authority police force concerns itself with public housing projects. It is not typically required to balance the interests of the business community, middle-class homeowners, and residents of the housing projects. Rather, this special-purpose police agency gives priority to the needs and concerns of the housing project's residents. It has a more focused and more limited jurisdiction than the general-purpose agency.

Both special-purpose police and private police have a different organizational relationship with policy makers than does the general-purpose local police. Particularly with regard to private policing, the employer sets the priorities and limits of the agency instead of imposing an ambiguous goal such as to "serve and protect the citizens and property of the city." Further, because of the limited jurisdiction and interests of these types of police, they frequently have more authority over the public than do general-purpose police.

Private security officers, for example, are fully within their powers to "eject" someone from private property, such as a factory. The public police, on the other hand, cannot "eject" a citizen from public space (although they might "remove" a person by arrest or order someone to "move along"). Consider the differences between airport police and the average city police officer. No matter how useful the practice might be, it is highly unlikely that the local police will be allowed to operate magnetic scanners on passersby. In airports around the nation, however, all persons entering the airport concourse are subjected to such searches.

In the past, passenger searches and much airport security was provided by private security firms working under contract with the airlines. Airport police, where they existed (usually at larger airports), had general policing responsibilities (traffic, order maintenance, etc.) in the airport area. In the wake of the terrorist attacks on September 11, 2001, changes in airport security and airport policing are underway. In the future, airport police will be federal, creating a new federal police agency. At present, military personnel have been assigned to increase security at airports, and to oversee the operations of private security companies.

Both private police and special-purpose police add an increment of police service to areas in which they operate. As Ostrom, Parks, and Whitaker (1978) discovered, there is little duplication of police service in practice. Thus, if there is a housing authority police, the general-purpose police are not normally going to patrol the area. If there is a private police agency operating in an industrial facility, the public police are not as likely to be called for service. Private and special-purpose police reduce the workload and demands on the general-purpose police, allowing the local police to concentrate more heavily in those areas lacking private or special police.

The disadvantages of private and special policing also reflect their specialized functions and organization. A frequent criticism of these agencies is that they add to the confusion and fragmentation of policing in America.

Consider the situation in which someone steals valuables from a shipment delivered to the docks of a city's harbor. This miscreant then makes an escape via public transportation. Upon arriving at the desired station, the offender next travels several blocks on foot before entering the public housing project in which she resides. Now assume there is not only a general-purpose police agency in the city, but also a harbor police, transit police, and housing authority police. Four distinct police organizations can become involved in this crime, and their efforts at pursuit and apprehension must be coordinated. From an administrative perspective, the case would be more easily managed if only one police agency were involved.

Just as the specialized functions and organization of these private and special-purpose police make their task easier by narrowing its scope and area, they also limit what these agencies can do. Although police themselves, if the cause of a problem exists outside their jurisdiction (say, for example, an unruly crowd gathered outside the gates of the park, factory, or other protected area), these police must now "call the cops" for help.

The existence of private and special-purpose police agencies also raises difficult issues of fairness and accountability. The exact conception of these issues depends on the viewpoint of the observer, but in general it can be asked if private and special-purpose police are fair. How do these types of police affect the distribution of police protection among the citizenry, and how do they affect the general-purpose police?

Given that the police, at least government police, have a **commonweal obligation** (Lundman, 1980) to serve and protect the general citizenry, these agencies represent the provision of special, additional police protection to certain people that is not available to all. Housing authority police can be said to give added service to those living in government-operated housing as compared with those residing in private residences. Transit authority police similarly protect public transit riders and not those taking taxis and other forms of private transit. Is it fair that these people receive additional policing?

This issue is even more apparent in the case of private police. As we saw, when needed, even the private police can themselves call the cops. Public, general-purpose police protection is available to all. In addition, those who can afford it, and choose to do so, can pay for additional police service through the private sector. In this instance, the difference in levels of protection is based on economic status rather than some definition of special government interest. With private police, "the rich get richer" in terms of police protection, while the less well-to-do must get by with whatever level of protection is generally afforded by the public police. Is it fair that levels of police protection are based, at least in part, on ability to pay?

When it comes to the impact of these private and special-purpose agencies on the public, general-purpose police, we may raise similar issues. If a special-purpose police agency is responsible for an airport, does that mean that the general-purpose police are excused from serving this area? Can we excuse the police from responsibility in some areas and yet hold them accountable for crime, order, and service throughout the general jurisdiction?

If the special-purpose police or private police are deficient in some fashion, to whom do the citizens complain? Can we and should we have a situation where we expect the general-purpose agency not only to police the community but also to regulate these other agencies? The dedication of resources (money, personnel, and equipment) to special and private police may detract from a community's ability or desire to invest in its general-purpose police. That is, does the existence of several police agencies mean that they compete with each other for public interest and support? What then is, or ought to be, the pecking order among these agencies? Which of them should be given first consideration in the competition for resources?

The fact that special-purpose police have developed, and that the private police component of the American police industry is experiencing such tremendous growth, indicates that people perceive a need for these additional focused police services. What is as yet unclear is whether the creation of limited-interest policing agencies is the most effective means of responding to the needs peculiar to special places.

SOME CORRELATES OF PRIVATE AND SPECIAL-PURPOSE POLICE

As with federal and state police, the differences between private and special-purpose police as compared with general-purpose local police correlate with the people, organizations, and communities involved. The major distinctions between private and special police versus their general-purpose local counterparts can be summed under the heading of *jurisdiction*. The jurisdiction of the agency reflects both its geographic and functional limits. These limits stem from the people and communities served and are reflected in the organizations of the agencies themselves.

People

The "people" component of private and special police is composed of agency personnel, clients, and audience, just as the people component of other aspects of the police industry is defined. However, here we have two different types of agencies; one a government agency and the other not. This difference in structure affects the people involved.

Private policing has long suffered an image problem related to its personnel. The traditional view (and depiction) of the private police officer was that of an uneducated, undertrained, underpaid, and out-of-shape rent-a-cop. Although having changed somewhat for the better, this image still haunts private police personnel (Bennett, 1987). Moore (1987:25) reports that educational, experience, and training requirements for private police are relatively rare, concluding,

Personnel involved in private security come into regular contact with the public. In fulfilling their responsibilities to their employers to protect person and property private security personnel exercise very substantial powers. These powers include the use of deadly force. Yet, as we have seen, there are few requirements that these private security personnel know when and how to use the power that is bestowed upon them.

Maahs and Hemmens (1998:123) report an earlier stereotype of the typical private security officer as, "an aging white male who is poorly educated and poorly paid." Since then, they suggest, today's security personnel are younger and slightly better educated than in the past, but still poorly paid. The low pay hinders recruitment of officers, and helps explain a personnel turnover rate estimated to be between 121 and 300 percent per year (Maahs and Hemmens, 1998:123). In a study of private security officers working at a Canadian college, Anthony Micucci (1998) identifies three distinct types of officers. "Crime fighters" were younger, better-educated officers whose professional identity and aspirations were linked to careers in public policing. "Bureaucratic cops" were administrators whose personal attitudes supported crime fighting, but whose professional responsibilities (the demands of the employer) required them to support service and loss-control activities. The "guards" were older, least educated, and had the longest tenure in the present job. These officers saw their service and loss-prevention activities as being most important, and sought to avoid using force or formal law-enforcement actions. The orientations of the "guards" were most in line with the desires of the college concerning how the security service should act. Micucci (1998:49) observes, "There is some irony in the observation that the much maligned and numerically shrinking guard segment best exemplified in its operational style . . . the employer's goals and performance requirements. . . ."

Often, in the case of private police, powers are granted to the agency with little attention paid to the employees of the agency. In this regard, private police are frequently expected to behave in a ministerial manner. That is, the policies and procedures of the organization, coupled with supervision by superiors, are expected to make the work of private police officers more or less routine. To the degree that private police perform a routine ministerial function, of course, their status as professionals (a title that assumes expertise and the exercise of judgment) is limited. The relative freedom of private police administrators to fire or otherwise discipline employees, together with their narrower, client-defined role, enables these agencies to perform in a more ministerial fashion than general-purpose public police agencies.

Special-purpose police typically are required to meet any state standards for law-enforcement officers. As we saw in regard to Ohio, the officers of housing authority, transit authority, and campus police departments in that state must meet the minimum training standards for general-purpose police officers. In many ways, the personnel of special-purpose agencies mirror the characteristics of those in general-purpose local police departments. They tend to be hired from the same personnel pools and be subjected to similar training and civil service personnel policies.

The clients of these two types of agencies differ from those of the general-purpose local police. For private police, the primary client is the person or organization that pays for the service. For special-purpose police, the clients are those who use or frequent the facilities or institutions they are directed to protect. In both cases, these agencies serve a more homogeneous clientele than does the general-purpose local police department. On the one hand, private and special-purpose police agencies interact with a broad spectrum of the population. On the other hand, there is typically a focus or common interest in that population that binds them together. Those who deal with the transit authority police, for example, tend to be users of public transportation and tend to share a desire to travel between points A and B in an orderly and efficient manner. The local police, on the other hand, interact with a variety of people who will infrequently share a common interest.

This important difference in clientele is also related to differences in audience. The general-purpose local police serve several audiences, but their general audience is an amorphous collective of the general public. The employers of these police—city and other municipal governments—serve the general public. As an audience, the general public represents a diversity of interests and expectations; it provides little direct guidance as to how a police agency should behave or what it should do.

Private and special-purpose police, in comparison, serve a smaller slice of this pie. To be sure, a rude or ineffectual private police agency will generate public resentment. However, it is less important how the general public reacts to the private police than it is how the employees or customers of the contractor react. Special-purpose police, too, must attend to the relevance of their audiences, since the general public as a collective is relatively unimportant. They deal with people within the confines of a small part of their lives. Park police, for example, need not be overly concerned with whether or not someone is disorderly or drunk in general, but only with how that person behaves in the park. This limited focus enables special-purpose police to be more flexible in their response to people and, in turn, means that their behavior is less routine and ministerial.

We can use a television broadcasting analogy to illustrate these differences in audience. A national network is similar to the general-purpose police. To be successful (to attract advertising revenue) the network must appeal to a large, broad audience. It must provide many programs for many people, from news through documentaries and dramas, to situation comedies and soap operas and Saturday morning cartoons. The special-purpose police are like a public broadcasting station. Public broadcasting stations attempt to provide many types of programs, but these stations must first and foremost provide educational programs for schools. Although the support of a broad, general audience is helpful and desirable to them (especially around pledge time), service to the educational audience is mandatory. The private police can be likened to a pay cable station. If one person is willing to pay enough, a separate station could be created to broadcast nothing but documentaries on the lives of insects, The Insect Channel. It is not important if most people (even if that means everyone else) are repelled by the idea. As long as that viewer is willing and able to support the channel, it can operate. Private police might like a broader audience (or a wider market share of the private police business), but a private agency must serve the wants of the people who pay for it first.

Organization

Private and special police display a variety of organizational structures reflecting the diversity of their functions. Most special police agencies are organized along bureaucratic lines, with a chief administrator and functional units for patrol, inspection, investigation, and the like. The height of this structure is related to organizational size, so that large agencies tend to have more special units and more levels of command than do smaller agencies. As public, general-purpose police agencies, these organizations resemble the local general-purpose police organizations. The primary distinction between the two is that the special-purpose police are limited in geographic jurisdiction and mandate to a specific government function or set of related functions.

Private police, too, run the gamut of organizational structures from small, undifferentiated offices to large, multinational corporations. They tend to include not only an administrative structure for the management and delivery of police services, but additional units or responsibilities for marketing and other business concerns. In some cases, individual officers are dispatched from a home office to serve a defined area. In other cases, a typical police agency is developed to provide police services to an institution such as a hospital or amusement park.

The more general the responsibilities of officers for order maintenance, service provision, and law enforcement, and the larger the geographic area covered, the less able the organization is to define and control employee behavior. Thus, a large private policing entity could administer the provision of well-structured police service of a ministerial sort to a variety of customers by tightly defining the role and responsibility of officers and providing closer supervision. One company, for example, could be organized to supply virtually interchangeable officers to provide security at banks in general. In other circumstances, the variety of tasks performed by employees requires that broader discretion be granted to officers.

Primarily what distinguishes these police agencies is the functions they specialize in serving. For example, large-contract private police agencies are likely to have a specialized marketing unit, but municipal, special-district, and even proprietary private police agencies are not likely to have such a unit. The municipal police may have a burglary investigation unit, but it is unlikely a similar unit will be found in a park police agency. The distinction, therefore, is not based on organizational structure but rather on functional specialization.

Special-purpose and private police organizations tend to differ from general-purpose police in terms of the types of tasks they routinely perform. These agencies, especially private police, have a much clearer mandate and therefore are much better able to tell their employees what to do. They can be more directive than the typical municipal police department. In contrast, officers in general-purpose departments, who must be all things to all people, are more self-controlled. Thus, administrators of special and private police agencies can more easily control employee actions than can those of municipal police agencies.

Community

As indicated in our earlier discussion of agency audiences, private and special police tend to serve a more narrowly defined community than general-purpose local police organizations. Although they must attend to the interests of the communities they serve, the definition of their jurisdictions as transportation facilities, factories, commercial establishments, or college campuses includes within it a community of interest among the people they serve. This community of interest plays a significant role in shaping agency practice.

For example, proprietary private police will be likely to base intervention decisions on the status of the individual. Accordingly, customers will be treated as customers, employees will be handled circumspectly and deferentially, and trespassers will be dealt with firmly. In all cases, participants are likely to understand the rules. In many ways, the existence of rules and their general understanding by all concerned make order maintenance, in particular, an easier task for these police.

Visitors at a park, for example, probably know if alcohol consumption in the park is banned. Seizure of a contraband six-pack by the park police is less troublesome than a similar intervention by city police involving someone drinking in a vacant lot in the city. The shared understanding of what is allowed and what is not informs both the citizen and the police about how and when to intervene. As long as these shared understandings are maintained, interactions among citizens and officers are less troublesome.

While the general-purpose local police can be considered to have an overriding goal of maintaining the peace, this goal does little to help direct their activities or inform the public of what to expect from them. Private and special-purpose police have narrower and more broadly agreed-on overriding goals. The public is at liberty to avoid the intervention of these police simply by not entering their jurisdiction. Thus, the relationships between private and special-purpose police are more predictable and more circumspect.

For example, if one does not wish to deal with transit authority police, one need only avoid public transportation (admittedly, this is sometimes easier said than done). If one wishes to avoid the local police, however, one is forced to avoid the city. Although private and special-purpose police perform enforcement and regulatory duties, most of the people with whom they come into contact recognize that they do so as a service. The private officer at the amusement park may make sure that we do not cut ahead in line, but we recognize that he or she is also protecting our place in line. For a variety of reasons, this same understanding of the general-purpose police is often lacking. People in general tend to be more accommodating and accepting of private and special-purpose police. There is a greater sense of community, or shared interest, and of these agencies as providing a community service than exists with the general-purpose police.

CONCLUSION

Just as perceived shortcomings or limitations in the ability or willingness of the local, general-purpose police to meet specific needs led to the creation of federal and state police, it has supported the development of private and special-purpose policing. These agencies typically have a narrower definition of jurisdiction, which includes either a restricted geographic area, narrower function, or both.

The development of private and special-police agencies was influenced by the same forces that shaped the development of other American police, but with important differences. Local general-purpose police developed in keeping with our traditional distrust of government power. Private and special-purpose police had the advantage of being more directly under the control of their creators, or having the local police serve as forerunners to them.

Instead of having a multipurpose, broad-based social-control mandate, these agencies are restricted in terms of place and interest. Their narrower jurisdictions make them both avoidable and more understandable to the public they serve. Although they often perform the same duties as the local police, they do so in a context qualitatively different because of their particular combination of jurisdiction and community.

CHAPTER CHECKUP

1. How are private and special-purpose police similar to state and local police?
2. What types of organizations and individuals are included under the heading of private police?

3. Describe the origins and development of private policing in the United States.

4. Distinguish between *proprietary* and *contract* security structures in private policing.

5. Identify four types of special-purpose police.

6. Distinguish between *special-purpose* police and *general-purpose* local police agencies.

7. What are the major advantages and disadvantages of private versus special-purpose police?

8. What are the major correlates of private and special-purpose police?

REFERENCES

Becker, D. C. (1995) "Private security," in W. G. Bailey (ed.) *The encyclopedia of police science,* 2nd ed. (New York: Garland):650–656.

Bennett, G. (1987) *Crimewarps: The future of crime in America.* (New York: Anchor Press).

Bureau of Justice Statistics (1983) *Report to the nation on crime and justice: The data.* (Washington, DC: U.S. Department of Justice).

Colby, P. W. (1995) "Contract police," in W. G. Bailey (ed.) *The encyclopedia of police science,* 2nd ed. (New York: Garland):120–122.

Cunningham, W. C. and T. H. Taylor (1984) *The growth of private security.* (Washington, DC: U.S. Government Printing Office).

Cunningham, W. C. and T. H. Taylor (1985) "Ten years of growth in law enforcement and private security relationships: A summary of the Hallcrest report," *In Criminal Justice 85/86,* edited by J. Sullivan and J. Vicron (1987) (Guildord, CT: Dushkin): 100–103.

Gallati, R. R. J. (1983) *Introduction to private security.* (Englewood Cliffs, NJ: Prentice-Hall).

Germann, A. C., F. D. Day, and R. R. J. Gallati (1988) *Introduction to law enforcement and criminal justice.* (Springfield, IL: Charles C. Thomas).

Governor's Office of Criminal Justice Services (1987) *The state of crime and criminal justice in Ohio.* (Columbus, OH: Author).

Hou, C. and C. J. Sheu (1994) "A study of the determinants of participation in a private security system among Taiwanese enterprises," *Police Studies* 17(1):13–24.

Kakalik, J. S. and S. Wildhorn (1972) *Private police in the United States: Findings and recommendations.* (Washington, DC: U.S. Government Printing Office).

Klockars, C. B. (1985) *The idea of police.* (Beverly Hills, CA: Sage).

Lundman, R. (1980) *Police and policing: An introduction.* (New York: Holt, Rinehart & Winston).

Maahs, J. R. and C. Hemmens (1998) "Guarding the public: A statutory analysis of state regulation of security guards," *Journal of Crime and Justice* 21(1):119–134.

Micucci, A. (1998) "A typology of private policing operational styles," *Journal of Criminal Justice* 26(1):41–51.

Moore, E. (1995) "Special function police," in W. G. Bailey (ed.) *The encyclopedia of police science,* 2nd ed. (New York: Garland):722–724.

Moore, R. H. (1987) "Licensing and the regulation of private security," *Journal of Security Administration* 10(1):10–28.

Morn, F. (1995) "Allan Pinkerton," in W. G. Bailey (ed.) *The encyclopedia of police science,* 2nd ed. (New York: Garland):525–527.

Murphy, B. (1995) "Transit police, New York City," in W. G. Bailey (ed.) *The encyclopedia of police science,* 2nd ed. (New York: Garland):782–783.

Ostrom, E., R. B. Parks, and G. P. Whitaker (1978) *Patterns of metropolitan policing.* (Cambridge, MA: Ballinger).

Petersen, D. M. and D. C. Bordner (1995) "Campus police," in W. G. Bailey (ed.) *The encyclopedia of police science,* 2nd ed. (New York: Garland):50–52.

Reaves, B. (1989) *Police departments in large cities, 1987.* (Washington, DC: Bureau of Justice Statistics).

Reaves, B. (1996) *Local police departments,* 1993. (Washington, DC: Bureau of Justice Statistics).

Reaves, B. and A. Goldberg (1996) *Campus law enforcement agencies, 1995.* (Washington, DC: Bureau of Justice Statistics).

Reaves, B. and A. Goldberg (1998) *Census of state and local law enforcement agencies, 1996.* (Washington, DC: Bureau of Justice Statistics).

Reiss, A. J. (1988) *Private employment of public police.* (Washington, DC: National Institute of Justice).

Shearing, C. D. and P. C. Stenning (1983) "Private security: Implications for social control," *Social Problems* 30:493–506.

Walsh, W. F. and E. J. Donovan (1989) "Private security and community policing: Evaluation and comment," *Journal of Criminal Justice* 17(3):187–197.

MUNICIPAL AND LOCAL POLICE

The officers employed by county, city, village, and township police agencies are the most numerous and most visible of government police. General-purpose police are organized at all levels of municipal government, including sheriffs, county, and metropolitan police. The Bureau of Justice Statistics (Hickman and Reaves, 2001; Reaves and Hickman, 2001) estimates that there are over 16,000 municipal and local police agencies in America, employing almost 850,000 persons full time. Of all local and municipal police employees, about three-quarters are sworn peace officers.

These police organizations range in size from those having only one employee to those employing thousands of officers. The nation's largest local police agency, the New York City Police Department, employs over 39,000 sworn officers. In 1988 local governments in cities of more than 10,000 population spent an average of $80.03 per resident for police protection (Hoetmer, 1989). Of the total estimated expenditure for police services, local governments paid nearly 72 percent of all costs for public police in 1998, over $33 billion (Maguire and Pastore, 1999:4).

A recruit class for the New York City police, a local police organization but the largest agency in America. (Courtesy of *New York's Finest* magazine)

We have already reviewed the development of local police in America in Part I. The purpose of this chapter is to describe local policing in America today and to examine the position and role of municipal and local agencies in the context of what we have called the police industry. To this end, we will begin with a description of local policing in America, move through a discussion of the variety of local policing that exists, and conclude with an examination of some correlates of that variety in local policing.

UNDERSTANDING LOCAL POLICE

In some sense, the local police are "all things to all people." They have an ambiguous role that includes law enforcement, service delivery, and peacekeeping, all of which must be performed at all times throughout the entire local jurisdiction. Further, local police agencies exhibit a variety of shapes and sizes, at all levels of local government. In all probability, whatever we say about the local police is not accurate for all local police. The purpose of this section is to describe the distribution of local police in the United States.

The descriptive task is complicated by a lack of information about the most common local police, relatively small organizations. Samuel

Walker (1983) warns about what he calls a "big-city focus" in our thinking about the local police. Ralph Weisheit, David Falcone, and L. Edward Wells studied rural police. They observe (1994:1), "Most studies of variations in police behavior have been conducted in urban settings. . . . It is evident, however, that rural environments are distinct from urban environments in ways that affect policing, crime, and public policy." As Charts 7.1 and 7.2 show, over three-quarters of local police organizations in America employ fewer than 25 sworn personnel. In fact, the New York City Police Department employs more sworn officers than the total for more than 7,000 of the smallest departments in America. Yet, the majority of what we know about the local police has been learned from the larger agencies.

Large police agencies are attractive as research sites for a number of reasons. First, obviously they are found in larger cities, which makes them more available to researchers since most Americans live in cities. Second, the very nature of their large size means that they encounter more situations about which researchers might be interested. For example, a small police agency in a rural area might not receive any homicide complaints in an entire year, whereas homicides occur daily in our large cities.

Further, if the object of research interest is police officers, large agencies are favored subjects because they employ the majority of police officers in America. It is easier to interview hundreds of police officers in one large department than to travel to many different localities in order to interview officers working in smaller departments. Finally, large police departments are more likely to maintain computerized databases, which make the task of gathering information much easier for police researchers.

For all of these reasons, and others we have not mentioned, most of the literature written about the police in America is based on the experiences and practices of larger agencies. As Walker (1983:36) observes, "Most of the research on American law enforcement pertains to big-city departments.

CHART 7.1
Number of local police departments by number of sworn personnel, 1999. (Source: M. Hickman and B. Reaves (2001) *Local Police Departments, 1999*. Washington, DC: Bureau of Justice Statistics, p. 2.)

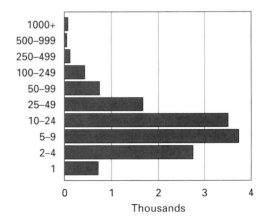

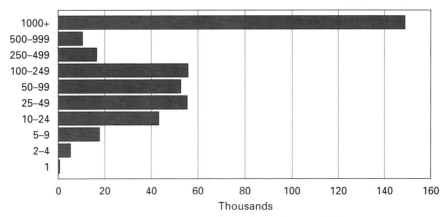

CHART 7.2 Number of full-time sworn police officers by size of department, 1999.
(Source: M. Hickman and B. Reaves (2001), *Local Police Departments, 1999.*
Washington, DC: Bureau of Justice Statistics, p. 2.)

And even that body of data is heavily skewed toward a few departments." Not only are we normally limited to descriptions and analyses of large agencies in big cities, often we are limited to studies of fewer than 20 of America's 16,000 local and municipal police organizations.

A growing body of evidence indicates that the size of the police organization is an important correlate of policing in practice. Decker and Ward (1989), for instance, contend that there are important differences between the police of smaller, rural agencies and those of large, big-city organizations. Both the personnel and practices of these two types of agencies appear to differ in important ways. Langworthy (1985a) observes that police agency size correlates with differences in organizational structure, although it seems that size alone does not determine how a police agency is organized. Galliher (1975) and co-workers conclude that the role of the police in smaller agencies is substantially different from that of the big-city police. Other researchers have also found that the size of the police agency correlates with various attitudes of police officers (Travis and Vukovich, 1990; Witte, Travis, and Langworthy, 1990; Regoli et al., 1987).

What research is available indicates that small police agencies are different from larger ones, and that the practice of policing correlates with agency size. Wendy Christensen and John Crank (2001) report an in-depth study of a rural sheriff's department, comparing policing there with what is known from studies of large, urban police settings. They conclude that while the themes of policing were similar (e.g., patrol, community relations, officer safety, etc.), the meanings or explanations given by deputies in their study were different from those of urban officers. The deputies they studied reported better relations with community members, less concern about danger on the job, and more independence in dealing with problems and citizens. In general, smaller agencies tend to be more "personal," both in terms of the relations among officers and the relations between officers and the community. Small agencies tend to receive disproportionately fewer calls for criminal matters and more for regulatory and peacekeeping functions than do large agencies (Weisheit, Falcone, and Wells, 1994). Size of agency is related to the size of the community served, as well as to the characteristics of those communities and of their police organizations (Crank, 1990). Larger agencies tend to serve a more diverse population in terms of socioeconomic and demographic characteristics than do smaller ones.

What all of this means for us is that police organization size appears to be related to differences in the correlates of policing we identified in Chapter 1. Yet, despite this, most of what has been learned about the police in America is based on information taken from only the largest police organizations. Thus, as we attempt to describe and understand the municipal and local component of the American police industry, we must be careful to remember that our knowledge base reflects a "big-city" bias.

Local Police Organizations

Municipal police organizations are structured in several different ways. They include general-purpose police agencies operated by county, city, township, and village government entities. More than 90 percent of municipalities with populations of 2,500 or more have their own police forces (Bureau of Justice Statistics, 1989b). In these local police organizations, the primary technology for the provision of police services is the use of patrol (Parks and Ostrom, 1984; Kelling et al., 1974; Kirkham and Wollan, 1980). Indeed, patrol is considered to be the backbone of policing (Wilson and McLaren, 1977). Between 60 and 80 percent of sworn officers in local police agencies have assigned duties that include responding to calls for service, or patrol. The uniformed patrol officer in many ways is the defining characteristic of local policing.

In addition to the provision of patrol, local police agencies are also responsible for the investigation of reported crimes. Thus, an additional fixture of the local police organization is frequently the detective or detective bureau. Unlike the patrol officer, these personnel tend not to be uniformed and generally are freed from the responsibility of routine patrol and its attendant service-delivery and peacekeeping functions. All local police organizations must provide for patrol, but the smallest cannot afford to assign officers full-time to work as detectives.

The first *Law Enforcement Management and Administrative Statistics Survey* (LEMAS), published in 1989, described the range of local policing (Bureau of Justice Statistics, 1989b). This survey identified 11,989 local and 3,080 sheriff's agencies. These organizations ranged in size from nearly

1,000 that employed only one sworn officer to 34 agencies that employed more than 1,000 sworn officers each. Budgets were similarly varied, ranging from less than $78,000 per year for those agencies serving populations under 2,500 to nearly $340 million per year for those agencies serving populations in excess of 1 million persons.

The second survey, released in 1992 (Bureau of Justice Statistics, 1992) reported similar findings. The survey estimated that there were 12,288 local police and 3,093 sheriff's agencies across the country. They ranged in size from 959 employing one sworn officer to 38 employing more than 1,000 sworn personnel. Average operating expenses for these agencies ranged from about $115,000 per year for departments serving populations of under 2,500 to over $334 million per year for those serving populations in excess of 1 million.

In 1996 the results of the third survey were published (Reaves, 1996). This study revealed 12,361 local police and 3,084 sheriff's agencies across the country, ranging in size from 851 that employed one sworn officer to 38 that employed more than 1,000 sworn personnel. Average operating expenses for these agencies totaled about $35 billion and ranged from an average of $107,000 for those serving populations of less than 2,500 to $427 million for those serving populations of 1 million or more. By 1999, (Reaves and Hickman, 2001; Hickman and Reaves, 2001) there were 13,524 local police and 3,088 sheriff's agencies with 792 employing one sworn officer and 63 employing over 1,000 sworn personnel.

In all local police agencies, however, the overwhelming majority of sworn personnel were assigned to **field operations** composed primarily of patrol and investigation. Nearly 90 percent of sworn officers were involved in patrol, investigations, traffic enforcement, and special operations. Only 10 percent were assigned to technical support, jail or court operations, and administration. Of those assigned to field operations, 75 percent were uniformed officers responsible for responding to calls for service—that is, patrol officers (Chart 7.3).

Officers working in smaller agencies are less likely to have specialized assignments. Those working in the thousands of police agencies with few personnel (especially agencies with fewer than five officers) are likely to perform their own technical-support and administrative duties in addition to field operations. Indeed, in the smallest departments even the police chief engages in patrol activities.

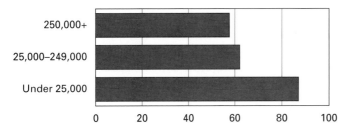

CHART 7.3
Percentage of officers assigned to respond to calls for service by jurisdiction population size. (Source: M. Hickman and B. Reaves (2001) *Local Police Departments, 1999.* Washington, DC: Bureau of Justice Statistics, p. 3.)

As the size of a police organization increases, its ability to specialize also increases. Larger police agencies are therefore able to field not only patrol and detective units, but to specialize further within these functions. Thus, a very large agency may have different types of patrol and detective units, as well as a number of different support and service units. The more specialized the agency is, the more complex its organizational structure. We will examine the nature of police organizational structures in more detail in the next section. For now it is sufficient to note that one effect of the big-city focus is a narrow description of local police organizations.

Sam Souryal (1989) describes a basic police organizational structure based on certain principles of organizational theory: authority, division of labor, unity of command, specialization, delegation of authority, and formal communication. While Souryal recognizes that the exact structure of any police organization is the product of many forces, including the size of the agency, he describes a typical police organization as a bureaucratic pyramid, with the chief of police at the top and patrol officers at the bottom. Beneath the chief are a number of middle-management positions that reflect functional specialization. Chart 7.4 shows an organizational chart of a municipal police agency.

A bureaucratic pyramid organization is the accepted model of police organization in most literature on the police. Yet, as we now know, it would be impossible for 25 percent of local police agencies to even come close to this model. It is difficult to have multiple divisions and specialized bureaus when one has a total of fewer than five employees! The general acceptance of this type of organizational structure as representing the typical police agency is a result of our traditional focus on large agencies in big cities.

The importance of this model lies more in what it tells us about our conceptions of the local police than in what it tells us about those police themselves. Police agencies often try to emulate this structure (in design if not in practice) in their own agencies. For this reason, the sole employee of a one-officer police organization is often granted the title of chief of police. In agencies with only a few personnel, this chief typically also performs routine patrol activities similar to those of the lowest-ranking member of the department.

Despite an apparent desire to develop a bureaucratic, paramilitary command structure with functional specialization, most small police agencies, even those that have developed some sort of command hierarchy and limited task specialization, have a different organization from that found in larger departments. In these agencies, all personnel are often **generalists** who perform all police functions, from patrol through investigation, as part of their daily activities. The smaller department also means that the personal relationships among the officers are at least as important as are the formal organizational roles in determining how they interact.

In short, then, there is no typical organizational structure found among municipal and local police agencies. Rather, these organizations display a number of different structures reflecting varying degrees of size, jurisdiction, and relative emphasis on police functions. The one common

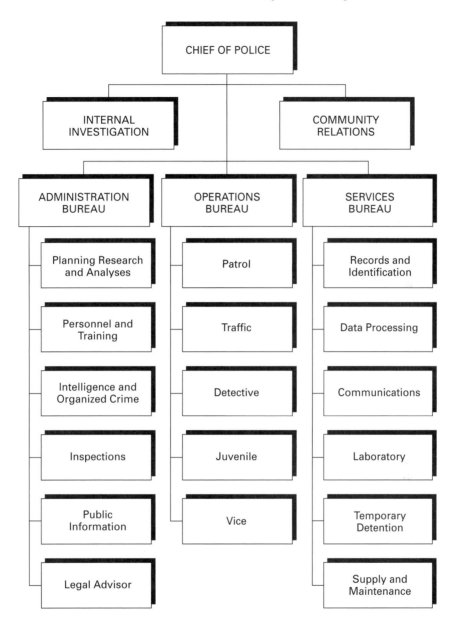

CHART 7.4 A model police department organizational chart. (Source: President's Commission on Law Enforcement and Administration of Justice (1967) *Task Force Report: The Police.* Washington, DC: U.S. Government Printing Office, p. 47.)

denominator of these thousands of organizations is the general-purpose police services that they provide within their communities.

Types of Local Police

As we saw with other components of the police industry, one of the central
defining characteristics of a police organization is its jurisdiction. Municipal and local police agencies include those with relatively small geographic
limits as well as those responsible for policing large areas. Although some
city police agencies have jurisdiction over large areas—for example, the
Houston Police Department serves an area of over 564 square miles—the
average local police agency serves a much smaller area.

One local police agency, the county sheriff's office, however, routinely
has jurisdiction over large areas. A descendant of the English shire reeve,
the American county sheriff is charged with the operation of a county jail,
civil functions such as service of eviction notices and other court orders, and
police responsibility. The Bureau of Justice Statistics (Reaves and Hickman, 2001) estimates that 3,088 sheriff's agencies exist in America, ranging in size from those employing less than five sworn officers to 17 agencies
having 1,000 or more officers (Chart 7.5). The largest sheriff's office, the Los
Angeles County Sheriff's Department, employs over 8,100 full-time, sworn
officers. The 1999 LEMAS survey reported 21 sheriff's offices that employed only one sworn officer. The average sheriff's department employs
about 60 sworn personnel (Reaves and Hickman, 2001) (Chart 7.6).

It is common for officers of a sheriff's department (commonly called
deputies) to provide routine police service to the unincorporated areas
within the county. Thus, each incorporated city, village or township in the
county might have its own police, with the sheriff's deputies patrolling and
serving residents in the unincorporated areas between them (Chart 7.7).
However, in many places the sheriff's office provides police service to incor-

CHART 7.5
Number of sheriffs offices by number
of sworn officers, 1999. (Source: B.
Reaves and M. Hickman (2001) *Sheriffs
Offices, 1999*. Washington, DC: U.S.
Bureau of Justice Statistics, p. 2.)

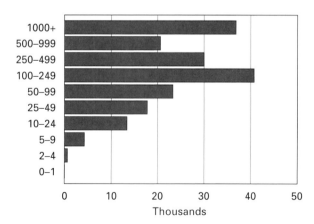

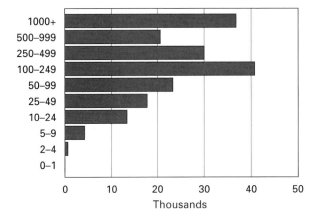

CHART 7.6
Number of full-time sworn sheriff's officers by size of sheriff's office. (Source: B. Reaves and M. Hickman (2001) *Sheriffs Offices, 1999.* Washington, DC: U.S. Bureau of Justice Statistics, p. 2.)

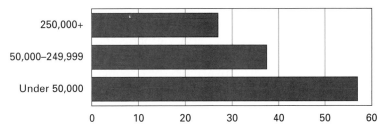

CHART 7.7 Percentage of sworn sheriff's officers assigned to respond to calls for service by population size of jurisdiction. (Source: B. Reaves and M. Hickman (2001) *Sheriffs Offices, 1999.* Washington, DC: U.S. Bureau of Justice Statistics, p. 3.)

porated municipalities through contracts. For example, the Los Angeles County Sheriff's Department not only serves the functions of operating the jails and supporting the county courts, it also provides police patrol for unincorporated areas and, under contracts with local government, provides police service to 40 percent of the municipalities in the county as well.

Other models of police organization have resulted in local police agencies with relatively large geographic boundaries. In an effort to achieve coordinated service and economies of scale, a number of places have consolidated police services into a single umbrella agency that covers several cooperating municipalities. These include county police, such as Prince George's County in Maryland, and metro police, such as the Miami-Dade Metropolitan Police.

The most common local police organization, and the one with which most Americans are more familiar, is the city police. City police have jurisdiction over matters that occur within the boundaries of the incorporated municipality. In theory, like the town marshal of Western movies, their interests end at the city line. It is common, especially in more densely populated areas, for scores of independent local police agencies to operate within a single county. In some places, which agency responds to a call for assistance depends on the side of the street the call originates from.

In practice, neighboring municipal police agencies develop working relations. In some cases interagency cooperation is guided by formal agreements called **mutual-aid pacts.** These agreements define the circumstances under which the agencies will cooperate and often grant jurisdiction to officers from the neighboring municipality under certain conditions. In other cases, the agencies, or at least the officers working in them, develop informal rules for cooperation.

In an important study of the delivery of local police services, Ostrom, Parks, and Whitaker (1978) conclude that cooperation among agencies is the norm. In the 80 metropolitan areas they studied, a total of 1,827 separate agencies were identified. Rather than the high degree of duplication and overlap that critics of police fragmentation have suggested, Ostrom and her colleagues report little inefficiency resulting from the existence of a large number of agencies. Instead, smaller agencies are able to use specialized services of larger ones, such as crime labs, and formal and informal agreements prevent duplication of service among agencies.

It appears, then, that though there are a large number of independent local and municipal police organizations, they cooperate in the delivery of police services. In studying a local police agency, it is important to consider the size and structure of the organization, but it is also necessary to examine its position within the network of local police in the area. A small agency need not maintain a detective unit if it can rely on others—for example, the county sheriff—for investigative duties. Municipal and local police share a common jurisdiction in terms of interests—law enforcement, service delivery, and peacekeeping. Their differences in geographic limits on jurisdiction are less important than might be first thought, given cooperation.

In some ways municipal and local policing represents what Ostrom and her colleagues (1978) call a "multi-firm industry." It must be understood as the product of a number of separate firms (agencies) that provide the same basic services (share interests) within a broader area than a single municipality. The total range of local police service is perhaps best understood as a regional phenomenon rather than a local one in places where a large number of agencies operate within a relatively small area. Where these separate agencies are likely to diverge, however, is in how they *do* policing. It is to this topic that we must now turn.

Varieties of Local Policing

We have seen that local police is a term that covers a range of agencies and organizations in the United States, from very large, urban police bureaucracies through one-person, rural police operations. Despite the differences in size, population, and organizational complexity, all of these local police agencies are required to serve and protect their citizens. Clearly, they cannot all do so in the same fashion. Let us now examine differences in the ways that local police agencies perform the job of policing.

In 1968 James Q. Wilson published *Varieties of Police Behavior,* a comparison of policing practices in eight American cities. This book established that local police agencies differ among themselves in terms of the ways in which they approach the tasks of policing. Wilson identified three distinct styles of local policing: legalistic, service, and watchman.

Regardless of the dominant style of policing in an agency, it must serve the three functions of law enforcement, service delivery, and peacekeeping. The differences among these styles are not based on the tasks performed by the agency but on the ways in which officers employed by the agency approach police problems. Thus, the distinction is not so much what they do but how they do it. After a brief description of the three styles, we will see how they differ.

Legalistic Style.   In Wilson's (1968) view, officers in a legalistic agency intervene frequently and formally with citizens. That is, employees of a legalistic-style agency rely on the formal criminal law as a definition of not only what situations merit their attention but in what manner they should intervene. A legalistic police department is characterized by officers who issue many citations and make many arrests. The kind of citizen complaint that requires police intervention is defined according to the criminal law.

If the complaint contains evidence that a law (any law) has been violated, the police are obligated to intervene. The form of their intervention is to enforce the law, either by issuing a citation or making an arrest. Thus, the legalistic style of policing stresses the law-enforcement function. The primary work of the officers is the enforcement of all codes, and their manner of interacting with citizens is that of ministers of the law.

Service Style.   Officers in agencies displaying a service style also intervene frequently in the lives of citizens, but they do so informally. The service-style police agency is characterized by many police–citizen contacts but relatively few citations and arrests. The officers in a service-style department treat every citizen complaint as requiring a police response; they do not rely on the criminal code to define police issues. As the label suggests, these agencies stress the service-delivery function of policing.

If a citizen contacts the police, the police in service-style agencies are required to respond. Officers determine the nature of the problem and suggest solutions. When appropriate, the officers will handle the problem directly, but usually without invoking the law. Thus, officers in service-style agencies are more likely to issue warnings and advice than to issue citations or make arrests. The primary task of these officers is the resolution of citizen problems, and their manner of interacting with citizens is that of a government gatekeeper. As gatekeepers, the officers frequently serve to direct citizens to available services.

Watchman Style.   Unlike officers in the legalistic and service styles, officers in watchman-style agencies do not intervene frequently with citizens. Although officers in all three styles of agency respond to citizen calls, watchman agencies are characterized by fewer officer-initiated contacts with citizens. These officers exercise considerable discretion in defining whether a situation requires police response and what that response should be. Watchman-style agencies emphasize the peacekeeping function of policing.

In answering a citizen's call for assistance, officers in a watchman department will assess the importance of the situation and determine an appropriate response. Depending on their assessment of the circumstances, they might intervene formally, by arrest or citation, or informally, by warning, referring, or even threatening the parties involved. The primary task of these officers is peacekeeping, and they perform this task by taking the role of arbiter, or judge.

A Comparison of Styles.   To indicate how these three styles result in different actions by the police officers involved, consider three situations:

1. The officer observes a motorist traveling at 40 miles per hour in a 30-mile-per-hour speed zone.
2. The police respond to a citizen complaint about an automobile that is illegally parked in front of a home.
3. A storekeeper calls police to report a robbery.

In the first example the officers will tend to react to the speeding motorist in ways consistent with their policing style. That is, officers from a legalistic agency would pursue and stop the driver, issuing a citation for violation of the speed limit. Officers from a service-style agency would probably also pursue and stop the motorist but would likely issue only a warning. The watchman-style agency basically leaves the intervention decision up to the officer. In this case, it is likely that the officer would simply ignore the speeder, absent some other reason for intervention, such as heavy traffic or a nearby school.

Officers from police departments characterized by these three different styles are also likely to respond differently to citizen complaints about regulatory ordinances such as those concerning parking (our second example—an illegally parked vehicle in front of a home). A legalistic agency's officers are likely to cite the vehicle for illegal parking. Police from a service-style agency might try to locate the car's owner and advise her or him to move the car rather than issue a citation. Those from a watchman-style agency might inform the complaining citizen that "there's nothing we can do about it," or that they would like to wait a few hours to see if the owner will move it.

In our final example, the robbery, regardless of agency style, officers will react in a similar fashion. That is, there are some instances (typically those involving serious crime) in which police are police, and differences in

A primary task of the police is to maintain order. (Akos Szilvasi/Stock Boston)

style do not apply. It is inconceivable that the watchman-style department would ignore a felony complaint or that the service-style department would merely issue a warning to the robber "not to do this again in the future."

Explaining Agency Styles.   Wilson (1968) arrived at his typology of police departments based on studies of officer activities in each of the eight local police agencies he reviewed. The initial focus of the research effort was on the exercise of discretionary authority by police officers on patrol. As we have seen, officers may react to specific situations in any of a number of ways, ranging from ignoring the problem to applying the criminal law. A review of the reported activity of officers in the departments revealed that they differed in terms of the circumstances under which they applied the criminal law and the number of times they intervened with citizens.

Thus, the initial focus of the research was on the decisions of individual officers. Patterns of decision making were revealed that differed according to the organization. Wilson concluded that the organization influenced the behavior of the officers on patrol, and his attention turned to explaining the differences in police organizations that correlated with the different patterns of officer decision making.

In sum, Wilson observed that the style of a police agency appeared to be related to the characteristics of the population of the community in which it operated and also to the form of government existing within the city. He found that service-style agencies were most likely to be found in areas with a homogeneous population. More mixed populations usually had either legalistic or watchman styles of policing. He also noted that a partisan political tradition linked to a mayor-council form of city government was associated with watchman-style policing, while more professional government (the city manager type, for example) and less partisanship in local politics were associated with legalistic or service-style policing.

Wilson concluded that the political culture of a community constrained the actions or style of the police. A later empirical test of this theory by Langworthy (1985b) replicated Wilson's observations of differences but raised questions about his theory. Wilson had suggested that the characteristics of a local community put limits on what would be acceptable behavior by the police. Thus, a "good-government" city would not tolerate watchman-style police behavior. Langworthy notes that within classes of political culture, police agencies display considerable variety in styles of policing. His reformulation of Wilson's theory is that political culture does not constrain police behavior. Rather, the characteristics of a community are related to police styles in that, on average, the practice of policing by officers is associated with the political culture of the community. That is, good-government cities tend to display legalistic policing, and traditional-politics cities tend to display watchman policing. Yet some good-government cities may have watchman-style policing, and some traditional-politics cities may have legalistic policing. In short, Langworthy suggests that Wilson's explanation is not enough for understanding local police agency style. Douglas Davenport (1999) studied the organizational environment, characteristics of the communities in which police organizations exist. He found that factors such as turbulence (changes in population, mobility), resource capacity (poverty, size of minority population), and complexity (number of police agencies in the area, population density) affect police organizational operations.

From these three studies we can draw three important conclusions about local policing. First, local police agencies differ among themselves in terms of policing styles. Second, the style of policing found in a community is related to its political culture. Finally, local police styles are the product of a variety of forces beyond the demographic and political structure of the communities in which they operate. In large agencies, specific squads or units may have different styles, so that even in one city, the police of different city areas may display different styles.

If we are to understand municipal and local policing, then, we must identify and analyze the relevant forces that influence the styles of policing adopted by these agencies. We will turn to that task by describing the factors that appear to be associated with different styles of local policing.

SOME CORRELATES OF LOCAL POLICING

Wilson's (1968) analysis of policing in eight cities serves as an example of the correlates of policing. His observations led him to conclude that the political culture of a community constrained the police organization in ways that influenced the work of individual police officers. He hypothesized a correlation between community and people (police officers) that worked through the organization. Langworthy's (1985b) test of this hypothesis illustrated that political culture and agency style are correlated, but that political culture is not sufficient to explain agency style. Davenport's (1999) study indicated that the organizational environment (the broader community in which the police organization exists) influences police structure and operations. The people, organizations, and communities involved in local policing all have their effect.

People

Peter Blau and Richard Scott (1962) explain that all formal organizations interact with four types of people: members of the organization, owners of the organization, clients of the organization, and the general public. Further, each of these groups holds somewhat different expectations of the organization, and these expectations are frequently in conflict.

Richard Lundman (1980) contends that styles of policing correlate with these four groups, or types of police audience. Each police organization, especially the general-purpose government police, has an obligation to serve the general public; that is one audience. Additionally, however, the organization has obligations to each of the three remaining audiences: members, owners, and clients. Lundman suggests that the style of policing in a local police organization reflects the relative importance of these audiences.

Members. The members of an organization are those who "belong" to the agency as employees and administrators. In the case of the local police, the membership includes the officers, any civilian employees, and those persons in command positions within the agency. That is, the people who work in the police agency are its members.

Members expect the organization to which they belong to meet several of their needs. In the case of an employing organization, members would expect adequate pay, good working conditions, and support from the organization. That is, they expect that the organization will take care of them, or at a minimum, care for them. This means that members expect both material and moral support from the organization.

Material support takes the form of salary, fringe benefits, and equipment. An agency that pays little, provides minimum health, retirement, and medical benefits, operates old and inadequate equipment, and otherwise fails to consider the material needs of its officers and staff violates the expectations of members. Similarly, an agency that applies tight controls on officers, through close supervision and numerous regulations, displays a lack of faith in its personnel from the perspective of those personnel. Therefore, an agency that does not provide training or that takes the word of complaining citizens over that of officers is viewed by the members as not providing enough moral support.

Members expect to be valued by the organization. Their value to the organization is proved (in their view) by the payment, trust, respect, and assistance they are given. The agency that pays well and shows faith in the judgment of its officers is perhaps the ideal organization from the perspective of its members.

Owners.   The owners of the police organization are represented by the municipal government that operates the agency. Owners expect organizations to meet needs as they define them and to do so in an efficient manner. In the case of local policing, the owners are those elected and appointed officials who ultimately control the budget and funding of the police agency.

Government officials, as owners of the local police, expect the agency to enforce the laws, control crime, and avoid creating problems for the owners. These problems would include citizen complaints about police corruption, brutality, and discrimination. Ideally, the organization will generate revenue (perhaps through the issuance of citations). At a minimum, it is expected to show a return on the investment of resources such (1) that the crime rate remains at an acceptable level and (2) the police respond to calls for assistance within a reasonable period of time.

A police agency that consistently receives complaints of brutality or corruption violates the expectations of its owners. In the same vein, if a community experiences an increase in its crime rate or faces a recurrent crime problem such as a series of unsolved murders or rapes, pressure mounts from the owners for that agency to "do something." That these pressures exist, and that local police agencies respond to them, indicates the importance of the owner audience.

Clients.   The clients of an organization are those people with whom the organization comes into contact in the course of doing business. For the local police, clients include complaining citizens, suspected offenders, and members of other organizations with which the police interact, such as public works departments, the courts, and related public and private institutions.

Clients expect that the organization will be responsive to their concerns and needs. While the clientele of the local police are members of the general public, they have a more particular and vested interest in police practices than the rest of that general audience. Local business leaders, for example, may desire increased (or decreased) police presence in a retail area. Neighborhood associations may want the police to regulate parking, littering, and other sources of disorder. Criminal suspects expect the police to treat them as innocent until proven guilty and protect their constitutional rights. A police agency that is not responsive to the demands or desires of clients violates the expectations of these audiences.

Styles and Audiences

As the descriptions of these three audiences of the police illustrate, they each expect different things of the police organization. Often their expectations are in conflict, such as those of owners wishing efficient (preferably cheap) policing versus officers, who expect substantial material reward. Similarly, many police clients expect the officers to be responsive to them regardless of the nature of their requests for service, whereas the members may view themselves as law-enforcement professionals and be reluctant to spend much time on matters not related to that function. The business leaders might want foot patrol while the officers prefer the work environment of the air-conditioned (or heated, depending on the season) squad car.

Each of the three styles of policing identified by Wilson (1968) is better adapted to meeting the expectations of one or another of these audiences. That is, the style of policing adopted by a local agency probably correlates with the relative influence of these audiences. For example, when the owners dominate, a legalistic style can be expected. Where the clients' interests dominate, a service style may be more common. Finally, in places where the members are most influential, a watchman style of police is to be expected.

Given that owners expect impartial and formal law enforcement by the police organization, the legalistic style is preferable. In a legalistic style, the police intervene in situations based on the criminal law, and their behavior is circumscribed by legal procedures. This style can prevent complaints of corruption and discrimination because the police enforce all laws equally against all people, and the procedures to be followed are explicit. The legalistic style also generates revenue through the issuance of relatively large numbers of citations, so the police are busy earning their keep through enforcement activities.

A service-style agency places the police at the service of their clients. Should a citizen call the police, an officer will respond and take the matter seriously. However, in most instances, the officer will act in nonpunitive and nonjudgmental ways to solve the citizen's problem. Consequently, citizens need not worry that calling the police might get them in trouble. Rather, they can (and do) expect the police to solve all sorts of problems and to be on call for problem solving at any time.

In a watchman-style department, the individual officer has tremendous freedom to define situations as deserving police intervention and to determine how to intervene. Thus, the officer is free to avoid work (for example, by ignoring a traffic violation) and to define his or her work situationally. The officer may decide to arrest, warn, or ignore persons who commit minor law violations. This kind of organization has recognized the officer's ability to make these judgments and generally does not interfere with the officer's daily activities. That recognition is usually not formally stated but rather is understood by the members, because the organization places fewer controls and demands on them.

The people involved in local policing thus exert an influence on the conduct of police business. Every organization must minimally satisfy the desires of the general public and each of the other three audiences, but usually one or another of these audiences will be more influential than the rest. Differences in the ability of members, owners, or clients to shape the role of the police organization correlate with the style of policing adopted by the police agency. You should recall that this style itself indicates a tendency of people—officers in the local police agency—to behave in a predictable pattern. A cycle of correlation exists in which people influence the organization, and the organization, in turn influences people.

There is ample research evidence to support the notion that the people involved in policing have an effect on how policing is done. Brooks, Piquero, and Cronin (1993) report that officer characteristics are related to support for organizational efforts to implement community policing. Officers who have a service orientation to their jobs are more supportive than those who are more crime-control in outlook. Similarly, Travis and Winston (1998) report that officer support for agency style (and satisfaction with the organization) is partly a matter of agreement between the officers' views of how policing should be done and the style of policing practiced by the organization.

Organization

That Wilson and others have been able to typify police organizations with reference to how their members perform the job indicates that the organization correlates with policing style. In this example, however, the organization appears to be a conduit, or filter, through which external forces (audiences or political culture) affect police practice. While it is impossible to fully separate the effects of external forces on both police organizations and police practice, there is evidence that the organization is itself an independent correlate of policing.

We have seen how local police agencies differ in terms of size and specialization. These differences affect the personal relationships among employees as well as the ability of the organization to direct and control the behavior of officers. Small, relatively undifferentiated agencies develop shared understandings among officers, and all officers routinely engage in the same activities. On the other hand, the sheer variety of tasks performed by officers makes it difficult to control their behavior. Larger, more specialized agencies tend to lose the personal touch among employees, but the activities of those officers in the most specialized functions are much more amenable to administrative control. The organization, then, has some effect on the behavior of its members.

Mastrofski, Ritti, and Hoffmaster (1987) examined the impact of police organizational characteristics on the discretionary decisions by police officers in cases of driving under the influence (DUI). They found that the likelihood of officers making arrests for driving under the influence of alcohol was related to the size and structure of the police agency. Officers in larger, more bureaucratic organizations were less likely to arrest DUI violators than were officers in smaller, less elaborately bureaucratic agencies.

Mastrofski and his colleagues offer two plausible explanations for their findings. One would expect larger, more bureaucratic and rule-bound organizations to produce more consistent officer behavior (in this case, arrests), but in fact the opposite occurred. One possible reason that officers in smaller agencies were more likely to apply the law and make arrests was that related to the fact that their agencies were small groups. As a result, the behavior of the officers was subject to more direct scrutiny by management. Mastrofski and colleagues explain, "The chief can have detailed knowledge of subordinates' behavior and can selectively reward compliance."

A second explanation is that the smaller agencies needed to show professionalism and impartiality in enforcement in order to maintain public support and to present a professional image. The larger department had "larger fish to fry," including many more serious criminal instances. Thus, DUI enforcement was not as critical to organizational survival and image of these larger agencies.

Both of these explanations point to ways in which the organization itself is an independent correlate of policing practice. Large agencies are more specialized and ostensibly more bureaucratic, but the officers within them are more anonymous. Small agencies are more closely knit and more personal, but their officers are also more observable. In this regard, police in small organizations work in a "small-town" atmosphere where everyone knows them and sees what they do. Police in large agencies work in a big-city atmosphere where they are part of the crowd and can hide.

In a later study on the same topic, Mastrofski and Ritti (1996) report that officer involvement in DUI enforcement is largely a function of organizational support for those efforts. If officers feel that DUI offenses are viewed as important by the organization, they are more likely to enforce DUI laws. They also note that the provision of specific training for DUI enforcement contributes to the sense that the organization is concerned with these offenses.

The need of the organization to retain public support can also affect how police behave. That is, the organization itself places demands on officers for its maintenance through requirements of reporting and the like. The reward structure within a police agency, for example, correlates with officer behavior. The organization rewards those behaviors that serve its interests. The authors know of one local police agency in Ohio, where officers tended not to serve court orders and subpoenas. The chief of police then began to include a count of such papers served in the officers' annual reviews, thus making this type of work count in performance appraisal. Thereafter, the backlog of unserved papers disappeared, and officers sought out opportunities to act as process servers.

The reward structure in the agency helps to ensure that tasks important to the agency are completed, such as serving court papers, which helps the agency maintain a good relationship with judges and prosecutors. On occasion, however, the existing reward structure hinders the agency. For example, an effort to get police officers to meet and interact with citizens in Cincinnati failed in part because officers were not rewarded for this effort.

The traditional reward structure in this Cincinnati agency based assessments of the officer's performance on the numbers of arrests made, citations issued, and the like. When instructed to make contact with citizens to help develop a better relationship with the public, many officers did. A problem arose when it came time for performance appraisals. Every hour spent talking with the citizens was an hour not available for writing citations, investigating suspicious persons, and other enforcement activities. The officers most involved in developing community relations were the ones who scored lowest on performance appraisals. Not surprisingly, few officers continued to try to meet the public. Rather, most returned to issuing citations and doing other work that counted in the reward structure.

In these ways, the organization represents an entity that places or defines restraints on police practice. Size of agency is a correlate of organizational structure, but its exact influence needs to be carefully considered. A large police organization may be decentralized into a number of smaller, relatively autonomous agencies. The use of team policing, neighborhood policing, and other decentralization models might make officers in large agencies experience a working environment similar to that of a small agency. Thus, it is not size alone, but the structure of the organization that correlates with police practice. Agency style, size, organizational structure, and demands for organizational needs (maintaining public support, required paperwork, etc.) are organizational correlates of local policing.

Community

Municipal and local police operate within communities, and the style of policing correlates with those communities. One of Wilson's (1968) correlates of police styles was the demographic composition of the community. He found that watchman and legalistic styles were associated with heterogeneous, or mixed, populations. They were most commonly found in cities that had populations showing a variety of racial, educational, economic, and ethnic characteristics. The service style was more typical of communities that had a more uniform population in terms of social class. Davenport (1999) found similar relationships between community characteristics and police operations.

Sherman (1980) reports on three community characteristics that appear to correlate with police behavior in terms of arrest rates and police use of deadly force. His review of the research indicates that the political, economic, and demographic structure of a community correlate with police actions. In short, although the data present mixed findings, police behavior appears to reflect community standards (Riksheim and Chermak, 1993).

Returning to our discussion of audiences, the relationships become a little clearer. In communities where government leaders (including the chief of police) hold and enforce identifiable expectations of the police, police behavior reflects these expectations. Thus, the police have been found to be unlikely to use aggressive detection and intervention practices in African American neighborhoods in communities where the relevant government officials are sympathetic to the aspirations of African Americans (Rossi, Berk, and Eidson, 1974). The likelihood of arrest or police use of deadly force is greatest in situations involving defendants or suspects who come from lower economic status. Also, the stability of the population (in terms of how few people have moved in a given time period) is associated with overall arrest rates. The more stable the population is, the lower the rate of either traffic citations or arrests (Jacobs and Britt, 1979).

From these and similar research findings, we can develop some hypotheses about how community characteristics correlate with police styles.

Torrance County, New Mexico, sheriff and deputies. (Mimi Forsyth/Monkmeyer Press Photos)

First, in communities with a stable and homogeneous population, a service style of policing is likely to emerge. The shared characteristics of the population, coupled with its stability, are likely to result in a shared understanding and expectation of the police among the public. Further, this public develops a long-term relationship with the police by virtue of their stable residence. Thus, the people who call the police, and with whom the police interact, are an important fixture in the work of the police organization. The organization, then, attends to the needs of its clients.

In communities with heterogeneous populations, or where there is little population stability, the clients of the police do not develop into a strong influence on the organization. The variety of clients, each with their separate needs and expectations of the police, coupled with the relatively transitory nature of their relationship with the police, makes the clients less important to the police organization (Webb and Katz, 1997). In these types of communities, either a legalistic or watchman style of policing is likely.

Which of these two styles dominates within a community correlates with the structure of the local government. In communities with strong and involved local governments, the owners can influence the police organization. Active and directing mayors, chiefs of police, city managers, and other officials are likely to emphasize a legalistic style as the best means of serving the needs of a diverse citizenry. In communities where active political direction or leadership of the police is absent, the members of the organization are free to define their work. In these circumstances, a watchman style of policing is likely.

Where the members of a community can agree in some fashion on what they want of the police, and where conflicts among citizens are less common because of common traits, a service style can develop. Where the population of the community is diverse, and agreement among subgroups does not emerge, the lack of agreement makes service problematic. In these diverse communities, there is also likely to be a fair amount of intergroup conflict among different social classes and ethnic or racial groups. If the government officials wish, they can constrain the police to enforcing the laws equally on all groups. But if this direction is lacking, the police members are free to define their jobs.

James Wilson (1963) suggests that the degree to which community leaders control the police agency is related to their view of political behavior. One view of political behavior, called the **Yankee-Protestant style,** rests on the belief that government (including the police) should be run according to general laws and principles, separate from personal relationships and needs. The alternative view, or **immigrant tradition,** views government as a product of personal relations, family needs, and personal loyalties. Legalistic-style agencies reflect this Yankee-Protestant approach to the impersonal provision of directed government services. Watchman-style agencies occur where political and community leaders allow the officers to discern and act on personal relationships and obligations among citizens.

In addition to the correlations of these community traits with styles of policing, other aspects of the community also correlate with police in more general ways. The number of police officers employed, for example, has been found to be associated with racial heterogeneity and distribution of wealth, and it also has been found to be a product of rational choice in responding to levels of crime (Chamlin, 1990; Liska & Chamlin, 1984; Nalla, Lynch, and Leiber, 1997; Sever, 2001). As a government entity, the police depend on general tax revenues for financial support. Wealthier communities are likely to pay higher salaries, have more modern and better equipment, and provide better training than poorer communities. Urban and more densely populated communities are likely to have a greater crime problem than rural and less densely populated communities. Communities with many employment opportunities for citizens will experience more difficulty in recruiting police personnel than those where employment opportunities are limited. These and other characteristics of the communities in which local police operate help explain the way municipal and local police agencies work.

CONCLUSION

There are thousands of municipal and local police agencies in the United States, employing hundreds of thousands of officers. Most of these agencies are small, and nearly every incorporated municipality has its own police force. Though only one component of the police industry in America, local policing is itself a complex topic. These local agencies differ in important ways, including the ways in which they practice policing.

Local police agencies demonstrate a variety of organizational structures. Although all such agencies have a general jurisdiction that covers the geographic area of the municipal government involved and the functions of law enforcement, service delivery, and order maintenance, they differ in terms of the relative emphasis each places on these functions. Styles of policing have been identified, and these styles appear to correlate with community characteristics. Thus, community factors correlate with the goals and structure of the police organization. The organization, in turn, correlates with how police officers tend to define and approach their work.

CHAPTER CHECKUP

1. What is the big-city focus in policing research, and why does it exist?
2. Why is there no typical organizational structure among municipal and local police agencies?
3. What are some types of local and municipal police?
4. Identify the three styles of policing described by James Wilson (1968).
5. Compare and contrast Wilson's styles of policing.

6. Why do local police in America display different styles?
7. What are some correlates of local policing?
8. Who are the "audiences" of local police agencies, and how do they affect the development of a style of policing?

REFERENCES

Blau, P. M. and W. R. Scott (1962) *Formal organizations.* (San Francisco: Chandler).

Brooks, L., A. Piquero, and J. Cronin (1993) "Police officer attitudes concerning their communities and their roles: A comparison of two suburban police departments," *American Journal of Police* 12(3):115–139.

Bureau of Justice Statistics (1989a) *Police employment and expenditure trends.* (Washington, DC: U.S. Department of Justice).

Bureau of Justice Statistics (1989b) *Profile of state and local law enforcement agencies, 1987.* (Washington, DC: U.S. Department of Justice).

Bureau of Justice Statistics (1992a) *Sheriffs' departments, 1990.* (Washington, DC: U.S. Department of Justice).

Bureau of Justice Statistics (1992b) *State and local police departments, 1990.* (Washington, D.C.: U.S. Department of Justice).

Chamlin, M. (1990) "Determinants of police expenditures in Chicago," 1904–1958, *Sociological Quarterly* 31:485–494.

Christensen, W. and J. Crank (2001) "Police work and culture in a nonurban setting: An ethnographic analysis," *Police Quarterly* 4(1): 69–98.

Crank, J. (1990) "The influence of environmental and organizational factors on police style in urban and rural environments," *Journal of Research in Crime and Delinquency* 27(1):166–189.

Davenport, D. (1999) "Environmental constraints and organizational outcomes: Modeling communities of municipal police departments," *Police Quarterly* 2(2):174–200.

Decker, S. H. and S. M. Ward (1989) "Rural law enforcement," in W. G. Bailey (ed.) *The encyclopedia of police science.* (New York: Garland):561–564.

Galliher, J. F. (1975) "Small-town police: Troubles, tasks and publics," *Journal of Police Science and Administration* 3(March):19–28.

Hickman, M. and B. Reaves (2001) *Local police departments, 1999.* (Washington, DC: Bureau of Justice Statistics).

Hoetmer, G. (1989) "Police, fire, and refuse collection, 1988," in *The municipal yearbook* (Washington, DC: International City Management Association): 179–234.

Jacobs, D. and D. Britt (1979) "Inequality and police use of deadly force: An empirical assessment of a conflict hypothesis," *Social Problems* 26(4):403–412.

Kelling, G. L., T. Pate, D. Dieckman, and C. E. Brown (1974) *The Kansas City preventive patrol experiment: A summary report.* (Washington, DC: The Police Foundation).

Kirkham, G. L. and L. A. Wollan (1980) *Introduction to law enforcement.* (New York: Harper & Row).

Langworthy, R. H. (1985a) "Police department size and agency structure," *Journal of Criminal Justice* 13(1):15–27.

Langworthy, R. H. (1985b) "Research note: Wilson's theory of police behavior: A replication of the constraint theory," *Justice Quarterly* 2(1):89–98.

Liska, A. and M. Chamlin (1984) "Social structure and crime control among macrosocial units," *American Journal of Sociology* 90:383–395.

Lundman, R. (1980) *Police and policing: An introduction.* (New York: Holt, Rinehart & Winston).

Maguire, K. and A. Pastore (1999) *Sourcebook of criminal justice statistics—1998.* (Washington, DC: Bureau of Justice Statistics).

Mastrofski, S. D., R. R. Ritti, and D. Hoffmaster (1987) "Organizational determinants of police discretion: The case of drinking-driving," *Journal of Criminal Justice* 15(5):387–402.

Mastrofski, S. and R. Ritti (1996) "Police training and the effects of organization on drunk driving enforcement," *Justice Quarterly* 13(2):291–320.

Nalla, M., M. Lynch, and M. Leiber (1997) "Determinants of police growth in Phoenix, 1950–1988," *Justice Quarterly* 14(1):115–143.

Ostrom, E., R. B. Parks, and G. P. Whitaker (1978) *Patterns of metropolitan policing.* (Cambridge, MA: Ballinger).

Parks, R. B. and E. Ostrom (1984) "Policing as a multi-firm industry," in G. P. Whitaker (ed.) *Understanding police agency performance.* (Washington, DC: U.S. Department of Justice):7–22.

Reaves, B. (1996) *Local police departments, 1993.* (Washington, DC: Bureau of Justice Statistics).

Reaves, B. and M. Hickman (2001) *Sheriffs' Offices, 1999.* (Washington, DC: Bureau of Justice Statistics).

Regoli, R. M., J. P. Crank, R. G. Culbertson, and E. D. Poole (1987) "Police cynicism: Theory development and construction," *Justice Quarterly* 2(2):281–286.

Riksheim, E. and S. Chermak (1993) "Causes of police behavior revisited," *Journal of Criminal Justice* 21(4):353–382.

Rossi, P. H., R. A. Berk, and B. K. Eidson (1974) *The roots of urban discontent: Public policy, municipal institutions, and the ghetto.* (New York: John Wiley).

Sever, B. (2001) "The relationship between minority populations and police force strength: Expanding our knowledge," *Police Quarterly* 4(1):28–68.

Sherman, L. (1980) "Causes of police behavior: The current state of the quantitative research," *Journal of Research in Crime and Delinquency* (January 1980):69–101.

Souryal, S. S. (1989) "Organizational structure," in W. G. Bailey (ed.) *The encyclopedia of police science.* (New York: Garland): 368–374.

Travis, L. and C. Winston (1998) "Dissension in the ranks: Officer resistance to community policing, cynicism, and support for the organization," *Journal of Crime and Justice* (forthcoming).

Travis, L. F. and R. J. Vukovich (1990) "Cynicism and job satisfaction in policing: Muddying the waters," *American Journal of Criminal Justice* 15(1):90–104.

Walker, S. (1983) *The police in America: An introduction.* (New York: McGraw-Hill).

Webb, V. and C. Katz (1997) Citizen ratings of the importance of community policing activities. *Policing* 20(1):7–23.

Weisheit, R., D. Falcone, and L. E. Wells (1994) *Rural crime and rural policing.* (Washington, D.C.: National Institute of Justice).

Wilson, J. Q. (1963) "The police and their problems: A theory," *Public Policy* 12:189–216.

Wilson, J. Q. (1968) *Varieties of police behavior: Law enforcement in eight American communities.* (Cambridge, MA: Harvard University Press).

Wilson, O. W. and R. C. McLaren (1977) *Police administration.* (New York: McGraw-Hill).

Witte, J., L. F. Travis, and R. H. Langworthy (1990) "Participatory management in law enforcement: Police officer, supervisor and administrator perceptions," *American Journal of Police* 9(4):1–23.

PART THREE

CORRELATES OF POLICING: ORGANIZATIONS, OFFICERS, AND COMMUNITIES

A large number of offices, agencies, and organizations perform police functions in the United States, and these organizations employ hundreds of thousands of individual police and law-enforcement personnel. In Part II we described some of the variety in policing in America by examining, in general terms, the types of police organizations and their government-level affiliations. In this section we will focus primarily on local policing.

The purpose of this section is to develop an analytical, or theoretical, framework to help us (1) understand the variety of structure in police organizations and (2) appreciate how they serve as intermediaries between the expectations of the community and those of the employees. The forces that appear to explain different types of organization are described and related to different priorities among the primary functions of policing—law enforcement, service delivery, and order maintenance. These different emphases are then linked to differently qualified personnel and differently successful leadership or management styles. A summary of what is known (or suspected) about local police officers and the effect of the individual officer on police practice is presented. Finally, we describe a variety of communities and how they influence the operations of police organizations and officers.

In Chapter 8 we describe police organizations as "tools" that can be used to meet a variety of police functions, and discuss these functions in terms of the police organization's goals. The importance of those goals to an understanding of organizational structure and practice is established. From this start, the chapter goes on to show how police organizations derive their structures from their goals and how the different structures make better or worse matches with an organization's environment.

In Chapter 9 we take up the question of police personnel in a general sense. Just as different organizational forms and structures are better suited to some goals than others, different personnel are better suited to some organizations than others. The chapter examines theoretical characteristics and qualifications of both police officers and police supervisors within the context of organizational constraints. We explore the question of which officers, under which supervisors, in which organizational forms, are most likely to successfully accomplish which organizational goals. This coverage includes an assessment of the consequences of mismatch among officers, supervisors, organizations, and goals.

In Chapter 10 we turn from the more theoretical material describing the nature of police organizations and the intersections of organizations, goals, and personnel to a description of contemporary police officers. Unfortunately, little research has been completed on the links among organizational goals, forms, and personnel in policing. There is, however, a relatively large body of work that examines and describes police officers. This literature is summarized and presented in the chapter. The chapter progresses from a description of police officers to a discussion of the social-psychological characteristics of police officers and their effects on police decision making and practice. It concludes with a review of the literature about police officer types. We must await further research before we can adequately link that research with characteristics of organizations.

In Chapter 11 we explore the concept of community and how characteristics of communities can be related to the social-control needs and types of policing that emerge. This chapter examines the link between the police and the community and illustrates that the concept of *a community* is difficult to apply in practice. A variety of communities may exist even within particular neighborhoods. To the extent that the police are expected to uphold and enforce community norms, values, and expectations, the characteristics of the community are important correlates of policing.

chapter *8*

POLICE ORGANIZATIONS

CHAPTER OUTLINE

What Police Do
  The Concept of an
     Organizational Goal
How the Police Do What
  They Do
  Law Enforcement
  Peacekeeping
  Public Service
Police Organizational Designs

Balance of Forces
  Organizational Goal
  Organizational Knowledge
  Police Methodology
  Interaction of Goal,
    Knowledge, and
    Methodology

Since the beginning of the twentieth century, if not midway through the nineteenth, policing in America has been the job of formal organizations. At that time, it stopped being legal to take up the hue and cry or to form a posse, and it became a civic obligation to call an organization to respond to problems. This organization, the police, is supposed to take care of the problem for you. There are a host of explanations for this development: Lundman (1980) suggests the shift to a functionally differentiated society, with each special division "doing its own thing," as the explanation. Bittner (1970) suggests a trend toward pacification of the society, leaving force to a select bureaucracy. Others suggest that the emergence of vocational policing is tied to a desire to control disorder, control crime, control classes, or to urban dispersion (Roberg, Crank, and Kuykendall, 2000). But the fact remains that individuals are no longer empowered to act on

187

their own. Rather, they must call an organization—the police.[1] The fact that policing in the United States is done by organizations requires that students of policing have an understanding of organizations. To this end, this chapter relates organizational structure to what police organizations are supposed to do and the ways they may elect to do those things.

Walker (1983) cites Perrow, noting that organizations are "tools" used to achieve objectives. The tool analogy is useful, and elaboration of the analogy will more clearly bring the notion home. Whereas tools are clearly means to ends, it is also clear that different tools are used to achieve different ends. That is, different tools (even within the same family of tools) serve different purposes, and are designed and administered differently.

For example, consider different types of hammers: a framing hammer, a tack hammer, and a jack hammer. Clearly, all three types of hammers are used to focus a great deal of energy in a small place. Equally clear is the fact that each of the hammers is designed and administered to focus this energy for different purposes. A good 24-ounce framing hammer is designed to be used in rough construction, a tack hammer is typically used to finish upholstering furniture, a jack hammer is typically used to break up pavements (roads, sidewalks, etc.). It is also apparent that the designs of these tools and their administration are shaped by these alternative ends. The framing hammer is designed to drive a larger spike than a tack hammer, and a jack hammer is designed to break up substantial materials. You would not wish your home to be framed with a jack hammer nor your street to be serviced with a tack hammer.

Just as the design is influenced by the goal, so is the administration of the tool influenced by its design. The imagery of someone misapplying a jack hammer, for example is equally as comical as that of someone misapplying a framing hammer. Picture someone swinging a vibrating jack hammer back and forth against the concrete or just holding a framing hammer against it, waiting for it to start jumping up and down.

Policing is done by organizations, that is, by tools that are designed and administered to achieve ends. Just as there are different hammers to achieve different ends, there are different police organizations designed and administered to achieve different police organizational goals. This chapter addresses the concept of the police organizational goal with the purpose of determining whether one type of police tool is all that is needed or whether police organizations might vary as do other tools. The next chapter correlates staffing and administration issues, a related subject. What will become apparent is that considerable variation exists in what people believe police organizations ought to do, and that each of these variants has organizational design, staffing, and administration implications.

---

[1]It is also noteworthy that personal victimization is not viewed as the subject of policing; societal victimization is. That is, violation of the law consists in what the state prosecutes for, not an individual's victimization. For the person to be "made whole," that person must act in a civil court.

WHAT POLICE DO

In 1973 the American Bar Association (ABA) (1973:53) published the results of a study of the urban police function in which 11 police organization objectives were identified:

1. To identify criminal offenders and criminal activity and, where appropriate, apprehend offenders and participate in subsequent court proceedings.
2. To reduce the opportunities for the commission of some crimes through preventive patrol and other measures.
3. To aid individuals who are in danger of physical harm.
4. To protect constitutional guarantees.
5. To facilitate the movement of peoples and vehicles.
6. To assist those who cannot care for themselves.
7. To resolve conflict.
8. To identify problems that could become serious law-enforcement or government problems.
9. To create and maintain a feeling of security in the community.
10. To promote and preserve civil order.
11. To provide other services on an emergency basis.

The ABA was principally concerned with police tasks that were a service to the society at large. Lundman (1980:44–45) notes that though the society at large is the intended primary beneficiary of police services, the public is not the sole constituent. Indeed, Lundman suggests that a number of other constituents must be served, if not placated, to sustain the organizational coalition. Toward that end, police departments provide for the needs of specific clients, local government, and labor (Duffee, Fluellen, and Roscoe, 1999; Freidman and Clark, 1999; Scheingold, 1999). Thus, we can add to the list of police organization concerns the management of public appearance, political relations, and labor relations.

It is clear that the police perform many tasks as agents of local government (Moore and Poethig, 1999); it is also clear that the organizational goal is complex and often consists of contradictory elements. Some have sought to reduce the complexity of the organizational goal by suggesting a prioritization of its elements. The National Advisory Commission on Criminal Justice Standards and Goals (1973), for example, suggests that although police should ensure that constitutional guarantees are protected and that the law is enforced, they also should

acknowledge that the basic purpose of the police is the maintenance of public order and the control of conduct legislatively defined as crime.

The basic purpose may not limit the police role, but should be central to its full definition. (1973:12)

This is in addition to providing various other services to the community.

Others who focus more closely on what police "actually" do suggest a different prioritization. Mastrofski (1988), for example, addressing the shift to community policing, notes that at the heart of the community policing reform effort stands a different prioritization:

[Community policing] redresses the previous reform era's obsession with law enforcement as the core of the police role. It [community policing] attempts to direct more administrative attention to the order-maintenance activities comprising a major but latent part of the police work. (1988:64)

Although it seems plausible that there is substantial agreement concerning the substance of police work, it is also evident that there is considerable debate concerning which elements of the job should receive emphasis. Our effort to better understand the police task can be improved by a systematic examination of organizational goals. To this end, the following discussion focuses on theories of the organizational goal.

The Concept of an Organizational Goal

Most discussions of an organization's *raison d'être* present the organizational goal as a fairly straightforward, known, agreed-on, and singular entity. These discussions suggest that organizations exist to achieve some particular objective. The problem with these presentations is they oversimplify the organizational purpose and create an illusion of simplicity where complexity is generally the case. This is particularly troublesome when we assume, as we do here, that organizations are designed to work toward the accomplishment of an organizational goal. If we overlook the complexity of the organizational goal, we risk bad organizational design and application. That is, we risk trying to upholster a chair with a jack hammer.

The following discussion is designed to help us better understand the complexity of organizational goals. We hope that such a better understanding will lead to a greater likelihood that readers will select the proper tool for getting the work of the organization done. There has been considerable discussion of the concept of an *organizational goal* in both the sociological and political science literature. This literature reveals a picture of the organizational goal as a complex, multidimensional construct that must be served in its entirety. The following discussion focuses on the work of Simon (1964), who addresses the organizational goal as a complex foundation for decision making, and Mohr (1973), who proposes a taxonomy of goals.

Simon (1964:20), in an effort to describe an organizational goal, concludes that "it is doubtful whether decisions are generally directed toward achieving a goal." He suggests that instead, organizational decisions are directed at "satisfying a whole set of constraints." He arrives at this conclusion after reasoning that the organizational goal is really a multidimensional construct that both guides and restricts organizational action. Simon also notes that not all elements of the constraint set are equally valued, and these inequalities ultimately result in organizational variety.

Simon's argument is based on the observation that organizations must satisfy a host of demands simultaneously. It can be argued, for example, that successful police organizations must simultaneously ensure a minimum level of community peace, a minimum level of law enforcement, and a minimum level of public service. If the police department fails to minimally satisfy any of these "constraints," it fails to achieve its organizational goal. Thus, if enforcement of law so burdens the community that riots occur, the police are properly accused of failing. Likewise, if fear of causing riots leads to gross underenforcement of the law, the department can also be (and likely would be) accused of failing to do its job. In neither case is the organizational goal, in all of its dimensions, minimally satisfied, and so the organization has failed to achieve that goal. Yet, survival of the organization is determined by its ability to minimally satisfy the constraint set.

It is also apparent that successful organizations, even within the same industry, take many different forms. Simon's basic premise is that although the organizational goal is composed of a set of constraints, not all of the constraints within the set are equally valued. This value-based distinction permits us to explain organizational variety even within the same industry. Again, in reference to the police, let us assume that all police organizations serve the same constraint set: law enforcement, peacekeeping, and public service. We know from the preceding discussion that all three of these elements must be minimally satisfied if the organization is to be successful. However, we also know that there is considerable variation, both organizationally and behaviorally, across organizations in the police industry (Wilson, 1968; Langworthy, 1986; Crank, 1990; Crank and Wells, 1991; Zhao, 1996; Maguire, 1997, 2001; Sanders, 1997; King, 1998). Simon (1964) would argue that this variation is the cumulative product of organizational decisions that are themselves products of different value orientations held by the varied organizations in the industry. In some police organizations peacekeeping is the most valued element of the constraint set, in others law enforcement is most valued, while in still others public service is most valued.

These inequalities ensure industrial variety by their cumulative impact on organizational decision making. In police departments where law enforcement is the most valued constraint, department policy would be designed to further law-enforcement efforts. Likewise, in departments where peacekeeping is most valued, policy would favor community tranquillity. The values attached to each of the constraints has the effect, over time, of shaping the organization to reflect those values. Therefore, department "personality" is a product of the accumulated organizational decisions that reflect the valued constraints of the organizational goal.

Taken together, Simon's points permit us to account for organizational variety within an industry. Organizations within the same industry serve the same constraint set. However, organizational variety within an industry is a product of differential emphasis placed on constraints within the set. Thus, although all successful police departments minimally satisfy the same constraint set (law enforcement, peacekeeping, and public service) the organizations often differ from one another because of different organizational values.

Chart 8.1 illustrates several different conceptions of police organizational goals using the trio of constraints just discussed. Goal 1 emphasizes law enforcement, goal 2 peacekeeping, and goal 3 public service. Nonviable goals fail to address one or more of the elements of the constraint set (e.g., goal 4 does not provide public service). It should also be noted that an organizational goal is not viable if the emphasis is insufficient to ensure a minimum level of constraint satisfaction. For example, goal 5 might not be viable because 5 percent of organizational effort devoted to law enforcement might be insufficient to minimally satisfy that constraint.

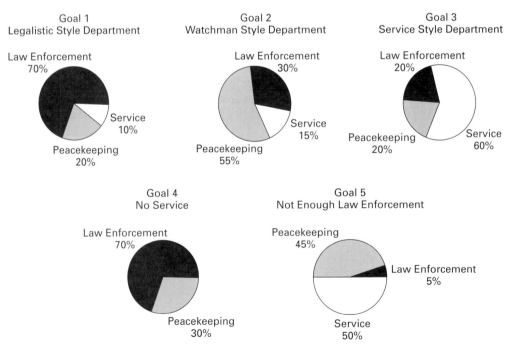

CHART 8.1 Conceptions of the police organizational goal.

Having established that there are different conceptions of the organizational goal possible, even within the same industry, the next stage of our analysis focuses on a taxonomy of constraints. In his discussion of the organizational goal, Mohr (1973) notes that there are several different types of elements. The first he notes are transitive elements, or intended environmental impacts. Our police example developed earlier focuses on the transitive elements of the constraint set.[2] The police organizational goal in that example reflects concerns with what the police are supposed to do for the community they serve. All agencies are to provide for the peace in a community, ensure a modicum of law enforcement, and provide some public services. These elements of the constraint set are the reasons communities have police organizations—so that the organization can affect the environment.

The second type of constraints within an organizational goal are reflexive. Reflexive constraints are characterized as focusing on organizational maintenance and survival. Mohr's definition of the reflexive elements focuses on "inducements . . . sufficient to evoke adequate contributions from all members of the organizational coalition" (1973:476). Police, as a publicly sustained element of local government, have a host of different constituencies that together form the police organizational coalition. If the organization is to survive in its current state, it must keep this coalition intact and is therefore constrained to at least minimally serve each constituent's concerns.

Blau and Scott (1962) have isolated four general organizational constituencies. Although they develop an organizational typology from their constituencies, they also note that organizations have primary and secondary beneficiaries. Thus, they recognize that an organization may serve more than one constituency. The four constituent groups are organizational members, owners (funders), clients (service recipients), and the society at large. Additional constituents might include "the profession" and the organizational network.

Reflexive elements in a police department's organizational goal would include concern with public appearance (society at large), local government (owners/funders), assistance to victims and constitutional rights of suspects (clients), relations to other social service agencies (the organizational networks), labor organizations (members), and the perceptions of the larger professional police community.

While reflexive and transitive elements of the constraint set are different, Simon's admonition that all elements of the organizational goal must be served is no less true. Mohr (1973:476) notes that the organization must survive if it is to address environmental needs. Therefore, reflexive constraints must be at least minimally satisfied if the organization is to be

---

[2]The ABA list of objectives are transitive elements.

successful. Likewise, there is no reason for the organization to survive if it does not address its intended environmental purpose. Therefore, the transitive constraints must also be minimally satisfied for the organization to be successful. Mohr concludes that transitive and reflexive elements of the constraint set are "co-equal" in importance and that an organization that does not minimally satisfy all transitive and reflexive elements will fail.

Chart 8.2 offers a more complete illustration of a police organizational goal, including both transitive and reflexive elements of the constraint set. The first goal conception (Chart 8.2a) indicates that the greatest emphasis is on the transitive elements—law enforcement first, then peacekeeping (this composition approaches Wilson's [1968] legalistic-style department).

CHART 8.2
Conceptions of the police organizational goal. (Both transitive and reflexive elements)

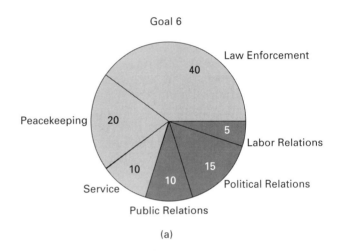

(a)

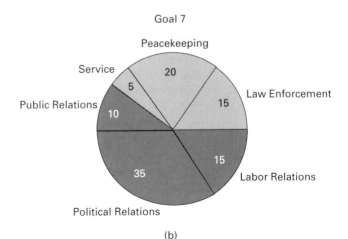

(b)

The second conception (Chart 8.2b) indicates that emphasis is placed on re-flexive elements generally (major emphasis on political relations), but that the peacekeeping transitive element is of secondary importance (this is similar to Wilson's watchman style). As noted before, all elements of the constraint set must be minimally satisfied if the organization is to survive.[3]

The foregoing discussion demonstrates that the organizational goal is quite a complex entity, composed of different types of constraints (transitive and reflexive), with different organizational value judgments attached to each of the constraints. So the police organizational goal is not a unitary construct but a multidimensional set of constraints that simultaneously guide and constrain the behavior of police departments. Further, it is ap-parent that the personality of a police department is determined by the values attached to each of the dimensions of the constraint set. However, the longevity of the department is determined by its capacity to satisfy the complete constraint set.

HOW THE POLICE DO WHAT THEY DO

Having established the fact that police address many different objectives, we shift our attention to how they do what they do. Our focus will be on the degree to which individual discretion is relied on. Although it is widely ac-knowledged that individual police officers exercise discretion, it is also ap-parent that not all officers in all departments exercise the same degree of discretion. Consequently, as with the organizational goal, the way police do what they do is also variable.

The preceding discussion has suggested that we can conceptualize the transitive element of the police organizational goal as composed of three ba-sic elements: law enforcement, peacekeeping, and public service. Though there are clearly different tasks assigned to the police, it is evident that each of these tasks can be left to the discretion of individual officers, or the agency can specify what is to be done, when it is to be done, and how it is to be done. That is, law enforcement, peacekeeping, and public service can be specified by the agency or left to the discretion of the street officer.

Law Enforcement

Police organizations can see to the enforcement of laws in a number of ways. Some agencies may instruct their officers to enforce all laws without fear or

---

[3]Where police departments (and many other public service agencies and utilities for that matter) are concerned, organizational failure is generally manifest as organizational reform rather than dissolution (Crank and Langworthy, 1992). That noted, research by King, Travis, and Langworthy (1997) suggests that failure of police departments may result in dissolution.

favor. That is, officers are to observe behavior, determine whether that behavior is legal, and if it is not, they are to invoke a specified criminal prosecution. In this type of department, officers are not encouraged to exercise discretion. First, officers are instructed in the law so that they can recognize violations when they see them. Second, officers are instructed in due process issues so that they know precisely what to do and how to do it, that is, effect an arrest, thereby engaging the process of prosecution. Third, officers are instructed to follow orders and administer the law as written and in accordance with established administrative principles. Departments of this type were called legalistic by J. Wilson (1968).

Police organizations may also elect to specify which laws they will enforce rather than enforce all laws on the books. In 1973 the National Advisory Commission on Criminal Justice Standard and Goals (1973:23) observes that one of the responsibilities of police executives as active policy makers is to establish enforcement priorities that are within the "budgetary, organizational, and legal constraints that limit his discretion." The commission further suggests that

> As long as those policies [enforcement priorities] are openly adopted, reduced to writing, and applied in a nondiscriminatory manner, the public and governing bodies are assured that the law enforcement is being administered properly by the police. (1973:23)

While this specification promotes discretion, it promotes it for the organization, not for the street police officer. In this organization, as in the legalistic police department just described, the street police officer is instructed in the law to recognize violations, informed as to how to intervene, and directed to follow orders, particularly regarding which laws to enforce. The only substantial difference between the legalistic department and this one is that this department enforces only some laws whereas the legalistic department enforces all laws. We call these departments **selective legalistic** police departments.

The two organizational styles outlined here seek to have police officers administer the law in accordance with the specification of legal behavior and in accordance with due process stipulated in law. Police organizations may also elect to let police officers "do their own thing" to further satisfy the law-enforcement goal. These departments are called **enforcement guilds,** composed of nominally independent craftspersons enforcing the law. Just as selective legalism has its advocates, so does this style of policing, which advocates that street police officers be permitted to exercise judgment in deciding which laws to enforce and when (in which situations) to enforce them. The same arguments that suggest the organization should be allowed to decide which laws to enforce and when to enforce them can be applied to street officers. That is, officers can be asked to consider legislative intent and community tolerance. The essential difference between the selective le-

The police enforce the law by taking a suspected drug offender into custody.
(Piet van Lier/Impact Visuals)

galistic style and the enforcement guild style of law enforcement is that in the latter the organization vests the officer with discretionary authority rather than retaining it.

Peacekeeping

Just as police departments can elect to enforce the law in a number of different ways, so can they elect to maintain peace and order in a number of different ways. The administrative mode is less directly apparent in support of peacekeeping and order-maintenance constraints than it is in law-enforcement constraints. Nevertheless, the organization can stipulate what constitutes peace and order and then follow that stipulation with action directives. This administrative ability to manage the peace is no more clearly demonstrated than in the Sherman-Berk (1984) domestic-violence study conducted in Minneapolis (1984) and replicated by others in Miami, Colorado Springs, Milwaukee, Omaha, and Charlotte (see Sherman, 1992, for discussion). The domestic-violence experiment sought to determine the effects of different police interventions in domestic-violence situations. During the study the police department had to variably define "simple (misdemeanor) domestic assaults" as (1) a violation of law to which a legalistic response (arrest) was required, (2) a violation of the peace to which a peacekeeping response (separate the parties) was required, and (3) a threat to the family requiring a service response (counseling). Thus, the department demonstrated the capacity to specify disorder (simple domestic assault) and specify the intervention—arrest, separation, or counseling. These police departments are called **peace agencies,** because they tell officers what the peace is and how to support it.

The principal distinction between administrative support of the peace versus law enforcement is that the number of laws to enforce is finite whereas definitions of peace are virtually infinite. It is possible to conceive of enforcing "all laws" but not to conceive of supporting all definitions of peace. That distinction noted, it is also necessary to acknowledge that it is a practical impossibility to enforce "all laws"—there are simply too many of them. What remains distinct, however, are the means by which law and peace are defined. Laws are defined by the state through a conventional process and routinely communicated to the police. Peace, by contrast, tends to be locally defined through an unspecified process and unevenly communicated to the police. Thus, the police organization may be less equipped to know what the "peace" is and therefore less able to maintain it.

Just as police organizations may elect to permit officers to determine which laws to enforce and how to enforce them, they can also allow individual police officers to define the peace and decide how to support it. The principal concern in such a department will be with what Rubenstein (1973) calls "keeping the beat quiet." The officer is considered to be doing a "good" job if the beat is quiet, meaning there are no complaints about loiterers or traffic flow, and commerce is supported. In such departments officers are expected to know how the beat defines peace and how the beat expects the peace to be supported. They are also expected to exercise judgment to ensure that the beat gets what is needed to keep it quiet.[4] Wilson (1968) calls this type of police department a watchman department.

Public Service

The 24-hour availability of the police to the public makes it likely that police will be asked to help in circumstances that do not necessarily relate to law enforcing or peacekeeping. As with their other responsibilities, the police organization has options concerning responses to requests for provision of public services.

J. Wilson's (1968) description of the service style of police organizational behavior details the practices of an organization that administers public services. Wilson (1968:201) notes that in some communities, typically homogeneous and affluent communities, the police organization re-

---

[4]Wilson (1968) refers to a "zone of indifference" that is essentially community tolerance. Tolerant communities have a wide zone of indifference, whereas intolerant communities have a narrow zone of indifference.

The police help keep the peace and maintain order by directing traffic when traffic lights fail. (Tom Kelly)

sponds to service demands by creating special units to ensure that the service is provided in a uniform manner. In this type of organization, the public services that the police organization will provide are specified, and officers are instructed regarding how those services are to be provided.[5] Police officers in the service-style organization do not decide what services are needed or how they should be provided; they simply follow instructions that specify the services to be provided and the way they are to be provided.

As was the case with the other police constraints, models that promote the exercise of discretion in the provision of public services are conceivable. The team policing model, more recently called **community-oriented policing,** was developed and tested expressly to provide police organizations with the ability to respond to a broad array of public service issues. In the team model, groups of police officers are assigned to provide marginally

---

[5]Wilson also suggests that public service is antithetical to the police role, as service is something desired and evaluated by the client. According to Wilson, the police role is defined at least in part by the fact that the client's wishes and assessments are not a central concern. Further, he argues that services can be provided by market forces—that is, if there is demand for service, it should be priced and sold.

The police providing emergency medical services at the site of a construction accident. (David A. Cantor)

defined police services to an area and left to decide, collectively, how to provide those services. The definition of area problems and the development of intervention strategies are left to the team. The team utilizes a collegial process by which a consensus is reached concerning community problems and the means of solving them. The team then executes its plan.

TABLE 8.1 Types of Police Organization by Elements of the Organizational Goal and Police Technology

Constraints	POLICE TECHNOLOGY (THE WAY THEY DO WHAT THEY DO)	
	Administer	Discretion
Law enforcement	*Legalistic:* Officers enforce all laws in accordance with federal, state, and local laws.	*Enforcement guild:* Officers decide which laws to enforce and when and how to enforce them.
	Selective legalistic: The agency sets enforcement priorities and determines intervention strategies. Officers do as they are told.	
Peacekeeping	*Peace agency:* The agency defines the peace and specifies how and on whom it will be maintained.	*Watchman:* Officers determine what the peace is and how to maintain it.
Public service	*Service:* The agency specifies the services it will provide and how it will provide those services to the jurisdiction.	*Team:* There is a collective assessment of community needs and development of interventions to address those needs.

Obviously, the way police organizations address their responsibilities are variable. The primary source of this variability is whether the organization specifies the work of the police or whether it leaves that specification to the officers on the street. Table 8.1 summarizes the distinction. Police organizational designs must therefore provide not only for different organizational goals but for a variety of ways of goal achievement.

POLICE ORGANIZATIONAL DESIGNS

Scholars concerned with police organization tend to overgeneralize their descriptions of police organizations. They summarily describe police organizations as highly centralized bureaucracies and then typically suggest that this model is inappropriate for doing the job of policing (Bittner, 1970:51; Angell, 1971:194; Sandler and Mintz, 1974:458; Cordner, 1978:34; Roberg, 1979:314; Kuykendall and Roberg, 1982:243; Manning, 1992:327; Moore, 1992:107). The problem with this generalization is that it causes students of police organization to overlook police organizational variety.

Critics of police bureaucracies suggest that their hierarchical structure and division of labor are antithetical to the nature of police work. The essence of this argument is that because police work is so varied, and because line police officers are vested with so much discretion, bureaucracies are inappropriate. Bureaucracies, by virtue of their hierarchical relations and functional division of labor, are best suited to dealing with routine duties and curbing the exercise of line-level discretion. What the critics fail to recognize, however, is that police organizations, like organizations in other industries, are more or less hierarchical and more or less divisive in task. However, Wilson (1968), Guyot (1979), and more recently Langworthy (1986), Slovak (1986), and Maguire (1997) have demonstrated empirically that it is inappropriate to characterize police organizations simply as bureaucracies, because, as Langworthy notes, police organizations "are neither bureaucratic nor democratic, but more or less so" (1986:28). It follows, then, that to the extent that organizational structure is related to task variability and discretion, some extant police organizations are better able to handle task variety and are more permissive of line-level discretion. Likewise, it follows that the nature of the job of policing, including both its ends and the means used to reach them, determines the appropriateness of the organizational form.

Contrary to the position held by many police organizational scholars, it is evident that police organizations are highly varied in hierarchal structure and in the degree to which they functionally divide the task of policing. The question to ask, then, is which organizational form is appropriate, and when? To answer that question, we must understand the varieties of organizational form.

In 1961 Burns and Stalker, as a product of their empirical study of British manufacturing firms, identified two basic types of organizational

A team of police officers discussing solutions to problems in their assigned area. (Courtesy of Dallas Police Department)

structures: mechanistic and organic. These two organizational forms are perfect opposites. *Mechanistic* organizations are characterized as hierarchically tall, functionally differentiated, and rule directed. *Organic* organizations, by contrast, are hierarchically flat, having a low degree of division of labor and little reliance on direction (rules) for the conduct of organizational business. Burns and Stalker (1961) suggest that the mechanistic form is appropriate when the organizational environment is stable, and the organic form is appropriate when the environment is unstable.

Others have sought to relate the mechanistic-organic typology to police organization. Cordner (1978) contrasts mechanistic and organic organizational forms, noting that each form can be viewed as appropriate for policing in 10 of 11 dimensions. Table 8.2 summarizes Cordner's descriptions of mechanistic and organic organizations. He suggests that the problem with the mechanistic organizational form for policing is that the form is inappropriate for tasks performed in unstable environments, and he contends that the environment of policing is unstable. The problem with the organic organizational form is that it is ends-oriented versus means-oriented in an industry where due process—that is, means—are a paramount concern. Thus, neither the mechanical nor the organic organizational form represents a panacea for police organizational design; each has an Achilles heel.

Langworthy (1992) adds another consideration to the discussion. He suggests that the propriety of the organizational form is conditioned by the way the job is done. That is, if the job of policing is viewed as ministerial, administering the law, or "the local will," then mechanistic organizational structures are appropriate. However, if the job of policing is viewed as being done by skilled craftspersons or professionals using their good judgment, then organic organizational structures are appropriate.

The foregoing discussion makes it apparent that alternative conceptions of organizations are differentially appropriate. Burns and Stalker (1961) suggest that environmental stability conditions the appropriateness of the organizational form. Cordner (1978) suggests that environmental stability is consequential for design but adds that for police, means–ends considerations are significant. Langworthy (1992) observes that the way the task of policing is to be done conditions the propriety of the organizational form. If police are to administer the law, the peace, or public service, a mechanistic organizational structure is indicated. However, if the job of policing is to be done by officers exercising their individual or collective judgment, organic organizational forms are more appropriate.

The question, then, becomes how can we reconcile these differences? A knowledge base, scientific or otherwise, makes it possible for an agency to fit these alternatives together. Knowledge reasonably permits organizations to specify how the job of the organization is to be done. If the organization (or industry, for that matter) knows enough about causes and effects that operate within the industry, it is in the position to direct operations in the industry. In the absence of this knowledge, it must rely on the experience

TABLE 8.2 Comparison of Mechanistic and Organic Organizations

MECHANISTIC	ORGANIC
1. Routine tasks occurring in stable conditions.	1. Nonroutine tasks occurring in unstable conditions.
2. Task specialization (i.e., a division of labor).	2. Specialized knowledge contributing to common tasks (specialized knowledge possessed by any one member of the organization may be applied profitably to a variety of tasks undertaken by various members of the organization).
3. Means (for the proper way to do a job) are emphasized.	3. End (or getting the job done) is emphasized.
4. Conflict within the organization is adjudicated from the top.	4. Conflict within the organization is adjusted by interaction with peers.
5. Responsibility (or what one is supposed to do, one's formal job description) is emphasized.	5. Shedding of responsibility (i.e., formal job descriptions are discarded in favor of all organization members contributing to all organizational problems) is emphasized.
6. One's primary sense of responsibility and loyalty are to the bureaucratic subunit to which one is assigned.	6. One's sense of responsibility and loyalty are to the organization as a whole.
7. The organization is perceived as a hierarchic structure (i.e., the organization looks like a pyramid).	7. The organization is perceived as a fluidic network structure (i.e., the organization looks like an amoeba).
8. Knowledge is inclusive only at the top of the hierarchy (i.e., only the chief executive knows everything).	8. Knowledge can be located anywhere in the organization (i.e., everybody knows something relevant about the organization, but no one, including the chief executive, knows everything).
9. Interaction among people in the organization tends to be vertical (i.e., one takes orders from above and transmits orders below).	9. Interaction among people in the organization tends to be horizontal as well as vertical.
10. The style of interaction is directed toward obedience, command, and clear superordinate–subordinate relationships.	10. The style of interaction is directed toward accomplishment, provides "advice" (rather than commands), and is characterized by a "myth of peerage," which envelops even the most obvious superordinate–subordinate relationships.
11. Loyalty and obedience to one's superiors and the organization generally are emphasized.	11. Task achievement and excellence of performance in accomplishing a task are emphasized.
12. Prestige is internalized; that is, personal status in the organization is determined largely by one's office and rank.	12. Prestige is externalized; that is, personal status in the organization is determined largely by one's professional ability and reputation.

Source: Adapted from Cordner, G. W. (1978) "Open and closed models of police organizations: Traditions, dilemmas, and practical considerations," *Journal of Police Science and Administration* 6(1):23, 24, 29.

and good judgment of its workers. According to Burns and Stalker (1961), the environmental stability constraint may indeed operate through the degree of understanding. If an organization finds itself operating in an unstable environment, it will likely require much more knowledge to manage that volatile environment than it would need for a stable one.

Knowledge also conditions the way a job can be done. If a great deal is known about how the job should be done, it is possible for the organization to instruct those who are to administer organizational tasks. It is also quite reasonable to expect that those instructions will be complied with. However, if knowledge is lacking, it is not feasible for the organization to tell its agents how to achieve the organizational goal. Neither is it reasonable to expect that those not instructed could comply with nonexistent instructions.

Because the concept of *knowledge* is pivotal to our understanding of the appropriateness of particular organizational structures, it warrants further discussion. For our purposes here, there is very little difference whether the knowledge is derived scientifically, drawn from the community, appears metaphysically, or is "written on a rock." This is so because what knowledge does is permit instruction, provide a reference when questions arise, and permit process evaluation of employee performance. Hence science-based knowledge is no more effective than policy or decree as a vehicle for instruction, problem solving, or evaluation. Knowledge, policy, and decree can all be taught, referred to, and used as evaluation criteria. The essential difference is in their derivation. Science-based knowledge is empirically derived through rigorous, systematic observation of the relations between causes and effects. Policy or decree, by contrast, tend to be fairly unsystematically derived opinions of those with the power to decide what will be done and how it will be done.

Perrow's (1970) comparison of two juvenile correctional institutions, Dick and Inland by name, provides an example. These two institutions operated under two different sets of assumptions. The staff at Dick believed that they completely understood juvenile behavior. Their theory was that children misbehave because they do not understand the rules. Therefore, the way to "fix" the problem was to simply establish clear rules and insist on their strict enforcement. The knowledge base at Dick permitted a mechanistic structure wherein workers administered rules in pursuit of the organizational goal.

The staff at Inland, in contrast, were not at all sure of the causes of juvenile misbehavior. They believed that each of their wards was unique and in need of a custom solution. At Inland knowledge on "how to fix kids" was presumed lacking, so staff were left to derive treatments protocols for each individual. The Inland organization was decidedly organic. It supported the need for staff to exercise considerable discretion to analyze unique problems and discover distinctive treatment regimens for each ward.

It should be apparent that what determined the organizational structure was not whether the knowledge base was true or false, but whether it existed. In the first organization, Dick, the knowledge base existed, and it made the job of the staff ministerial and the structure mechanistic. In the other institution, Inland, knowledge was lacking, rendering the job judgmental and the structure organic.

BALANCE OF FORCES

The forces that must be balanced in proper organizational design are the organizational goal, the organization's knowledge, and the police methodology. Examination of the interaction of these concepts provides a framework by which we can consider organizational alternatives.

Organizational Goal

Police organizations, like other organizations, are tools designed expressly to do work. The work is done to accomplish something—achieve an organizational goal. We have demonstrated that police organizational goals are quite complex entities composed of transitive and reflexive elements and that some of these elements will be emphasized rather than others. Transitive elements of the organizational goal are intended environmental impacts—the organization's reason for existing. Where police organizations are concerned, the transitive elements include law enforcement, peacekeeping, and service. Reflexive elements of the organizational goal focus on organizational maintenance or survival. Reflexive elements of the police organizational goal include public relations, political relations, and labor relations.

Finally, we have shown that organizational personality is determined by the cumulative effect of decisions that manifest elemental preferences. For example, legalistic police departments are distinguishable from watchman or service police departments because their decisions, over time, favor choices that emphasize law-enforcement elements rather than peacekeeping or provision of public services.

Little research has directly assessed the development and impact of the organizational goal on police organization structure and activity, but some recent studies support the thesis that reflexive elements of the organizational goal influence police agency activities. In a study of a midwestern police department, Katz (2001) reports that the agency created a specialized gang unit largely in response to external pressures from political and community interests. Once created, the unit evolved, taking on tasks and responsibilities aimed more at maintaining the survival of the unit than at the transitive goal of reducing and controlling the gang problem. Similarly, there is growing evidence that at least part of the explanation for the rapid spread of community policing across American law enforcement agencies has been the availability of federal support for such initiatives (Zhao, Lovrich, and Thurman, 1999; Maguire and Mastrofski, 2000).

Organizational Knowledge

The significance of knowledge for organizational design is that where knowledge exists, discretion can be restricted. However, if knowledge is lacking, organizations must rely on the "good judgment" of the organization's members. It has been suggested that knowledge can assume a number of different forms. It can be scientifically derived or specified in policy. Regardless of the form, knowledge permits organizations to specify how the job will be done. Thus, the organization can instruct subordinates and evaluate their performances.

There have been a few studies that indicate that police organizations can control the behaviors of officers through the promulgation of rules, policies, and directives. Geoffrey Alpert and John MacDonald (2001) examined the use of force by police officers in a national sample of American police departments. They report that agencies that require only involved officers to report use of force show higher rates of police use of force than those requiring supervisors or other officers to also report use of force incidents. It may be that an increased level of supervision or accountability reduces officer use of force. Policies on pursuit driving have also been found effective in controlling police officer discretion in high-speed pursuits. Policies that are most restrictive are associated not only with fewer high-speed pursuits, but also with reduced use of force by police after pursuits (Becknell, Mays, and Giever, 1999). Finally, research indicates that organizational climate is an important correlate of citizen complaints about police use of unnecessary force (Cao, Deng, and Barton, 2000).

Where knowledge exists, either presumed or actual, mechanistic organizations that restrict the exercise of discretion are appropriate. Where knowledge is lacking, again either presumed or actual, organic organizations that promote the exercise of discretion are appropriate.

Police Methodology

The last force we need to bring into the equation of police organization is police methodology. Our discussion of police methodology focused on ministerial versus discretionary styles. We demonstrated the possibility of addressing each of the transitive elements of the organizational goal from either a ministerial or discretionary style. How police work is to be done has significant implications for police organizations. If police are to minister (ministerial style) to the law, the peace, or to provide specified public services, then mechanistic police organizations are appropriate. However, if police are to exercise their individual or collective judgment (discretionary style) when enforcing the law, keeping the peace, or seeing to the provision of unspecified government services, then organic police organizations are appropriate.

Interaction of Goal, Knowledge, and Methodology

The question now turns to the interaction of these forces—the organizational goal, organizational knowledge, and police methodology—as they collectively determine the appropriateness of types of organizations. We have noted that mechanistic police organizations are appropriate when the police method is ministerial and when knowledge is present. It has also been noted that organic police organizations are appropriate when the police method relies on the exercise of discretion and when knowledge is lacking. What has yet to be incorporated is the organizational goal.

What we know about achieving the organizational goal conditions the degree to which it is possible to be ministerial and mechanistic or discretionary and organic. The question now is, which elements of the organizational goal do we know how to achieve? We have demonstrated that each of the transitive elements of the organizational goal can be addressed in a ministerial or discretionary manner. What distinguishes treatment of each of the transitive elements is the willingness of the organization to define situations and stipulate interventions.

Where legalism is the focus, government must stipulate the situations in law and specify interventions in due process if ministerial law enforcement is desired. If the locality elects to leave the decisions to the police officers (discretionary law enforcement), then officers will decide when laws are violated and when to intervene. When peacekeeping is the focus, government must be willing and able to define the peace and specify interventions if ministerial peacekeeping is desired. If government lacks the ability or is unwilling to define the peace (or more likely, to specify when the peace has been broken) and cannot or will not specify how to intervene, then keeping the peace will fall on the shoulders of individual police officers exercising their best judgment (discretionary peacekeeping).

Finally, if policing is to be viewed as ministerial public service, it will be necessary for government to specify which services the police will provide, when police will provide these services, and to whom they will provide the services. If government is unwilling to specify the services to be provided, then once again police officers will provide public services and assistance as they see fit, when they see fit, and for whom they wish.

Ultimately, the appropriate form of the police organization turns on the ability or willingness of the government to tell its subordinates what to do, how to do it, and when to do it. If the government is willing and able to tell subordinates what to do, how to do it, and when to do it, then ministerial police methods and mechanistic organizational structures are appropriate. If the government is unwilling or unable to tell its subordinates what to do, how to do it, and when to do it, then discretionary police methods and organic organizational structures are appropriate.

CHAPTER CHECKUP

1. What do we mean when we note that "policing in America is done by organizations"? How does this contrast with other means by which the police role might be accomplished?
2. How do values shape the organizational goal?
3. Within the context of our discussion of the organizational goal, why do municipal police departments differ across the country?
4. What are *transitive* elements of the organizational goal? Give examples.
5. What are *reflexive* elements of the organizational goal? Give examples.
6. What is the *organizational goal?*
7. We distinguish between *ministerial* policing and *discretionary* policing. What is the difference, and does it limit the police organizational goal?
8. What is the difference between *organic* and *mechanistic* organizational structures?
9. Are mechanistic structures appropriate in policing? When are organic structures appropriate? What role does knowledge play in determining the appropriateness of alternative structures?

REFERENCES

Alpert, G. and J. MacDonald (2001) "Police use of force: An analysis of organizational characteristics," *Justice Quarterly* 18(2):393–409.

American Bar Association (1973) *The urban police function.* (Chicago: American Bar Association).

Angell, J. (1971) "Toward an alternative to the classic police organizational arrangements," *Criminology* 9:185–207.

Becknell, C., G. Mays, and D. Giever (1999) "Policy restrictiveness and police pursuits," *Policing: An International Journal of Police Strategies and Management* 22(1):93–110.

Bittner, E. (1970) *The functions of the police in modern society.* (Washington, DC: National Institute of Mental Health).

Blau, P. and W. R. Scott (1962) *Formal organizations.* (San Francisco: Chandler).

Burns, T. and G. Stalker (1961) *The management of innovation.* (London: Tavistock).

Cao, L., X. Deng, and S. Barton (2000) "A test of Lundman's organizational product thesis with data on citizen complaints," *Policing: An International Journal of Police Strategies and Management* 23(3):356–373.

Cordner, G. (1978) "Open and closed models of police organization: Traditions, dilemmas, and practical considerations," *Journal of Police Science and Administration* 6:22–34.

Crank, J. (1990) "The influence of environmental and organizational factors on police style in urban and rural environments," *Journal of Research in Crime and Delinquency* 27:166–189.

Crank, J. and R. Langworthy (1992) "An institutional perspective of policing," *Journal of Criminal Law and Criminology* 83:338–363.

Crank, J. and L. Wells (1991) "The effects of size and urbanism on structure among Illinois police departments," *Justice Quarterly* 8:170–185.

Duffee, D., R. Fluellen, and T. Roscoe (1999) "Constituency building and urban community policing," in R. Langworthy (ed.) *Measuring what matters: Proceeding from the Policing Research Institute meetings.* (Washington, DC: USGPO):91–119.

Friedman, W. and M. Clark (1999) "Community policing: What is the community and what can it do?" in R. Langworthy (ed.) *Measuring what matters: Proceeding from the Policing Research Institute meetings.* (Washington, DC: USGPO):121–131.

Guyot, D. (1979) "Bending granite: Attempts to change the rank structure of American police departments," *Journal of Police Science and Administration* 7:253–284.

Katz, C. (2001) "The establishment of a police gang unit: An examination of organizational and environmental factors," *Criminology* 39(1):37–73.

King, W. (1998) *Innovativeness in American municipal police organizations.* Dissertation Abstracts International, 59(9A), 3654 (UMI No. AAG9905427)

King, W., L. Travis, and R. Langworthy (1997) "Police organizational death: Preliminary findings from a study of forty-three disbanded Ohio police departments." Paper presented at the annual meeting of the American Society of Criminology, November 1997.

Kuykendall, J. and R. Roberg (1982) "Mapping police organizational change: From a mechanistic toward an organic model," *Criminology* 20:241–256.

Langworthy, R. (1986) *The structure of police organizations.* (New York: Praeger).

Langworthy, R. (1992) "Organizational structure," in G. Cordner and D. Hale (eds.) *What works in policing?* (Cincinnati, OH: Anderson):87–105.

Lundman, R. (1980) *Police and policing.* (New York: Holt, Rinehart & Winston).

Maguire, E. (1997) "Structural change in large municipal police organizations during the community policing era," *Justice Quarterly* 14(3):547–576.

Maguire, E. (2001) *Context, complexity and control: Organizational structure in large police agencies.* (Albany, NY: SUNY Press).

Maguire, E. and S. Mastrofski (2000) "Patterns of community policing in the United States," *Police Quarterly* 3(1):4–45.

Manning, P. (1992) "Technological dramas and the police: Statement and counterstatement in organizational analysis," *Criminology* 30(3):327–346.

Mastrofski, S. (1988) "Community policing as reform: A cautionary tale," in J. Greene and S. Mastrofski (eds.) *Community Policing.* (New York: Praeger): 47–67.

Mohr, L. (1973) "The concept of organizational goal." *American Political Science Review* 67:470–481.

Moore, M. (1992) "Problem solving and community policing," in M. Tonry and N. Morris (eds.) *Modern policing.* (Chicago: University of Chicago Press):99–158.

Moore, M. and M. Poethig (1999) "The police as an agency of local government: Implications for measuring police effectiveness," in R. Langworthy (ed.) *Measuring what matters: Proceeding from the Policing Research Institute meetings.* (Washington, DC: USGPO):151–167.

National Advisory Commission of Criminal Justice Standards and Goals (1973) *Police.* (Washington, DC: U.S. Government Printing Office).

Perrow, C. B. (1970) *Organizational analysis: A sociological view.* (Monterey, CA: Brooks-Cole).

Roberg, R. (1979) *Police management and organizational behavior: A contingency approach.* (St. Paul, MN: West).

Roberg, R., J. Crank, and J. Kuykendall (2000) *Police and society,* 2nd ed. (Los Angeles: Roxbury).

Rubenstein, J. (1973) *City police.* (New York: Ballintine Books).

Sanders, B. (1997) *Variety in policing: Job activities of municipal police officers in Ohio.* Dissertation Abstracts International, 58(8A), 3316 (UMI No. AAG9804428).

Sandler, G. and E. Mintz (1974) "Police organizations: Their changing internal and external relationships," *Journal of Police Science and Administration* 2:458–463.

Scheingold, S. (1999) "Constituent expectations of the police and police expectations of constituents," in R. Langworthy (ed.) *Measuring what matters: Proceeding from the Policing Research Institute meetings.* (Washington, DC: USGPO):183–192.

Sherman, L. (1992) *Policing domestic violence: Experiments and dilemmas* (New York: Free Press).

Sherman, L. and R. Berk (1984) "The specific deterrent effects of arrest for domestic assault," *American Sociological Review,* 49:261–272.

Simon, H. (1964) "On the concept of organizational goal," *Administrative Science Quarterly* 9:1–22.

Slovak, J. (1986) *Style of urban policing.* (New York: New York University Press).

Walker, S. (1983) *The police in America.* (New York: McGraw-Hill).

Wilson, J. (1968) *Varieties of police behavior.* (Cambridge, MA: Harvard University Press).

Zhao, J. (1996) *Why police organizations change: A study of community-oriented policing.* (Washington, DC: Police Executive Research Forum).

Zhao, J., N. Lovrich, and Q. Thurman (1999) "The status of community policing in American cities: Facilitators and impediments revisited," *Policing: An International Journal of Police Strategies and Management* 22(1):74–92.

chapter *9*

INDIVIDUALS IN POLICING: OFFICERS AND SUPERVISORS

In the preceding chapter we explained that policing is done by organizations. We noted that it is no longer permissible for individuals to "take the law into their own hands" or to organize groups for law enforcement. Indeed, it has become the case that the functions of police are held as the sole province of government. We also noted that the function of policing is varied both in what different police agencies try to accomplish (organizational goal), and how they try to achieve those goals (technology). During these discussions we concluded that particular organizational forms are contingent on the nature of the organizational goal and what the organization knows or presumes to know (organizational knowledge) about how to achieve the organizational goal.

This chapter seeks to build on that contingency position by examining varieties of people and management styles. Our position is that as different organizational forms are appropriate to serve varied organizational goals, so are different types of people. Further, we suggest that different management styles are appropriate for supervising the work of different types of officers.

The first part of this chapter explores types of people, addressing the question of who is best suited to serve what kind of organizational goal and in what kind of organization. The discussion of "who" will focus on Maslow's (1954) hierarchy of needs and lead up to McGregor's (1960) characterization of human nature as Theory X or Theory Y type. This discussion will give us the foundation to logically relate different kinds of people to different conceptions of the police job and set the stage for a discussion of management styles.

Our discussion of management styles relies on Hersey, Blanchard, and Johnson's (1996) situational-leadership theory. The goal of this discussion is to link varied management styles to types of individuals and police tasks.

This chapter concludes as the last did, with a discussion of forces that must be balanced. The balancing of forces helps us determine what kinds of people police departments should hire and how they should be managed.

OFFICERS

We have determined that in "modern" society organizations rather than individuals carry out the task of policing. However, policing is not accomplished by the simple creation of police organizations; the work of police organizations is done by managed individuals. The first part of this chapter explores thoughts about human motivation on the way to developing alternative conceptions of human nature. At the conclusion of this discussion we will be able to logically relate types of people to different conceptions of the police job.

Motivation

We begin our discussion of individuals with a brief examination of motivation, or more particularly, a discussion of things that motivate persons. We will demonstrate the obvious—different people are motivated by different things. Maslow's famous needs hierarchy is addressed first, followed by Herzberg's motivation-hygiene theory.

In 1954 Maslow presented his **needs hierarchy.** His position was that people are motivated to behave in a manner that will satisfy pressing personal needs. He noted that persons in different situations and circumstances are motivated by different classes of needs, and that these classes of needs can be ranked hierarchically. Needs ranging from the most basic physiological needs through safety, social, self-esteem, and self-actualization needs will motivate a person sequentially.

People are first and foremost motivated to meet basic physiological survival needs (food, shelter, etc.). After immediate survival is assured, they are motivated to ensure continued survival; that is, safety becomes a paramount consideration. When a person feels safe and secure, efforts turn to gaining social acceptance and sustaining friendship. When social needs are satisfied, recognition and self-esteem needs will motivate. Finally, when an individual has been recognized, motivation will be for "doing your own thing," or self-actualization. Self-actualization is the apex of Maslow's hierarchy, as depicted in Figure 9.1.

Building on Maslow's work, Herzberg (1966) sought to distinguish what he called motivating factors from hygienic factors. **Motivating factors** are those that encourage individuals to perform beyond the minimum fulfillment of job assignments. **Hygienic factors** are those that go into making the workplace habitable and thus are essential to adequate worker performance but will not encourage workers to excel. These hygienic factors are extrinsic to the work—quality of supervision, pay, job security, safety in the workplace, and so on. If supervision or pay are poor, if the job is not secure or is unsafe, these poor working conditions will have to be improved before workers can be expected to perform adequately.

Herzberg's extrinsic, hygienic factors tend to address the more basic of Maslow's needs. For example, job safety and security are comparable to Maslow's safety needs in that they ensure continued satisfaction of basic physiological needs. The quality of supervision, another hygienic factor, is comparable to Maslow's social acceptance needs.

Motivating factors, by contrast, are intrinsic to the work. Herzberg proposed that the things that motivate an individual to excel are inherent

FIGURE 9.1
Maslow's hierarchy of needs.

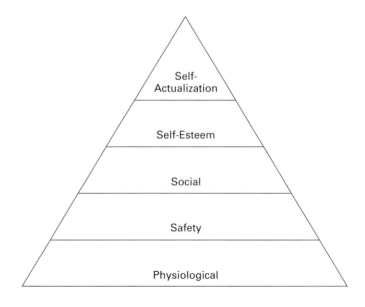

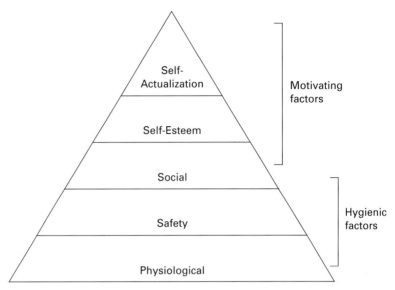

FIGURE 9.2 Herzberg's theory superimposed on Maslow's hierarchy of needs.

in the work itself—such as accomplishment, recognition, increased responsibility, and personal growth and development. These motivating, or intrinsic, factors address Maslow's higher-order needs of self-esteem and self-actualization. Figure 9.2 illustrates the overlap of Herzberg's and Maslow's theories.

Types of People: Theory X and Theory Y

Whereas Maslow and Herzberg both focus on things that motivate, another theorist, McGregor, focuses on the different conceptions of human nature implicit in motivational styles of managers. In his seminal treatise on the management of human resources, McGregor (1960) suggests that there are two competing conceptions of human nature: Theory X and Theory Y. Theory X holds that

1. The average human being has an inherent dislike of work and will avoid it if he can.
2. Because of this human characteristic of dislike of work, most people must be coerced, controlled, directed, and threatened with punishment to get them to put forth adequate effort toward the achievement of organizational objectives.
3. The average human being prefers to be directed, wishes to avoid responsibility, has relatively little ambition, and wants security above all (McGregor, 1960:33–34).

Theory Y holds the exact opposite view of humankind. The assumptions about human nature under Theory Y are

1. The expenditure of physical and mental effort in work is as natural as play or rest.
2. External control and the threat of punishment are not the only means for bringing about effort toward organizational objectives. People will exercise self-direction and self-control in the service of objectives to which they are committed.
3. Commitment to objectives is a function of the rewards associated with their achievement.
4. The average human being learns, under proper conditions, not only to accept but to seek responsibility.
5. The capacity to exercise a relatively high degree of imagination, ingenuity, and creativity in the solution of organizational problems is widely, not narrowly, distributed in the population.
6. Under the conditions of modern industrial life, the intellectual potentialities of the average human being are only partially utilized (McGregor, 1960:47–48).

McGregor's characterization of Theory X people suggests they will be motivated, in their jobs, by opportunities to satisfy safety and social needs. Theory X people, according to McGregor, seek job security, are not inclined to assume responsibility in the workplace, and prefer to be directed in their work. In sum, people who fit the Theory X profile simply want to do their jobs and go home—they do not "live for their work."[1]

Theory Y people, by contrast, are characterized as people who do "live for their work." Theory Y people are motivated by opportunities to satisfy self-esteem and self-actualization needs. That is, Theory Y people are self-directed and self-controlled, achievement oriented, and willing to assume responsibility if they are committed to the organization's objectives. In sum, people who fit the Theory Y profile enjoy the intrinsic nature of their work; they really like being involved in doing their jobs.

Mixing the theories of Maslow, Herzberg, and McGregor permits us to develop characterizations of different kinds of people that are motivated by

---

[1]It should be noted that McGregor contends that the Theory X conception of human kind implicit in the "traditional view" fails to accurately depict human nature (1960:42). He argues that the Theory Y view is "far more consistent with existing knowledge in the social sciences than are the assumptions of Theory X" (1960:49). However, Perrow reminds us that "one cannot explain organizations by explaining attitudes and behavior of individuals" (1986:114). The significance of this discussion is that regardless of which view of humankind is "truth," organizations may only need one type or the other. Thus, while individuals may not feel fulfilled, the organizational goal may nevertheless be addressed.

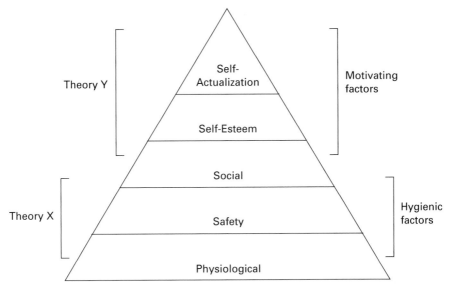

Figure 9.3 McGregor's and Herzberg's theories superimposed on Maslow's hierarchy of needs.

different things. Figure 9.3 highlights the relationship among Maslow, Herzberg, and McGregor's theories.

Staffing Police Departments

These competing conceptions of human nature have significant implications for the staffing of police agencies. If the work of a particular police agency is viewed as ministerial—that is, officers are to do what they are told to do, the way they are told to do it—then persons amenable to close supervision (Theory X people) are indicated. However, if the organization relies on the exercise of individual officer discretion, then the organization would be best served by hiring self-directed persons willing to assume responsibility (Theory Y people).

Legalistic police departments that focus on dispassionate enforcement of the law in accordance with due process will minister to the law. Officers working for these departments are expected to observe civil behavior, characterize that behavior as legal or not legal, and if it is not legal, intervene in a prescribed manner. Legalistic officers are expected to do what they are told to do (enforce the law) the way they are told to do it (with strict adherence to due process). Theory X persons, who are motivated by job security and are not eager to assume responsibility but prefer to be directed, seem well-suited to serving in legalistic police departments.

Peace-agency police departments define the peace and instruct officers in how to keep the peace. These departments are also best served by Theory X officers. Innovation, creativity, self-direction, and self-control are not important attributes in departments that are able to tell officers what peace to keep and how. Indeed, innovation and creativity could result in peacekeeping interventions that are not consistent with departmental desires and by means that are at odds with departmental procedures. Thus, if the police department knows what peace it wants to keep and how it wishes to keep it, then the department will be best served by employing the type of persons (Theory X persons) who will keep the designated peace in the designated manner.

Like legalistic police departments that enforce laws according to specified procedure, and peace-agency police departments that maintain a specified peace in a designated manner, service-style police departments that focus on providing specified services to the community in a specified manner also are best served by employing Theory X officers. Officers in these service-style departments are expected to provide services to the community as specified by the department, in the manner designated by the department. Thus, as in legalistic and peace-agency departments, officers in service departments are expected to do what they are told to do, in the manner they are told to do it.

Watchman-style police departments are of two types: law enforcement and peacekeeping. In contrast to the legalistic, peace-agency, and service departments just described, enforcement guilds and watchman depart-

Officers receiving the recognition of their peers by accepting awards. (Meredith Davenport)

ments rely on the judgment of individual officers to meet the organizational goal. In enforcement guild police departments, officers are expected to decide which laws to enforce and in what situations the laws are to be enforced. Likewise, in watchman police departments, officers are expected to define the peace and to use their individual judgment in deciding how to maintain or restore it. Enforcement guild and watchman police departments rely on individual officer discretion to get the work of policing done. These organizations need people who are willing to assume responsibility, who act with little or no direction, and who are capable of fashioning solutions to problems. These kinds of people fit the Theory Y profile.

In team police departments, or to use the more recent term, community-oriented police departments, the organizational goal is set in collaboration with the community. These departments work with communities to determine the kinds of services the department will provide and the means of provision. In this type of department officers are expected to "team up" with other officers and the community to detect problems, specify the problems, and participate in the development of solutions. Officers in team departments need to be creative collaborators willing to assume responsibility for participation. As with the enforcement guilds and watchman police departments, human characteristics consistent with the demands of team departments fit the Theory Y profile.

Consequences of Mismatch.

Consequences of Mismatch. The principal purpose of the foregoing discussion is to draw corollaries between the nature of police work, however conceived, and types of people. Implicit in this discussion is the idea that police departments that are properly staffed will be the "best," or most efficient, at what they are trying to do. So the question remains, what are the consequences of staffing police departments with the "wrong" kind of people? That is, what happens if we staff administrative police departments with Theory Y people or discretionary police departments with Theory X people?

The principal product of mismatch for individuals employed in police organizations is what Festinger (1957) terms **cognitive dissonance**. Festinger's thesis deals with perceptions individuals have of themselves and of their environment—in this case, the work environment. He posits that if perceptions of self and relevant aspects of the environment are consistent with one another, the relationship will be consonant. However, if perceptions of self and relevant aspects of the environment are inconsistent, the relationship will be dissonant. The importance of this observation is that consonant relationships are comfortable and stable. Dissonant relationships, by contrast, are stressful and require individuals to adapt to relieve the tension between perceptions of self and the environment.

In regard to policing, the most eloquent explanation of adaptation to dissonance is offered in Neiderhoffer's (1969) examination of police cynicism.

FIGURE 9.4
Niederhoffer's cynicism thesis. (Source: Adapted from A. Neiderhoffer (1967) *Behind the Shield.* Garden City, NY: Anchor Books, p. 103.)

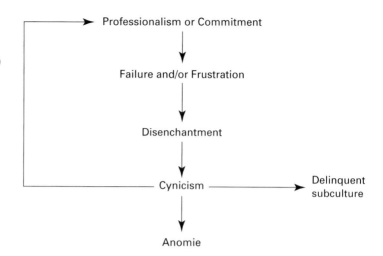

The thesis is that individuals come to policing with a professional ideal or, at a minimum, with a commitment to the training-academy view of policing. Early on in the new officers' career, they "goof up," or find out that they cannot accomplish the things they had hoped to accomplish, and a sense of failure or frustration sets in. This in turn leads to a disenchantment with "the job," and then to cynicism. Figure 9.4 illustrates the process that leads some police officers to become cynical and shows some of their adaptations to cynicism.

Cynicism is a dissonant state. So are its foregoing unstable states— naive professionalism, frustration, and disenchantment. Cynical individuals must adjust their perceptions of self or perceptions of their job in order to relieve the tension associated with their dissonant state. Neiderhoffer suggests three plausible adaptations: (1) integration into a delinquent subculture, (2) development of personal anomia, and (3) commitment to a more mature concept of police professionalism.

The adaptations an individual chooses has profound implications for the organization's capacity to manage its employees. If a cynical officer elects to relieve the stress of dissonance by accepting the precepts of a delinquent police subculture, then that officer's allegiance will be to the subculture's beliefs, which are at odds with the organizational ethos (ergo, delinquent). The organization will then be in the position of competing with the subculture as the vehicle for defining appropriate officer behavior.

The anomic officer, by contrast, is one who does not know which rules specify appropriate behavior and essentially gives up. These officers are controlled neither by the precepts of the organization nor by the ethos of a subculture. Indeed, it is fair to characterize anomic officers as being not controlled.[2] Organizations that employ anomic police officers are in the unenviable position of being unable to manage their employees.

Clearly, the best officer adaptation from an organizational perspective is commitment to a more mature, "professional" view of policing. If the cynical officer adapts to the stress of cynicism by recommitting to the organization, the organization and its ethics retain (or regain) control over the individual officer.

Mismatching police officers and police jobs clearly increases the likelihood of failure and frustration, disenchantment, and cynicism. The mismatched officers must adapt to these dissonant states. In the course of that adaptation, the organization may lose the capacity to manage its employees. Clearly, it is in the interest of the police organization to match the type of person they make a police officer to that organization's particular concept of the police role and the means of fulfilling that role.

SUPERVISORS

In the introduction to our discussion of officers, we noted that the work of police departments is done by managed individuals. This part of the chapter explores variations in managers and links those variations to the types of individuals that are managed. As in the preceding section, we address the consequences of mismatch.

Management literature is replete with management typologies, but we have elected to base this discussion on Hersey and Blanchard's (Hersey, Blanchard, and Johnson, 1996) situational-leadership theory.[3] We have chosen the situational-leadership theory because its two-dimensional foundation (task and relationship) accommodates the full range of police goals and methods previously identified. Single-dimension theories (typically focused on relationship) neglect or denigrate management styles that are appropriate for select police missions and methods.

Hersey and Blanchard (Hersey, Blanchard, and Johnson, 1996:134–135) base their typology of management styles on the interaction of what they describe as task behavior and relationship behavior. They define *task behavior* as the extent to which leaders are likely to organize and define the roles of the members of their group (followers), that is, to explain what activities each member is to do and when, where, and how tasks are to be

---

[2]It is important to note that many theorists, Merton (1957) most notably, contend that an anomic state is not a consonant condition but rather a state that must be adjusted to. Merton suggests five principal adaptations to the dissonance of anomie: conformity, innovation, ritualism, retreatism, and rebellion.

[3]For a summary description of theories of management styles see R. Roberg and J. Kuykendall (1997); Swanson, Territo, and Taylor (1993); and Southerland and Reuss-Ianni (1992).

accomplished. Task behavior is characterized by an attempt to establish well-defined patterns of organization, channels of communication, and ways of getting jobs accomplished.

Relationship behavior is defined as the extent to which leaders are likely to maintain personal relationships between themselves and members of their group (followers) by opening up channels of communication, providing socioemotional support, "psychological strokes," and facilitating behaviors.

Figure 9.5 illustrates the four basic leadership styles to emerge from the interaction of task and relationship behaviors. The first quadrant (S1) leadership style is called *telling.* Telling leaders rate high on task behavior and low on relationship behavior. Leaders employing a telling style are highly directive. That is, they engage in one-way communications telling subordinates what they are to do, how they are to do it, and when. Telling leaders are little concerned about their relationship with subordinates or with what subordinates think about the task at hand.

The second quadrant (S2) leadership style is called *selling.* Selling leaders rate high on both task and relationship behaviors. Unlike telling leaders, selling leaders discuss things, both personal and job-related, with their subordinates. Like telling leaders, though, they tell their subordinates what to do, when to do it, and how. On the job, selling leaders explain their decisions and answer questions, thereby convincing subordinates to do what they are instructed to do, the way they are instructed to do it, and when. Telling leaders, by contrast, render decisions and provide direction without discussion or explanations intended to convince or motivate subordinates; they simply give orders.

The third quadrant (S3) leadership style, called *participating,* emphasizes relationship behaviors rather than task behaviors. Participating leaders try to create a supportive environment that encourages participation in decision making. The emphasis on two-way discourse characteristic of relationship behaviors promotes fuller discussion of problems and actively en-

FIGURE 9.5
Hersey and Blanchard's leadership styles. (Source: Adapted from P. Hersey, K. Blanchard, and D. Johnson (1996) *Management of Organizational Behavior,* 7th ed. Englewood Cliffs, NJ: Prentice Hall, p. 193.)

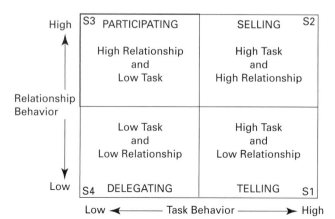

gages subordinates in the specification of task behaviors. Participating leaders tend to rely to a greater extent on subordinates for determining what should be done, how it should be done, and when. Therefore, they need be less directly concerned with task behaviors (what, how, and when).

The fourth quadrant (S4) leadership style, called *delegating,* is characterized as low on both relationship and task behaviors. Delegating leaders permit subordinates to do their job as they see fit. Delegating leaders rely on subordinates to decide what their job is, how it should be done, and when. Delegating leaders, unlike participating leaders, do not get involved in discussions of what should be done, how it should be done, or when it should be done. Instead, they rely on subordinates to exercise their own judgment in making those decisions.

Hersey and Blanchard (Hersey, Blanchard, and Johnson, 1996) specify the situational appropriateness of each leadership style by linking the style to subordinate readiness, which is described in terms of task and psychological readiness. *Task readiness* refers to an individual's ability to do the job. It varies from able to unable. Task-unable individuals do not know how to do the job; task-able individuals do. *Psychological readiness* refers to an individual's willingness to assume responsibility for "getting the job done." As with task readiness, psychologically unwilling individuals are reluctant to assume responsibility, whereas psychologically willing individuals will assume responsibility. A subordinate who is low on readiness is neither willing nor able to do the work. A subordinate who is high on readiness is both willing and able to do the job. Figure 9.6 illustrates the relationship between subordinate readiness and leadership styles.

The logic underlying Figure 9.6 is really quite simple: If you are dealing with low-readiness employees, the appropriate leadership style is directive, either telling or selling. However, as employees become more ready, the appropriate leadership style shifts work responsibilities (task) from leaders to subordinates. Leaders whose subordinates are moderately ready—that is, generally willing to assume responsibility for doing the work and possessing a substantial degree of job skill—lead best with a style that transfers a substantial degree of task behavior to the subordinate while maintaining a high level of relationship behaviors. By contrast, leaders of fully ready subordinates lead best by delegating task and responsibility to subordinates—that is, by getting out of the way.

Police Leaders

The foregoing leadership contingencies have implications for police departments. In our discussion of staffing police departments, we noted a relationship among the goal of the police department, the way it chooses to do police work, and the nature of people employed. We showed that ministerial

FIGURE 9.6
Leadership styles and employee readiness. (Source: P. Hersey, K. Blanchard, and D. Johnson (1996) *Management of Organizational Behavior,* 7th ed. Englewood Cliffs, NJ: Prentice Hall, p. 200.)

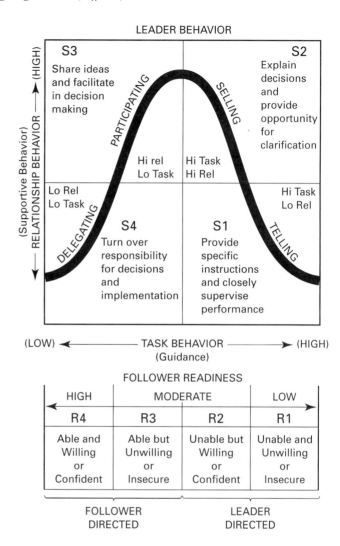

police departments are best staffed with Theory X people, that is, with people who prefer to be directed and instructed in their work. People fitting the Theory X profile are quite consistent with Hersey and Blanchard's depiction of relatively immature subordinates.[4]

Discretionary police departments, by contrast, are best staffed by Theory Y people, that is, with people who are willing to assume responsibility for do-

---

[4]A number of police organizational theorists have suggested that one of the principal problems associated with the bureaucratic police organization is that it tends to maintain or promote immaturity in subordinates (Angell, 1971).

Willie Williams, Chief of the Los Angeles Police Department and former Chief of the Philadelphia Police Department, addressing the media. (Ted Soqui/Impact Visuals)

ing the work of the organization and who, if given the opportunity, will develop a high degree of skill. People fitting the Theory Y profile are quite consistent with Hersey and Blanchard's depiction of relatively mature subordinates.

In our discussion of staffing issues, enforcement guild and watchman police departments were distinguished from team police departments based on the way discretion is exercised. In team police departments, discretion is the product of collective decision making. A participatory leadership style promotes collective decision making and would be indicated for team police departments.

The work of police in enforcement guild and watchman police departments is done through the exercise of individual discretion. This stands in sharp contrast to the collegial style of the team departments. In both of these types of police departments, individual officers use their experience, training, and intuition to do the work of the organization. In Hersey and Blanchard's terms, they must be mature. A delegating leadership style is indicated for the management of mature subordinates who will be taking responsibility for doing the work.

Table 9.1 summarizes the marriage of police organizational goals, technology, and leadership styles. Ministerial police departments (legalistic, peace-agency, and service) should rely on either telling or selling leadership styles. Conversely, discretionary policing is properly done either using a delegating

TABLE **9.1** Leadership Styles for Alternative Police Organizational Goals and Technologies

	TECHNOLOGY	
Organizational Goal	*Ministerial*	*Discretionary*
Law Enforcement	Telling/Selling	Delegating
Peacekeeping	Telling/Selling	Delegating
Public Service	Telling/Selling	Participating

leadership style (enforcement guild or watchman) or a participatory style (team), depending on whether discretion is individually or collectively held.

Consequences of Mismatch.   What happens if there is a mismatch of leadership styles to police objectives? In their presentation of situational-leadership theory, Hersey and Blanchard (Hersey, Blanchard, and Johnson, 1996) assess the effects of mismatch in terms of leader effectiveness. When leadership styles and situations (employee maturity) match, leadership is effective. When there is a mismatch, leadership is ineffective. What is unique in Hersey and Blanchard's discussion is their explicit recognition that there are varying degrees of mismatch and therefore varying degrees of ineffectiveness.

They note (see Figure 9.6) that when subordinates are least ready (S1), a telling style is most effective, selling next most effective, participative somewhat ineffective, and delegating least effective. If subordinates are somewhat more ready (S2), then a selling leadership style is most effective, telling and participating styles next most effective, and delegating least effective. If subordinates are modestly ready (S3), a participating style is most effective, delegating and selling styles next most effective, and telling least effective. Finally, if subordinates are ready (S4), then a delegating style is most effective, participating style next most effective, selling somewhat ineffective, and telling least effective.

Where police departments are concerned, it is also apparent that different organizational goals and technologies are best matched to differing leadership styles. Table 9.1 highlights the matches of goals and technology to leadership styles. Table 9.2 highlights the variable effectiveness of different leadership styles depending on the nature of the police technology and organizational goal.

Review of Table 9.2 shows that all ministerial police departments, regardless of organizational goal, exhibit the same pattern of variation of leadership style effectiveness. That is, all are most effective when either telling or selling leadership styles are used, next most effective when a participating style is used, and least effective when a delegating style is used.

TABLE 9.2 Effectiveness of Leadership Styles by Police Technology and Organizational Goal

TECHNOLOGY/GOAL	MOST EFFECTIVE	NEXT MOST EFFECTIVE	LEAST EFFECTIVE
Ministerial			
Law Enforcement	Tell/Sell	Participating	Delegating
Peacekeeping	Tell/Sell	Participating	Delegating
Public Service	Tell/Sell	Participating	Delegating
Discretionary			
Law Enforcement	Delegating	Participating	Tell/Sell
Peacekeeping	Delegating	Participating	Tell/Sell
Public Service	Participating	Delegating/Sell	Tell

It is apparent that the directive nature of ministerial police departments is best served by a leadership style that instructs subordinates (either tells them what to do or sells them on what to do), marginally served by a style that promotes discussion of alternative procedures (participatory discussion), and not served well at all by a leadership style that instructs officers to use their own judgment and intuition in the performance of organizational duties (delegation of decision making to officers).

In a discretionary police technology, on the other hand, there are two patterns of leadership style effectiveness. When discretion is reserved to the individual, as in the cases of discretionary law enforcement and peacekeeping, a delegating style is most effective, followed by a participating leadership style, and selling and telling styles are least effective. Thus, if a police department wishes to rely on the judgment and experience of officers to decide which laws to enforce or what peace to keep, it will be best served by a delegatory leadership style, marginally served by a style that promotes collective decision making (participatory discussions), and least well served by directive leadership styles (selling and telling).

As previously noted, discretionary public service tends to be a collaborative enterprise in policing. Team policing focuses on collaboration among police officers who have joint (team) responsibility for serving a relatively small geographic area. Newer, community-oriented policing strategies expand the collaboration to include influential elements of the community, but the focus on collaboration to derive solutions to community problems remains. Thus, for a police department that emphasizes discretionary public service, a participating leadership style that encourages collaboration is most effective. The next most effective leadership style is either a delegating style, which promotes discretion but not collaboration, or a selling style, which promotes two-way discourse but not discretion. The telling style would be least effective. A telling leadership style promotes neither two-way discourse (essential to collaboration) nor the exercise of discretion.

BALANCE OF FORCES

Early in this chapter we noted that police departments do policing through the efforts of managed individuals. Accordingly, the chapter has explored issues related to staffing and leading police departments. We have also carried forward the contingency perspective suggesting that variations in police technology and goals have implications for the kinds of people that police departments ought to employ and the way they ought to lead them. The forces to be balanced are the police organizational goals and technology, the natures of people, and the styles of leadership.

In the previous chapter we developed a framework that describes different kinds of police organizations in terms of the interaction of goals and technology (methodology). The typology suggests that there are three transitive police organizational goals (law enforcement, peacekeeping, and public service) and two technologies (ministerial and discretionary). We also noted that all three of the police organizational goals could be addressed by either of the technologies. Interactions of the three transitive goal elements and the two technologies, or methodologies, provides us with six types of police organizations.

At the conclusion of that chapter we noted that this variation has implications for organizational design. We concluded that organic organizational forms are appropriate for police departments that rely on discretion, and mechanistic forms are appropriate for departments that seek to do their work by administration. This chapter has demonstrated that police organizational goal-technology variation also has implications for staffing and leading police departments.

Police departments (or divisions within departments, for that matter) that seek to achieve their ends by relying on the good judgment of mature police officers (discretion) will be best served by hiring persons who fit McGregor's Theory Y profile, then encouraging their maturation and leading them in a manner consistent with whichever form of discretion the department encourages (individual or collective). Theory Y people are willing (if not eager) to exercise their judgment in the performance of duties—clearly a prerequisite to institution of a technology that relies on the exercise of discretion.

We have also noted that there are two forms of discretion prevalent in theories of policing: collective discretion in team, or community-oriented, policing models; and individual discretion in enforcement guild and watchman-style policing. This distinction is important because it conditions the appropriateness of particular leadership styles. In team, or community-oriented, policing modalities, collective discretion is encouraged such that teams of officers or officers and community members collaborate in developing solutions to police problems. Participative leadership styles are appropriate in collegial situations where members of the team are viewed as mature contributing members who interact with one another and are capable of assuming responsibility for task accomplishment.

In enforcement guild and watchman-style police departments, the organization relies on individual discretion to achieve the organizational goal. In contrast to the collaboration expected in team, or community-oriented, models, officers in watchman-style departments are expected to exercise their own good judgment. Delegation leadership styles are appropriate in these situations where mature officers are expected to rely on their experience and intuition to satisfactorily do the work of the police department.

Police departments that seek to achieve their ends through detailed instruction and close supervision of officers (i.e., ministerial police departments) will be best served by hiring persons who fit McGregor's Theory X profile and then leading them with directive styles. Theory X people prefer to be directed and are uncomfortable if placed in a position where they are expected to use their own judgment. This profile is consistent with the needs of ministerial police departments, where officers are expected to do what they are told to do and when and how they are told to do it. Directive leadership styles (Hersey and Blanchard's telling and selling) are appropriate in this situation because they rely on leaders who presumably know what they want done and how they want it to be done. Directive leaders tell, or sell, subordinates what, how, and when to do what they want done.

Recent research has shed some light on the links among organizational structure, supervisory style, and officer motivation. Robin Engel (2001) identifies four supervisory styles among sergeants and lieutenants in two police departments. She does not use the same classification scheme we have presented, but her results suggest that there are distinct supervisory styles, that these styles differ in terms of both relationships and task orientation, and that leadership styles are associated with different types of activities among officers. Some supervisors focus on supporting and protecting subordinates, some on closely directing and controlling the activities of subordinates, and some focus on developing the skills and abilities of their subordinates.

Studies of police officer turnover, job commitment, and desire to participate in the promotional process have provided data concerning the effects of mismatch among officer, supervisor, and organizational characteristics. The lack of promotional opportunities means that most officers seeking promotion will be unsuccessful, and this acts to reduce police officer interest in the promotional process (Scarborough et al., 1999). The monetary rewards of promotion do not serve as strong motivators for officers in agencies that provide good pay and benefits to patrol officers. Officers who do not "fit" the organizational goal, whose self-definition is at odds with the dominant organizational goal grow dissatisfied with the job and are most likely to

TABLE **9.3** Police Department Variety and Appropriate Organizational Designs, Staffing, and Leadership Styles

TECHNOLOGY/GOAL	ORGANIZATIONAL DESIGN	STAFF*	LEADER STYLE
Ministerial			
Law Enforcement	Mechanistic	X	Tell/Sell
Peacekeeping	Mechanistic	X	Tell/Sell
Public Service	Mechanistic	X	Tell/Sell
Discretionary			
Law Enforcement	Organic	Y	Delegating
Peacekeeping	Organic	Y	Delegating
Public Service	Organic	Y	Participating

Staffing refers to McGregor's alternative conceptions of human nature—Theory X and Theory Y.

leave police employment, or otherwise resist and disrupt organizational operations (Travis and Winston, 1998; Harris and Baldwin, 1999; McElroy, Morrow, and Wardlaw, 1999). Finally, David Hoath, Frank Schneider, and Meyer Starr (1998) report that police officer career orientations, how they viewed themselves and their careers, were an important correlate of job satisfaction. Officers who were committed to the policing career were most satisfied with and committed to their jobs. Officers whose self-definitions as service providers or independent "artisans" found their jobs rewarding, but only if their assignments were compatible with their self-definitions. Individuals who value independence and see themselves as problem solvers would thrive in a decentralized, community policing assignment. In contrast, they might grow quite dissatisfied if assigned to traffic enforcement.

Ultimately, the effectiveness of a particular police department depends on the degree to which the organizational design (Chapter 8), staffing, and leadership styles are consistent and appropriate to achievement of the police organizational goal. Table 9.3 highlights the relationships described in Chapters 8 and 9.

CHAPTER CHECKUP

1. Do organizations or people do policing?
2. How do Maslow (1954), Herzberg (1966), and McGregor's (1960) theories of things that motivate people relate to one another?
3. What kind of people should be hired to staff the different kinds of police departments (legalistic, peace agency, service, watchman, and team), and why?

4. What happens when there is a mismatch between type of police department and persons hired?

5. Describe Niederhoffer's (1969) thesis regarding police cynicism and tell why it is an example of adaptation to dissonance.

6. Within the context of the Hersey and Blanchard (Hersey, Blanchard, and Johnson, 1996) situational-leadership theory, what is *task behavior?* What is *relationship behavior?*

7. Hersey and Blanchard's typology isolates four types of leaders. What are they and when are they appropriate?

8. What kinds of leaders are appropriate for different types of police departments? What are the consequences of mismatch?

REFERENCES

Angell, J. (1971) "Toward an alternative to the classic police organizational arrangements: A democratic model," *Criminology* 9:185–206.

Engel, R. (2001) "Supervisory styles of patrol sergeants and lieutenants," *Journal of Criminal Justice* 29(4):341–355.

Festinger, L. (1957) *A theory of cognitive dissonance.* (Stanford, CA: Stanford University Press).

Harris, L. and J. Baldwin (1999) "Voluntary turnover of field operations officers: A test of confluency theory," *Journal of Criminal Justice* 27(6):483–493.

Hersey, P., K. Blanchard, and D. Johnson (1996) *Management of organizational behavior,* 7th ed. (Englewood Cliffs, NJ: Prentice-Hall).

Herzberg, F. (1966) *Work and the nature of man.* (New York: World Publishing).

Hoath, D., F. Schneider, and M. Starr (1998) "Police job satisfaction as a function of career orientation and position tenure: Implications for selection and community policing," *Journal of Criminal Justice* 26(4):337–347.

Maslow, A. (1954) *Motivation and personality.* (New York: Harper and Row).

McElroy, J., P. Morrow, and T. Wardlaw (1999) "A career stage analysis of police officer work commitment," *Journal of Criminal Justice* 27(6):507–516.

McGregor, D. (1960) *The human side of enterprise.* (New York: McGraw-Hill).

Merton, R. (1957) *Social theory and social structure,* 2nd ed. (Glencoe, IL: Free Press).

Neiderhoffer, A. (1969) *Behind the shield.* (Garden City, NY: Anchor Books).

Perrow, C. (1986) *Complex organizations: A critical essay,* 3rd ed. (New York: McGraw-Hill).

Roberg, R. and J. Kuykendall (1997) *Police management,* 2nd ed. (Los Angeles: Roxbury).

Scarborough, K., G. Van Tuberger, L. Gaines, and S. Whitlow (1999) "An examination of police officers' motivation to participate in the promotional process," *Police Quarterly* 2(3):302–320.

Southerland, M. and E. Reuss-Ianni (1992) "Leadership and management," in G. Cordner and D. Hale (eds.) *What works in policing.* (Cincinnati, OH: Anderson):157–177.

Swanson, C., L. Territo, and R. Taylor (1993) *Police administration,* 3rd ed. (New York: Macmillan).

Travis, L. and C. Winston (1998) "Dissension in the ranks: Officer resistance to community policing, cynicism, and support for the organization," *Journal of Crime and Justice* 21(2):139–156.

POLICE OFFICERS

CHAPTER OUTLINE

In this chapter we examine the characteristics of local police officers, focusing on sworn police officers in local police agencies in the United States. These people represent the members of police organizations, and these are the people who actually "do" policing. The purpose of this chapter is to identify many of the forces acting on police officers that appear to correlate with how they work.

State, local, and special police agencies in America employ over 930,000 persons full-time (Hickman and Reaves, 2001). Nearly 73 percent of these employees were sworn police officers, and over 90 percent worked for local police and sheriffs' departments.

Nearly half of all full-time local police officers were employed by fewer than 2 percent of local agencies. Over 90 percent of local police agencies employ fewer than 50 officers. Thus, while the typical local police agency is small, the "typical" local police officer works in a large agency. As we saw earlier, many agencies have specialized units and a bureaucratic structure. Most local police employees work within large police organizations that have a bureaucratic structure. As shown in Charts 10.1 and 10.2, officers in local police agencies are predominantly male (91 percent in 1997), and white, non-Hispanic (78.5 percent). In total, female and minority officers accounted for

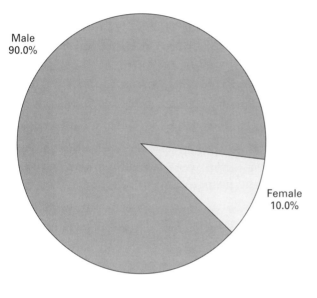

CHART 10.1
Gender of American police officers, 1997. (Source: B. Reaves and A. Goldberg (2000) *Local Police Departments, 1997.* Washington, DC: U.S. Bureau of Justice Statistics, p. 3.)

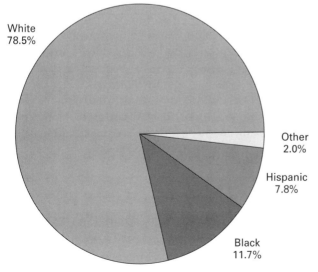

CHART 10.2
Ethnicity of American police officers, 1997. (Source: B. Reaves and A. Goldberg (2000) *Local Police Departments, 1997.* Washington, DC: U.S. Bureau of Justice Statistics, p. 3.)

less than 25 percent of all sworn personnel in local police agencies in 1997. Representation of women and minority officers is more likely in agencies serving larger populations than in those serving small communities. Nearly 85 percent of officers in local agencies serving communities of fewer than 50,000 people were white males, while fewer than 70 percent of officers in communities of over 250,000 were white males (Reaves and Goldberg, 2000:3).

It is evident that the police industry is large and the characteristics of police officers are varied, but the implications of these facts are complex. The literature of policing in America contains a variety of statements about the relative importance of agency and officer characteristics in explaining police behavior. This chapter is devoted to the debate about the correlates of officer/agency characteristics as they apply to policing in practice. We will examine the "police personality," its genesis, police officer demographic and work correlates, and police officer typologies.

Many observers of the police suggest that what happens in any case where the police interact with the public is partly a function of the police officer involved. That is, officers differ among themselves in how they deal with different circumstances. One outgrowth of this perspective is to seek to change policing by changing the characteristics of the persons employed as police officers. Others have argued that police officers, as a group, share common traits that are different from the civilian population in important ways. Individual police officer attitudes and behaviors are the products of becoming a police officer, according to both sets of observers.

ARE POLICE OFFICERS DIFFERENT?

Numerous observers have noted how different from other citizens police officers appear to be. These observations stem, in part, from the fact that the observers are looking for differences. Yet, when so many observers report the same thing in different places at different times, it becomes fairly safe to conclude that there is a police personality. William Westley (1970), Jerome Skolnick (1994), James Wilson (1968), Arthur Neiderhoffer (1967), John Crank (1998) and others have noted that police officers, as a group, tend to share attitudes and perceptions about the public, police work, and the role of policing. There appears to be support for the contention that police officers represent a special set of people who have values, beliefs, and attitudes that differ from nonpolice (Caldero and Larose, 2001).

The Police Personality

In 1966, Jerome Skolnick described what he called the "working personality" of police officers (1994:42). His thesis is that characteristics of the job of policing, including danger, authority, and isolation, led police officers to

Policing in America consists of different people working for different organizations in different places. (Tom Kelly)

develop a particular approach to their jobs. At work the police tend to be suspicious, authoritarian, and cynical, according to Skolnick. He sketches this **working personality** as follows (1994:43):

> The process by which this "personality" is developed may be summarized: the policeman's role contains two principal variables, danger and authority, which should be interpreted in the light of a "constant" pressure to appear efficient. The element of danger seems to make the policeman especially attentive to signs indicating a potential for violence and lawbreaking: As a result, the policeman is generally a "suspicious" person. Furthermore, the character of the policeman's work makes him less desirable as a friend, since norms of friendship implicate others in his work. Accordingly, the element of danger isolates the policeman socially from that segment of the citizenry which he regards as symbolically dangerous and also from the conventional citizenry with whom he identifies. The element of authority reinforces the element of danger in isolating the policeman. Typically, the policeman is required to enforce laws representing puritanical morality, such as those prohibiting drunkeness, and also laws regulating the flow of public activity,

such as traffic laws. In these situations, the policeman directs the citizenry, whose typical response denies recognition of his authority, and stresses his obligation to respond to danger. The kind of man who responds well to danger, however, does not normally subscribe to codes of puritanical morality. As a result, the policeman is unusually liable to the charge of hyprocrisy. That the whole civilian world is an audience for the policeman further promotes police isolation and, in consequence, solidarity. Finally, danger undermines the judicious use of authority.

Thus, the working personality of the police officer includes the characteristics of authoritarianism, suspicion, and isolation. Compared to others, the police officer is less trusting and more aloof. William Westley (1970) adds "secretive" to the list of police personality characteristics. He suggests that the police tend not to share their concerns and activities with the public for fear of negative public reactions and reprisals. This secrecy further isolates the police from the public. In interactions with the public, the officer expects obedience and acts in a domineering way. Neither the police officer nor the citizen is comfortable or natural in their interactions. The differences between the police and the average citizen are highlighted in each encounter. In this vein, Irving Piliavin (1973:25) argues that alienation of police from the citizens "is itself a consequence of structural conditions" of the role of police in society.

Others, however, have not found evidence of a distinctive police personality and therefore question its existence (Bayley and Mendelsohn, 1969; McNamara, 1967). More recently, Eugene Paoline, Stephanie Myers, and Robert Worden (2000) report that a survey of police officers in two cities indicated that the "police culture" is overstated because there is much variation in the attitudes and values held by police officers, even within the same police department. These findings have led some observers to conclude that even if police officer attitudes differ from citizens, the differences are probably not significant (Lefkowitz, 1975). Nonetheless, Senna and Siegel (1987:208) conclude, "the weight of existing evidence generally points to the existence of a unique police personality."

What Makes Cops Different?

James Q. Wilson (1968) writes, "It is not money, or organization, or training that defines the policeman's job, it is the job that defines the policeman." In a similar vein, Peter Manning and John Van Maanen (1978) observe, "Policing is more than a job, it is a way of life." George Kirkham (1976), in telling his experiences as a college professor turned police officer, comes to much the same conclusion. In short, what makes cops different from nonpolice is the job itself.

Six characteristics of the job of police officer seem particularly useful in explaining how police come to be different from civilians: the uniform, power, working hours, danger, suspicion, and dirty work involved in the police function. In total, these factors work to separate the police from the public and to accentuate any differences.

Uniform

You may recall that in the early years of American policing, officers resisted the idea of wearing distinctive uniforms as being un-American. The uniform sets the officer apart, and it clearly identifies his or her status as a police officer. It is not possible for the uniformed officer to be inconspicuous or to blend into the crowd. Wearing the uniform marks the officer as a member of a particular minority group—the police.

Power

The uniformed police officer, with the sidearm, baton, badge, handcuffs, citation book, and other trappings of office, is a walking symbol of government power. In most circumstances, both the officer and the citizen recognize that the officer is in a dominant position and can do things to reward or punish the citizen. The fact of power means that the officer is not an equal of the citizen on the street. It changes the nature of the relationship between the citizens and the police. For example, people frequently thank an officer after receiving a traffic citation. This polite, "Thank you, officer," symbolizes the citizen's recognition of the officer's power (and the desire not to antagonize the officer and receive any further "gifts").

Working Hours

In most places, police officers work rotating shifts. A typical pattern involves changing work hours every month, so that the officer works days for one month, afternoons for the next, nights for the third, and then days again. Police officers also work weekends, holidays, and frequently must put in overtime in emergency situations, and for court appearances. This type of work schedule makes it difficult, if not impossible, to maintain friendships with nonpolice, or even the semblance of a normal existence in their own personal lives.

Danger

The presence of danger in policing makes the job exciting (Storch and Panzarella, 1996; Brandl, 1996) but also serves to make police officers wary. Police officers must expect danger at almost any time and must be prepared to respond to threats. For this reason, when an officer stops a motorist, rather than approaching the driver from the front, the officer comes from the rear. The driver must screw her head around uncomfortably to speak with the officer. On the street, police typically violate expectations of personal space, either standing too close or too far away from the citizen. Similarly, while conversing with a citizen, police often do not maintain eye contact.

All of these are defensive tactics. Often without thinking, almost always without explanation, police officers violate social conventions about personal interactions so that they are positioned to defend themselves. The citizen who has no intention of harming the officer does not understand why the officer is too close, too far, standing behind him or not looking at him. All the citizen knows is that the officer is "strange"—does not behave "normally." In combination with the officer's power, this strange behavior makes interaction with the police uncomfortable for the citizen.

Suspicion

In part for reasons of self-defense, police officers tend to be suspicious of citizens. Beyond the concern for personal safety, however, police are trained, and learn, to be suspicious. Police officers are trained to notice details, such as someone wearing a heavy coat on a warm afternoon. Over time, police learn to question the motives of persons who seek their aid. For example, a person reporting a theft may in fact be attempting to defraud the insurance company. Thus, the police tend to be skeptical, at least initially, of the stories they hear from citizens.

This suspicion confuses and frustrates citizens who call the police, because they do not understand why the police don't believe them. Other citizens, such as the person wearing the heavy coat, are angered by police intervention when they "haven't done anything." In these circumstances, the police officers are distrustful, the citizens defensive, and the interaction between them uncomfortable.

Dirty Work

One of the most important things police do for society is its dirty work. Police officers are exposed to the seamier, seedier side of society. The police deal with dead bodies, crimes, crime victims, criminals, automobile wrecks, and other aspects of society that most of us would rather not think about. As dirty workers, police officers are untouchables, because we do not wish to be reminded of what they do. Fogelson (1977) suggests that the police occupation suffers from what he calls a "pariah complex." On a somewhat broader level, Egon Bittner (1975) suggests that because the police have become the repository of coercive force, they have what he terms a "tainted" occupation. That is, the police can (and do) use blatant force to accomplish their tasks in a society that has increasingly become pacified.

Dirty work separates police officers from other people in two important ways. First, they cannot share their experiences with nonpolice because those experiences are frightening and unpleasant. Thus, in social settings, officers frequently cannot discuss their jobs, a major topic of conversation at most gatherings. Second, for psychological survival, police officers must harden themselves to suffering.

Police officers often develop a black, gallows humor, making jokes about horrible situations. Confronted with a tragedy, police officers seem unconcerned. They have seen it all before, and they may even appear bored with the situation. To the upset citizen, the officer appears calloused, uncaring, cold, and inhuman. Police officers who insulate themselves against the normal emotional reactions to human suffering do not receive much empathy from civilians.

In these ways, if not others, the job of policing acts to isolate officers from citizens. The requirements and experiences of the job make police officers behave in ways that are alien to the average person and accentuate the differences between them. The working police officer does not "fit" into the world of the civilian, and the civilian does not understand the working world of the officer. Yet, it is precisely in this work role that the officer must interact with the citizen. Officers and citizens come together without a solid understanding of each other, often in times of crisis, and must somehow arrive at solutions to immediate problems.

Becoming a Police Officer

Developing an understanding of police officers raises the old issue of nature versus nurture. Another way of stating the issue is to ask, "Are police officers born, or are they made?" The answer to this question might be found in the recruitment, selection, and retention process for police personnel. On the one hand, it may be that the ways in which we choose and reward officers results in certain people being more likely to become police. On the other, it may be something about the role of policing that turns ordinary people into cops. The first task is to determine if police as police are somehow different from the rest of society who are not police officers. Those who say police are different point to the police personality as evidence that officers differ from civilians.

How this question is answered has profound implications for the staffing and structure of the police industry. If police officers are born, we must develop screening and selection measures that tell us which people among a pool of applicants will make good police officers. If we wish to change the nature of policing, we need only change the nature of police recruits. Conversely, if police officers are made, we must examine the ways we make them. If we wish to change the nature of policing we may be required to make major changes in the organizational culture, training, reward structures, and other aspects of the occupation.

John Van Maanen observed the tendency to think of the police as a unitary group. In 1973 (p. 407) he wrote,

> Whether one views the police as friend or foe, virtually everyone has a set of "cop stories" to relate to willing listeners. Although most stories dramatize personal encounters and are situation-specific, there is a common thread running through these frequently heard accounts. In such stories the police are almost always depicted as a homogeneous occupational grouping somehow quite different from most other men.

Van Maanen notes that occupational stereotyping is common (think of "used car salesman," for example), but what is unique about the police is that officers themselves recognize the implied differences.

Once researchers recognized that police officers might be different from nonpolice, they began to search for explanations. The first studies of the correlates of a police personality focused on the characteristics of people who became officers. As Richard Lundman (1980:73) notes, the initial explanations for a police personality were based on a belief that the people who became police officers were different before beginning the job.

From what we have seen already, we know that most police officers are white males. Assuming that white males as a group are different in important ways from females and minority group males, it seems logical that the characteristics of those who become officers might explain the personality. If we add to these demographic traits other considerations, such as local recruitment, minimal educational standards, and physical and psychological requirements, we can see that police officers may represent a select subgroup of the population. Yet, the initial studies of police recruit characteristics failed to reveal that policing attracted any distinct personality types (Neiderhoffer, 1967). The focus of the research then shifted to organizational and social correlates of the police personality.

If the people who chose to become police officers did not differ from those choosing not to be officers in terms of their backgrounds and attitudes, individual characteristics of recruits did not explain the observed working personality. New explanations for differences between the police and civilians were forged to explain the differences as a function of the nature of the job of policing. Police officers were not born different from nonpolice; they were made to be different by the job.

John Crank (1998) describes what he calls a "police culture." Like other cultures, the **police culture** is a way of understanding or thinking about things. Crank argues that there is a somewhat distinct worldview among police officers that structures how they view themselves and their jobs. This worldview, or culture, is a set of understandings about human nature, the ways people interact, and what kinds of behaviors and values are positive and what kinds are negative. In regard to the police culture, Crank (1998:5) suggests, "All areas of police work have meaning of some kind to cops, and as every reformer and chief who has sought to change any organization knows, these meanings tend to bind together in sentiments and values impossible to analytically separate and individually change."

Crank is arguing that what has often been described as a police personality is itself a product of police culture. Because police officers have a distinct way of understanding the world and their role in it, and because that understanding is shared among officers, they tend to view situations in the same way and act in similar ways. Even if different officers perceive situations slightly differently and react in somewhat different ways, they can understand and appreciate each other because they share a common culture.

Police Socialization

Socialization is the term sociologists use to refer to the process by which people acquire the knowledge, skills, values, and attitudes of their culture or society. The same general process of learning applies to the mastery of a profession or job. In this specific instance, the learning is called **occupational socialization.** Police socialization is the process by which police recruits learn the values, skills, knowledge, attitudes, and behaviors of policing. Socialization into the occupation of policing works to create a common culture and personality among the different people who become police officers.

Van Maanen (1973) identifies four stages in the socialization process of police officers: choice, introduction, encounter, and metamorphosis. The stages refer to important milestones in an officer's career. **Choice** includes the decision of both the officer and the agency that results in recruitment to policing. **Introduction** covers the period of academy training. **Encounter** relates to the early days on the job when the rookie officer is required to balance expectations of the work, which came from both personal beliefs and formal academy training, with its realities. The final stage, **metamorphosis,** refers to the change in the officer required to enable her or him to come to terms with being a police officer.

Choice. Van Maanen (1973) classifies the choice stage as "preentry," important because it indicates the types of people who seek employment in policing. The job of police officer is viewed by prospective applicants much like any other job. The relatively good pay, benefits, and job security offered to police officers are an attraction to many working-class, high-school-educated persons. As important, if not more important, however, is the motivation to do something important and meaningful. Most applicants, Van Maanen suggests, are motivated by a sincere desire to perform public service. To this motivation can be added the fact that policing is presented, at least in the popular entertainment media, as an exciting, nonroutine, outdoors job.

The applicant is not the only one who makes a choice at this stage, however. The police agency typically invokes a strict selection process. Minimum educational, health, and age requirements are common selection criteria for police officers. Traditionally, police agencies have also established minimum height, weight, strength, and residency requirements. Selection normally involves a test (usually a paper-and-pencil civil service examination), psychological and medical examinations, and background and character investigation. Thus, of all of those who apply for police officer positions, the departmental screening procedures winnow out the few who will be accepted and will proceed to the second stage of socialization—the introduction.

Introduction.   The police recruit is required to complete a training program before being assigned to duty in most departments. Many states now have legislation requiring minimum levels of police officer training. Van Maanen (1973) calls this formal academy training the "introduction" stage. Here the recruit is exposed to both formal and informal instruction, which aid the socialization process. The academy class member learns the fundamentals of the job, including report-taking procedure, criminal and traffic codes, regulations of the department, and similar knowledge and skills by means of formal lecture presentations and readings. Informal learning also occurs through "war stories," experiences with discipline, and other events that transmit the values of the occupation.

Encounter.   On completion of the academy training, the rookie police officer is assigned to working the street as a police officer. Traditionally, the rookie has been assigned to a field training officer who is an experienced officer, though not necessarily trained as an instructor. The typical process involved patrol officers training new patrol officers, and the peer group thus has the opportunity to reinforce and redefine what it considers to be appropriate values and behaviors. The literature is full of references to rookies being told, "Forget the academy; I'm here to teach you how to be a police officer," by their field training officers. During this period the rookie learns that working the street is not what he or she thought it would be. Hours of monotonous routine patrol, experiences that encourage disrespect and distrust of citizens, few opportunities to apprehend criminals, and other facets of real-life police work are at odds with what most rookies expected. Because this is a period when the rookie's expectations of the job are tested against the reality of police work, Van Maanen (1973) calls this stage the "encounter."

Metamorphosis. The final stage in police socialization, according to Van Maanen (1973), is "metamorphosis." Assuming the new officers successfully complete all aspects of their probationary period, it is time now for them to adapt to the reality of policing on a day-to-day basis if they are to be successful in the career. Thus, like most people, career police officers come to view policing as a job. In general, this means that they attempt to do what has to be done, but not to overwork themselves, and to maintain the support and friendship of other officers.

MAKING POLICE OFFICERS

Van Maanen (1973) suggests that this socialization process explains how it is that police officers share values, attitudes, and behaviors. Officers (at least within departments) are all products of the same socialization. The essential argument is that most people who are subjected to this selection and training process and who then work as police officers will respond to general social pressures so that their behavior will be similar to that of their peers. In short, the ways in which we select and train people for the position works to "make" police officers out of them.

Richard Lundman (1980) expands on Van Maanen's thesis by linking each of these stages to the development or reinforcement of particular characteristics of the police officer's working personality. His review of the stages reveals that powerful social and psychological forces are at work in the process to support group solidarity, authoritarianism, suspicion, dogmatism, conservatism, and cynicism. Like Van Maanen, Lundman concludes from his review of this traditional police recruitment process that police officers, complete with "working personalities," are not born but made.

Lundman observes that the traditional selection and recruitment standards for police officers produces what he calls a "remarkably homogeneous" group of high-school-educated, white males with blue-collar backgrounds who believe policing is the best job they are likely to find (1980:77). The police personality, however, is absent from this group at the beginning; it is something they learn in what Lundman calls the "quasi-military or stress academy" experience. That is, police departments start with a relatively homogeneous group of recruits and then subject them to a uniform training experience. This process is what produces the apparent uniformity in the police officers' working personalities.

The academy experience, according to Lundman, is as important for its informal training as it is for its formal content. The **formal content** includes

Despite individual differences, in many ways police officers are similar. (Phyllis Graber Jensen/Stock Boston)

departmental rules and regulations, criminal law, first aid, weapons training, and the like. The **informal content** is what conveys the attitudes and beliefs of the occupation. These are learned through experience in the academy and by the examples provided by instructors and their "war stories."

Through the enforcement of academy rules about attendance, note taking, appearance, and the like, recruits learn to rely on fellow recruits to prepare for inspections, share class notes, and "cover" for each other. The almost random assignment of discipline for rule violations leads recruits to see the rules as something to avoid rather than something to follow. The instructors, and by extension, departmental officials, come to be seen not as teachers and friends, but as bosses and punishers. Thus, the experience starts recruits along a path that leads to distrust, skepticism, or cynicism about the police organization, along with codependence and solidarity with fellow officers.

The emphasis of formal training on crime control and self-defense issues such as criminal law, defensive driving, and mostly about physical and weapons training implants the idea that policing is crime fighting and is dangerous (Birzer and Tannehill, 2001:236–238). Thus, recruits are prepared to define many police tasks that do not involve crime-control efforts as not being "real" police work. The war stories told by instructors typically convey to recruits the message that only other police can be trusted. Even when a citizen calls for assistance, the responding officer may end up the target of assault by that citizen. The decisions of courts, prosecutors, and the behavior of defense attorneys are criticized. These lessons lead the recruit to conclude that police are different from, and better than, other people. This attitude reinforces group solidarity and enhances alienation from the public and cynicism toward the public. The selection and training processes coupled with job experiences produce an occupational culture (Farkas and Manning, 1997).

When the recruit first takes to the street and is assigned to an experienced officer as a teacher, the informal lessons of the academy are reinforced. As we mentioned, the rookie is often told to "forget the academy" and to learn the craft of policing by watching his or her experienced partner. In this learning process, the rookie sees how to "do" policing—including using threats and other questionable tactics that go against academy training but that work on the street. These experiences reinforce cynicism, solidarity among street officers, and authoritarianism. Experiences with the public are primarily of the type that further alienate the officer and help create a we-versus-them mindset.

Lundman (1980) describes this training and its effect of creating the working personality of police officers as the "traditional practice." He suggests that changes in selection, training, and retention of officers have had the effect of changing the basic police personality, or at least of increasing the numbers of personalities available to police. The new procedures include recruiting a broader cross-section of the population, using a different emphasis and new procedures in training, and implementing improved field training programs for new recruits.

Making New Police Officers

In the early 1970s many police agencies began to change the ways in which they recruited, selected, and trained officers. Spurred in part by the requirements of the Civil Rights Act enforced by court decrees, police departments began an aggressive affirmative-action recruitment campaign that continues today. In short order, agencies found that minimum height and weight standards discriminated against women and certain ethnic groups. The reliance on a civil service test of general intelligence also had the effect of excluding many minority group members. Realization that the discriminatory effects of these tests and requirements were not related to qualifications for the police job led departments to alter their selection procedures. The tests were changed, tutorial programs were begun for recruits, and separate eligibility lists were established to increase the pool of minority applicants. The result has been a more heterogeneous group of recruits to policing.

Although the majority of recruits are still white and male, departments across the country are accepting increasing numbers of women and

CHART 10.3
Changes in officer characteristics, 1987–1997. (Source: B. Reaves and A. Goldberg (2000) *Local Police Departments, 1997.* Washington, DC: U.S. Bureau of Justice Statistics, p. 4.)

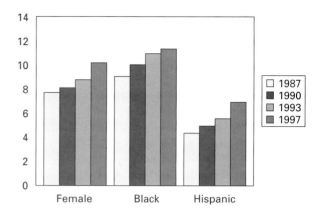

minority applicants, as can be seen in Chart 10.3. Their entry into the police occupation creates a greater diversity of attitudes, experience, and personality among police officers. Moreover, to this changed group of recruits, new training procedures have been applied.

Lundman (1980:97–99) contends that the combination of different people with different training has created different types of police personalities, or at least police officer styles. Training has shifted away from the police-dominated, quasi-military, stress academy to what he calls the "non-stress," or "college," academy. This academy experience is more relaxed and more issue oriented than the traditional training. Police instructors have been supplemented with nonpolice professionals in sessions on the use of discretion, police ethics, community relations, crisis intervention, and other subjects. Although rules still pervade the academy, their number and enforcement have both been changed to be more reasonable and predictable. Nonetheless, police training has not changed substantially in the past two decades, suggesting that as the role of policing changes, training may not be keeping pace with social and demographic changes (Bradford and Pynes, 1999).

The practice of assigning rookie police officers to a veteran patrol officer for on-the-job training continues, but today's **field training officers** are specialists who are selected and trained for the job of acclimating rookie officers to the new world of policing. Michael McCampbell (1986) reviewed the state of the art in police field training. He reports that nearly two-thirds of police departments responding to his survey indicated they used a field training program that involved specially trained and qualified trainers implementing a planned training program. Most of these programs were less than ten years old.

The typical program consists of different training phases operated by carefully selected and trained field training officers and supported by the agencies' chief executives. Phases might include an introduction in which the recruit observes the training officer, a practice phase in which the recruit takes responsibility under direct supervision by the training officer, and an evaluative phase in which the recruit works alone, with the training officer playing a passive observer role. In each phase the field training officer instructs and evaluates the performance of the rookie officer and provides feedback to the rookie to help her or him improve.

The goal of these newer field training programs is to reinforce the academy lessons and to allow the recruit to develop expertise and competence under the tutelage of an officer whom the agency considers to be an exemplary police officer. Street experience is often combined with additional classroom instruction, and weaknesses in the recruit's performance are addressed through remedial training. Rather than learning to "forget the academy," the purpose is to help the recruit apply the lessons learned there.

Lundman (1980) suggests that the newer procedures have improved police officer attitudes and skills. He believes that the traditional stress academy experience was one that almost invariably developed Skolnick's (1994) working personality in police recruits, whereas the non-stress academy, coupled with a diversity of recruits, has produced additional types of police personalities.

We will discuss types of officers in greater detail later, but for now the point of Lundman's argument is that the combination of police recruit traits and police training helps to explain police behavior. When recruits share common traits and undergo an authoritarian, dogmatic, and, in many ways, unrealistic training, they tend to share a common, narrow view of the job. On the other hand, when a variety of people are presented with a broader range of training experiences, divergent views of the job emerge.

CORRELATES OF OFFICER CHARACTERISTICS AND WORK

Laure Brooks (1989) reviewed the literature on discretionary decision making by police officers and concluded that a variety of forces appear to correlate with how officers do their jobs. These forces include situational characteristics (nature of the call, time of day, presence of witnesses, etc.), characteristics of the police organization, characteristics of the neighborhood, characteristics of the officer, and officer attitudes. Brooks considers attitudes to be a **"predisposition,"** preference, or tendency to act in certain ways.

These latter two, officer characteristics and attitudes, are our focus in this section. As Brooks (1989:126) observes, there is support for the contention that officer attitudes influence their decisions: "Predispositions supply the officer with a repertoire of possible behavior, and, from this collection, the officer selects an appropriate response to a specific situation." The existence of different predispositions helps explain different actions by different officers. But the relationship between attitudes and behavior is a complex one. As Worden (1989:705) observes, both occupational attitudes and behavioral patterns are developed at the same time on the job, and broader and deeper personal attitudes may compound any attempt to study the effects of officers' occupational attitudes.

Given the broad discretionary authority granted to police officers, differences in policing may be the product of differences in officers. There is ample evidence to suggest that officers vary among themselves in terms of how they do policing. This variation has been found in regard to the success of community-policing initiatives (Thurman, Bogen, and Giacomazzi, 1993). Similar differences in officer perceptions of stress (Violanti and Aron, 1995) and officer attitudes concerning pursuits (Falcone, 1994) have been reported. Officers also differ in how they deal with citizens and in terms of how qualified they are to use alternative methods of resolving citizen conflicts (Cooper, 1997). Thus, we have reason to believe that officer characteristics are an important correlate of policing in practice.

To the degree that different approaches to policing (or predispositions) correlate with characteristics of individual officers, the officer himself or herself becomes an important factor in understanding policing in America. Research to date has examined a number of officer traits as they relate to police decision making (Sherman, 1980; Riksheim and Chermak, 1993). These traits include gender, ethnicity, education, age, police experience, and officer attitudes.

Gender

Until relatively recently, policing in America was an almost exclusively male occupation (Lord, 1989; Golden, 1982; Horne, 1980; Martin, 1980). Since the 1970s, however, the number of women holding patrol officer positions in the United States has increased steadily. These female officers may bring a different set of predispositions to the job.

Most comparisons of male and female police officers report some differences in performance between the two groups (Milton, 1972; Townsey, 1982). In general, female officers appear to be better prepared to mediate disputes and less likely to make arrests than their male counterparts. Women officers may act to reduce the likelihood of violence in police–citizen encounters. In the case of domestic violence, Homant and Kennedy (1985) report that female officers viewed these situations as more serious and were more likely to involve themselves than were their male counterparts. Similarly, staff at shelter programs for battered women and the women victims themselves tend to see female officers as being more involved in the cases, devoting more time, providing more information, and being more caring than male officers (Kennedy and Homant, 1983).

For whatever reasons—personal experience, societal sex-role stereotyping, or other factors—female police officers appear to be more willing and more able to negotiate disputes. Consequently, they are less likely to resort to force or formal arrest. Abernathy and Cox (1994) report that female officers, in general, appear better able to manage their anger and are thus less likely to use force. In the case of domestic violence, as mentioned, female officers also appear to be more sympathetic to victims. Homant and Kennedy (1983:43) suggest several possible explanations for this:

> The entire correlation between policewomen and involvement could not be accounted for by education, feminism, and empathy, nor could it be accounted for by the combination of demographic variables on which the two samples (policewomen and policemen) differed. Therefore, we speculate that being a policewoman implies certain other predispositions, traits, or experiences.

It may also be, of course, that chivalry is not yet dead, and what these researchers have observed is a tendency on the part of citizens to defer to, or at least not challenge, the authority of the female officer. Michael Breci (1997) reports a survey of public attitudes that indicate the public, both males and females, hold sex-stereotypic views of police officers. Though agreeing that female officers are as effective as males, the public rates females as better suited to "traditional" female roles—the treatment of children and women, dealing with victims, and so on.

Thus, gender correlates in some situations at least, with police officer behavior. The question of how, if at all, gender affects officer effectiveness requires further study. Without addressing which is the most effective way of responding to a particular situation however, it is possible, as in the case of domestic violence, to see that gender correlates with different ways of responding. The Bureau of Justice Assistance recently reported (2001:2), "Research conducted both in the United States and internationally clearly demonstrates that women police officers use a style of policing that relies less on physical force. They are better at defusing and de-escalating potentially violent confrontations with citizens and are less likely to become involved in incidents of excessive force."

Ethnicity

Ethnic and racial minority group members, like women, are underrepresented on America's police forces and have been since the beginning of formal policing in the United States. As with females, members of minorities began to join the ranks of police departments in larger numbers after passage of the Equal Employment Opportunity Act of 1972. Prior to this law, and the lawsuits that resulted from it, test results, height and weight standards, and other selection criteria for policing worked to disqualify disproportionate numbers of female, black, Asian, and Hispanic applicants.

Sullivan (1989) notes that the police literature lacks studies comparing minority and white officers' performance in different situations. Still, anecdotal data, as well as "common-sense" understandings indicate that the ethnicity of an officer is correlated with officer behavior. Available data—for example, from Fyfe (1978) and Geller and Karales (1981)—show that black police are more likely to be involved in use of force and deadly-force situations than their white counterparts, but those researchers note that minority officers are usually assigned to patrol high-crime, minority ghettos. Friedrich (1977) reports that the patrol styles of black officers were more aggressive than that of white officers. In particular, black officers appear to be less lenient with black citizens than are white officers (Alex, 1976).

An important goal of increasing the numbers of black and other minority officers in policing was to improve community relations. Decker and Smith (1980) report that increased numbers of minority group officers alone do not appear to alter public appraisals of the police. Yet, some observers (Rossi, Berk, and Eidson, 1974; Buzawa, 1981; Berg, Truce, and Gertz, 1984) report that minority group officers were less antagonistic to the public and displayed greater ties to the community than their white colleagues. Others, such as Larry Stokes and James Scott (1996) support increased representation of women and minorities in policing as a means to make the police agency more similar to the public it serves.

The increase in the numbers of women and ethnic minority officers in policing has produced its own problems within the police organization. Haar (1997) reports that female officers experience specific job stress related to the traditional male domination of police work. Further, despite efforts on the part of police organizations and administrators to integrate the patrol force, female and ethnic minority officers tend not to interact freely or frequently with white male officers. The interaction patterns in police patrol units remain somewhat segregated (Morash and Haar, 1995). Additionally, affirmative-action efforts in policing may have increased dissatisfaction among both minority and majority officers. For minority officers, advancement seems too slow, in part because of requirements of time in rank for promotions (Polk, 1995). For majority officers, the presence of affirmative-action programs leads to a questioning of the fairness of the promotional process and dissatisfaction with the police organization (Seltzer, Alone, and Howard, 1996).

The lack of empirical data notwithstanding, and the confusion of assertions that abound on the impact of ethnicity on police officers, it seems safe to say that ethnic background correlates with police behavior. Minority officers may be harsher in dealing with minority offenders, or they may be more accepting or trusting of the public than white officers. Whatever the specific correlation in a particular situation, the ethnicity of officers may contribute to our ability to understand police behavior.

Education

For several decades now a concerted effort has been made to raise the educational level of police recruits. The National Advisory Commission on Criminal Justice Standards and Goals (1973), for instance, recommended that a bachelor's degree be the minimum educational standard for police recruits by 1982. Support for better-educated police stems from a variety of sources, including a desire on the part of police themselves to gain recognition as a profession (Swanson, 1977; Sherman, 1978).

Those who support higher educational standards suggest that college education will generally improve police officers in terms of their sensitivity to citizens, ability to communicate, and general demeanor. Further, college education is expected to assist officers in gaining a clearer perspective on the job, on their role within society, and on the ethical and moral problems they face. In short, the college-educated officer is likely to be more fair, more honest, and more effective (Lynch, 1976; Hoover, 1995). The data, however, are inconclusive.

A few studies comparing college-educated officers with those possessing only a high school diploma appear to support the call for more education (Witte, 1969; Cohen and Chaiken, 1972; Smith, Locke, and Walker, 1968; Sanderson, 1977; Trojanowicz and Nicholson, 1976; Cascio, 1977; Carter, Sapp, and Stephens, 1988). In sum, these authors conclude that college-educated officers are less likely to receive citizen complaints, use fewer sick days, are less likely to be assaulted on the job, and present fewer disciplinary problems to the organization. Bowker (1980), in particular, notes that college-educated officers display greater tolerance of minority group members, less authoritarianism, less dogmatism, and more tolerance than their less educated colleagues.

Other observers have questioned the value of higher education for the police. Charles Swanson (1977) notes that higher education may be both unnecessary and actually harmful for police officers. Describing the police job as largely routine, within a paramilitary bureaucracy, Swanson suggests that more highly educated employees have a greater likelihood of being dissatisfied with the job and understimulated by the work, and may suffer related morale problems. He predicts that better-educated officers could result in greater job turnover and that the more highly educated officers would disrupt the traditional command structure of the police agency. Regoli (1976) found that college-educated officers were more cynical than their less educated colleagues.

Whatever the effects of higher education on the satisfaction officers feel in their jobs, the available evidence suggests that educational attainment does matter in police behavior (Hoover, 1995). Education correlates with differing levels of citizen complaints, use of force, and officer injury, indicating that better-educated officers are less likely to be involved in violent encounters. Whether it is because they have learned sensitivity,

CHART **10.4**
Educational requirements for
police recruits by departments.
(Source: B. Reaves and A.
Goldberg (2000) *Local Police
Departments, 1997*.
Washington, DC: U.S. Bureau of
Justice Statistics, p. 5.)

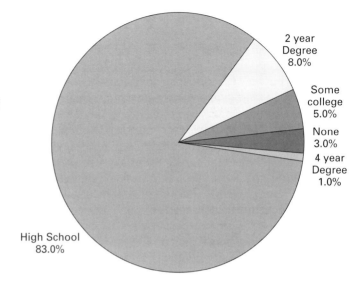

2 year
Degree
8.0%

Some
college
5.0%

None
3.0%

4 year
Degree
1.0%

High School
83.0%

negotiation skills, or something else, more highly educated officers appear to encounter less friction in their dealings with the public.

Age

Sherman (1980) notes that the commonplace understanding is that younger officers are more aggressive and also more likely to make mistakes than are older officers. Perhaps as a result of youthful vigor, perhaps as a function of inexperience, younger officers are expected to feel the "weight of the badge" more acutely than are older officers. Cohen and Chaiken (1972) report that officers who were older at the time of appointment to the force were less likely to have citizen complaints filed against them and less likely to become victims of assault. Hale and Wilson (1974) found that, on average, younger officers were more apt to become engaged in violent encounters with civilians than were older officers.

Ferdinand (1980) points out that as age increases, the officer's level of enthusiasm and effort decreases. He also reports an inverse correlation between officer attitudes supporting a crime-control philosophy and punitiveness with age. Other observers report no differences in the quantity or quality of arrests related to officer age (Forst, Lucianovic, and Cox, 1977).

In summary, Sherman (1980) concludes that the available research indicates that older officers may be somewhat more service oriented in their work than younger officers. He reports that age does not seem to be an important factor in detection of crime, number of arrests, or incidence of violence on the part of officers. Importantly, Sherman notes the near-impossibility of separating out the effects of officer age from those of job experience. Any correlates between officer age and officer behavior may actually be the product of job experience. It is to this subject that we now turn.

Police Experience

As with age, police officer behavior seems to vary in relation to the length of service. Officers with more years of experience are reported to do less work in terms of taking reports, making aggressive patrol, initiating citizen contacts, and the like (Friedrich, 1977; Forst, Lucianovic, and Cox, 1977; Hou et al., 1983; Neiderhoffer, 1967). On the other hand, experienced officers are more successful in obtaining convictions of offenders they arrest. These officers also tend to be more friendly with civilians and less aggressive in their dealings with the public.

Neiderhoffer's (1967) theory that police officer cynicism changes with years of experience anticipates these relationships. New officers, out to change the world, work very hard. In time, they become dissatisfied with their efforts and become cynical. Still later, they come to terms with their cynicism and recommit themselves to the job. More experienced officers are likely to have learned the "tricks of the trade" so that they can be as effective, or more effective, with less effort than can newer officers. Further, studies of police cynicism and other police officer attitudes indicate that both cynicism and job satisfaction among police correlate, to a modest degree, with length of service, so that officers with much experience tend to be both less cynical and more satisfied with their work (Cannizzo and Liu, 1995; Travis and Vukovich, 1990; Buzawa, 1984; Sparger and Giacopassi, 1983; Poole and Regoli, 1979).

Officer age and length of police experience exhibit a common relationship with police behavior. Although the available literature is still relatively scant, it appears that older and more experienced officers are less aggressive, are more sympathetic to and tolerant of the public, and are the subject of fewer citizen complaints. Younger officers, on the other hand, display more individual initiative, expend more effort and energy, and are more likely to make arrests of citizens. Both age and length of experience correlate with how officers approach the job of policing. Future research must control for the overlap between age and years of experience to allow us to better understand how each of these characteristics correlates with police behavior.

Officer Attitudes

In 1967 Black and Reiss concluded that the racial attitudes of officers they studied were unrelated to how officers treated black citizens. A reanalysis by Friedrich (1977), however, revealed that when specific officers were the unit of analysis, the more the officers disliked blacks, the more likely they were to arrest black suspects. Moreover, aside from the arrest decision, the reanalysis found that officers with negative attitudes toward blacks were likely to treat black citizens with disrespect. Thus, the officer's attitudes were related both to arrest decisions and to how the officers approached and dealt with citizens in nonarrest circumstances. Black himself recognized this limitation of the original analysis in a later edition of his work.

There has been interest in the relationship between individual officer attitudes and police behavior for several decades. Unfortunately, as Brooks (1989) notes, very little research has been conducted in this area. Brooks (1986) observes that differences in officer attitudes (which she aggregates into orientations) correlate with differences in police officer behavior. For example, she notes that officers having a crime-control orientation are more likely to use force but less likely to make arrests than are those not having that orientation. Other writers have drawn similar conclusions, arguing that officer behavior is the product of officer perceptions. For example, Westley's (1970) finding that police officers justify the use of force in order to maintain public "respect for the police" supports the conclusion that officer attitudes toward the public channel their decisions to use force (Brooks, 1989).

Much research in policing has focused on police attitudes, particularly in the area of police cynicism (Regoli, et al., 1987; Langworthy, 1987). Others have studied police officer attitudes toward professional ethics (Felkenes, 1984), and other topics. In general, however, none of this research has linked officer attitudes to officer behavior.

On an intuitive level, it makes sense that a bigoted white officer will treat minority group members differently than she or he treats whites. Similarly, a sexist male officer will react differently to women than a nonsexist officer will. Unfortunately, most of the research on police attitudes has been either descriptive (reporting the attitudes of police officers) or has compared police officers to nonpolice. More comprehensive analyses of the influence of police attitudes on behavior has not revealed very strong or direct relationships (Worden, 1989). Future research in this area may benefit from the use of analytic techniques that reveal how attitudes affect decision making by police officers (Worden and Brandl, 1990).

The Influence of Officer Characteristics

To date, the available research suggests that some personal characteristics of police officers are related to how officers perform their jobs. Much of the research has produced mixed findings, suggesting that officer characteristics are not solely responsible for officer behavior. As Paoline, Myers, and Worden (2000) observe, "Overall it appears that the variation in officers' outlooks is patterned only to a limited extent by officers' characteristics." While individual differences may be important, how those differences translate into different practices is conditioned by the organization, community, and specific situation in which an officer is working. Regarding the use of violent force by police, Jeffrey Cancino compared contemporary police with those studied in the 1950s by William Westley. He found that little had changed in officer reports of when and why they would feel justified in using violence. His explanation for the lack of change suggests the limited impact of changing officer characteristics on reforming police practice. He writes (2001:156), "While recruitment, selection, training, and demographic characteristics of officers has changed over the past 50 years, the reality of police work has remained the same."

In one area scholars have attempted to link differences in officer attitudes with differences in officer behavior. A number of observers have developed schemes for classifying officers into different types of police. Each police type is characterized as more or less fully supporting a set of attitudes and beliefs relevant to policing. Differences among officers in terms of these beliefs and attitudes are then correlated with differences in how officers approach the job of policing. Thus, while there are little data on how specific officer attitudes may affect officer behavior, several models suggest sets of attitudes that direct officer decision making. We now turn to the subject of police officer typologies.

TYPOLOGIES OF POLICE OFFICERS

Ellen Hochstedler (1981) examined the literature on police officer types and then tested the existence of types with data from a survey of Dallas, Texas, police officers. Her analysis failed to confirm the existence of distinctive types of police officers. Her conclusion was based on a criterion of practical utility for the types, that is, the ability of police officer type to predict how police officers would behave in specific situations.

This is a stringent test, as Hochstedler herself notes (1981:465): "Given that typologies seek to simplify, and that attitudes and behaviors of police officers are not simple phenomena, perhaps the most that can be expected of a typology of attitudes and behaviors of police officers is that it will provide conceptual clarity, or a greatly simplified, even distorted, perception of reality." Therefore, the value of a typology of police officers must be assessed in light of the purposes for which it is to be used.

Types and Styles

People use **typologies** and **stereotypes** in their thinking to categorize and organize their interactions. For example, college students tend to typify each other in terms of their majors (a typology, or categorizing by type of major). If someone is a business major, engineering major, premedicine, or art major, clear images come to mind (a stereotype of the average business major, for example). That stereotypical student with a major in engineering may not actually exist; still, the typology remains because it is useful in determining, quite quickly, what a person is likely to be interested in, and how

A grieving member of the force at the funeral of a police officer shows that beneath the badge, police are human. (Bill Mitchel/*Fort Pierce/Port St. Lucie Tribune*/ Silver Image)

she or he is likely to view the world. Social scientists often use typologies for the same purpose.

Faced with a large amount of information about a number of individuals, the typing process is a means of categorizing and organizing data. If you think about a police officer, whom do you envision? Your vision of the officer is a form of **typification.** It allows you to think about (and understand) the term police officer despite the fact that there are over 750,000 individual officers.

For scientists who observe differences in police behavior, one way to organize these differences is to generalize the characteristics of the officers involved so that thousands of observations can be classified into a few types. These types can then be distinguished from each other on the basis of a small set of characteristics. In this way, while prone to error in specific instances, a general understanding of police officers is developed.

Thus, one officer may be classified as a "law enforcer" while another is classified as a "social worker." In a particular instance, the officers may or may not be true to type. That is, the law enforcer may be caught counseling the parties to a family disturbance while the social worker arrests them. In general, however, the typology is useful if the law enforcer tends to arrest and the social worker tends to counsel in their interactions with the public.

A perfect typology might very accurately predict how an officer responds in specific situations. In large measure, this was the criterion for testing types that Hochstedler (1981) used. Thus, law-enforcer types would arrest in the vast majority of situations, and social-worker types would counsel in the vast majority of situations. In the ideal, social scientific typologies seek to attain this level of predictive accuracy. It is this accuracy that Hochstedler says is missing in existing police officer typologies.

Officer types are only one of several forces at work in determining police behavior. For instance, what would the typology predict in the case where the law-enforcer type and the social-worker type were partners? How would officer type affect the outcome of the family disturbance call if the department had a policy of nonarrest (or arrest) for these situations? As discussed in Chapter 9, the link between officer orientation and agency "style" or goals is also an important consideration. Amy Halsted, Max Bromley, and John Cochran (2000) report a study of sheriff's deputies that showed deputies whose orientation toward policing (crime control versus service delivery) was consistent with agency policy were most satisfied with their jobs. Still, different types of officers were found to be working in the same sheriff's office. It should not be surprising that, in reality, police officer type alone does not prove to be a powerful predictor of police behavior.

Rather, to use Laure Brooks's (1989) term from our earlier discussion of police officer attitudes, officer type is a shorthand summary of officer predispositions. That is, other things being equal, different types of officers will behave in ways consistent with their type. Unfortunately for easy explanation, other things (offense seriousness, presence of witnesses, departmental policy, etc.) are usually not equal. Yet another difficulty arises when we acknowledge that officers may have two or more sets of predispositions. That is, an officer may primarily be a social worker but quickly turn into a law enforcer when citizens reject attempts at mediation or counseling. Even so, typologies are useful for understanding officer behavior and interpretations of situations. Accordingly, a number of police officer typologies have been developed that seek to classify officers in terms of how they view or do the job of policing.

Police Officer Types

Coates (1972) applied Wilson's police department typology to individual officers and concluded that three distinct types of officers exist. The *legalistic abusive officer* defines the police role as defender or protector of the community and community values. This officer is quick to use coercive power and very authoritarian in action. The *task officer* does the job without applying his or her own values to the laws; this officer merely enforces the law. The *community-service officer* does not emphasize the law and law enforcement; rather, this officer tries to help people and solve problems. As we saw in the previous chapters, if police organizations can accurately fit these types, they would be best advised to hire and retain officers whose personal styles match that of the organization.

Broderick (1977) suggests that police officers differ among themselves in terms of how much they emphasize the due-process rights of suspects and how much they emphasize the need for social order. He develops a four-class typology based on whether officers place a high or low emphasis on each of these goals. Those who stress both due process and social order he calls *idealists*. Those who stress neither value he terms *realists*. Between these two are those who stress social order, the *enforcers,* and those who stress due process, the *optimists.*

White (1972) also develops four types of officers based on whether the officers are universalistic or particularistic in either their *values* or their application of *techniques.* She uses the terms *universalistic* to describe tendencies to ignore individual suspect characteristics and *particularistic* to describe officers who note and react to individual differences. White names those officers who are universalistic in both their values and techniques *crime fighters* and those who are particularistic *tough cops.* The middle ground includes the *rule applier,* who uses universalistic techniques based on particularistic values, and the *problem solver,* who uses particularistic techniques to support universalistic values.

O'Neill (1974) applies a similar set of dimensions to police officers. He suggests that officers differ in terms of their perceptions of activity and formalism. *Activity* (like White's "values") refers to the range of situations the officer feels warrant police intervention. *Formalism* (like White's "techniques") relates to the range of intervention options the officer perceives to be available. Officers who perceive a wide range of activity and a high degree of formalism (few intervention options) are *law enforcers,* while those who perceive few activities as deserving police intervention and a low degree of formalism (many intervention options) are *watchmen.* Those officers who see a range of activities needing police intervention and a low degree of formalism are *social agents. Crime fighters,* by contrast, define the police role narrowly to include only responding to crime and perceive the police intervention as highly formal.

Finally, Muir (1977) applies two dimensions to police officer role perception and arrives at four types of officers. On the one hand, officers may differ in their ability to empathize or understand people, or their *perspective.* In addition, they may differ in terms of their degree of comfort in using coercive force against people, or their *passion.* Officers who can understand others, and are also comfortable with the use of force, Muir calls *professionals.* Those who have neither perspective, nor comfort in their ability to use coercive power, are *avoiders.* In the middle are those who have perspective but are uncomfortable with coercive force, the *reciprocators,* and those who are comfortable with force but lack perspective, the *enforcers.*

When Hochstedler (1981) tested police types, she specifically looked at these five models. She notes that on closer examination, the five different models all supported the descriptions of four basic police officer types:

1. *Ideal:* This officer recognizes and manages the complexity of the police role. Able and willing to use authority when necessary, the officer is sensitive to the needs and desires of citizens and suspects as well as to the rights of citizens. In short, this is "supercop," labeled by Muir as the *professional.*

2. *Lawman:* This officer narrowly defines the police role as that of crime fighting and the police function as detection, investigation, and arrest. Other activities of police are not "real" police work. Broderick and Muir call this officer the *enforcer,* White and O'Neill apply the label *crime fighter.*

3. *Servant:* This officer is generally uncomfortable with the enforcement side of policing and the authority that goes with the job. He or she would prefer to help people in trouble and solve problems. Muir uses the term *reciprocator* to describe the officer who sees her or his primary function as "doing good." The other typologists each use a specific term, but perhaps Coates's *community-service officer* best exemplifies this type.

4. *Shirker:* Hochstedler notes that only O'Neill and Muir identify a type of officer who generally prefers to do nothing—the *shirker.* Muir calls this officer the *avoider,* while O'Neill uses the term *watchman.* This type of officer is perhaps self-motivated or even unmotivated. This officer prefers to do nothing rather than enforce the law or help people.

Despite problems with applying any particular typology to actual practice and being able to predict actual officer behavior, police officer typologies are powerful analytic tools. That there are so many, and that they identify similar types of officers at different points in time and in different departments, supports the theory that officers do fall into different categories.

Further, it is important to note that though not a controlling influence in all circumstances, officer type is likely to be an important correlate of officer behavior. When attempting to understand the balance of forces explaining a particular police action, one must remember that an important force is the police officer himself or herself.

CONCLUSION

This chapter described the characteristics of police officers and examined the literature concerning police officer characteristics and police behavior. Many earlier observers report the existence of a common police personality that colors the behavior of police officers toward citizens. There is some evidence that this observed "working personality" is a product of the nature of the police job and the ambiguous role of the police in American society. These factors combine to create what is known as an occupational (or police) culture, which is a way of interpreting and understanding the world that is shared among a set of people.

The possibility that working police officers exhibit more than one personality was raised during a discussion of various typologies of police officers and their policing styles. In total, the existing literature indicates that police officers view their jobs and the situations they encounter on the job differently than do the citizens with whom they interact. In addition, police officers differ among themselves in terms of their preferences, or predispositions, for interventions and their manner of intervening with citizens.

More research must be conducted in this area to determine if different types of police officers actually exist and, if so, what factors lead different officers to adopt different styles. This research is hampered by inadequate measures of police attitudes and of the relationship between attitudes and behavior. It is also limited by the likelihood that attitudes and behavioral preferences change over time. Finally, although the influence of officer characteristics and types is a potentially important explanation for police behavior, it is but one factor in a larger mix of forces that may explain that behavior.

CHAPTER CHECKUP

1. Are police officers different from citizens?
2. What is the police officer *working personality?*
3. What characteristics of the job of policing appear to make police officers develop the working personality?
4. What is *occupational socialization,* and how does it work to produce a police personality?
5. What is an *occupational culture?*
6. How do contemporary police selection and training procedures differ from those of the past?
7. In what ways do characteristics of officers correlate with the ways in which they do their jobs?
8. What is a *typology,* and what are some examples of typologies of police officers?
9. Why have police officer typologies been created?

REFERENCES

Abernathy, A. and C. Cox (1994) "Anger management training for law enforcement personnel," *Journal of Criminal Justice* 22(5):459–466.

Alex, N. (1976) *Black in blue: A study of the Negro policeman.* (New York: Appleton-Century-Crofts).

Bayley, D. and H. Mendelsohn (1969) *Minorities and the police.* (New York: Free Press).

Berg, B., E. Truce, and M. Gertz (1984) "Police, riots, and alienation," *Journal of Police Science and Administration* 12(1):186–190.

Birzer, M. and R. Tannehill (2001) "A more effective training approach for contemporary policing," *Police Quarterly* 4(2):233–252.

Bittner, E. (1975) *The functions of the police in modern society.* (Washington, DC: National Institute of Mental Health).

Black, D. (1980) *The manners and customs of the police.* (New York: Academic Press).

Black, D. and A. Reiss (1967) "Patterns of behavior in police and citizen transactions," in President's Commission on Law Enforcement and Administration of Justice, *Studies in crime and law enforcement in major metropolitan areas.* Field Surveys III, Volume 2. (Washington, DC: U.S. Government Printing Office):1–139.

Bowker, L. (1980) "A theory of educational needs of law enforcement officers," *Journal of Contemporary Criminal Justice* 1(1):17–24.

Bradford, D. and J. Pynes (1999) "Police academy training: Why hasn't it kept up with practice?" *Police Quarterly* 2(3):283–301.

Brandl, S. (1996) "In the line of duty: A descriptive analysis of police assaults and accidents," *Journal of Criminal Justice* 24(3):255–264.

Breci, M. (1997) "Female officer on patrol: Public perceptions in the 1990s," *Journal of Crime and Justice* 20(2):153–165.

Broderick, J. (1977) *Police in a time of change.* (Morristown, NJ: General Learning Press).

Brooks, L. (1986) "Determinants of police officer orientations and their impact on police discretionary behavior." Unpublished Ph.D. dissertation, Institute of Criminal Justice and Criminology, University of Maryland.

Brooks, L. (1989) "Police discretionary behavior: A study of style," in R. Dunham and G. Alpert (eds.) *Critical issues in policing: Contemporary readings.* (Prospect Heights, IL: Waveland):121–145.

Bureau of Justice Assistance (2001) *Recruiting and retaining women: A self-assessment guide for law enforcement.* (Washington, DC: Bureau of Justice Assistance).

Buzawa, E. (1981) "The role of race in predicting job attitudes of patrol officers," *Journal of Criminal Justice* 9(1):63–78.

Buzawa, E. (1984) "Determining patrol officer job satisfaction," *Criminology* 22(1):61–82.

Caldero, M. and A. Larose (2001) "Value consistency within the police: The lack of a gap," *Policing: An International Journal of Police Strategies and Management* 24(2):162–180.

Cancino, J. (2001) "Walking among giants 50 years later: An exploratory analysis of patrol officer use of violence," *Policing: An International Journal of Police Strategies and Management* 24(2):144–161.

Canizzo, T. and P. Liu (1995) "The relationship between levels of perceived burnout and career stage among sworn police officers," *Police Studies* 18(3–4):53–68.

Carter, D., A. Sapp, and D. Stephens (1988) *The state of police education: Policy direction for the 21st century.* (Washington, DC: Police Executive Research Forum).

Cascio, W. (1977) "Formal education and police officer performance," *Journal of Police Science and Administration* 5(1):89–96.

Coates, R. (1972) "The dimensions of police-citizen interaction: A social psychological analysis." Unpublished Ph.D. dissertation, University of Michigan.

Cohen, B. and J. Chaiken (1972) *Police background characteristics and performance: Summary.* (New York: RAND Institute).

Cooper, C. (1997) "Patrol police officer conflict resolution processes," *Journal of Criminal Justice* 25(2):87–102.

Crank, J. (1998) *Understanding police culture.* (Cincinnati, OH: Anderson Publishing).

Decker, S. and R. Smith (1980) "Police minority recruitment: A note on its effectiveness in improving black evaluations of the police," *Journal of Criminal Justice* 8(6):387–393.

Falcone, D. (1994) "Police pursuits and officer attitudes: Myths and realities," *American Journal of Police* 18(1):143–155.

Farkas, M. and P. Manning (1997) "The occupational culture of corrections and police officers," *Journal of Crime and Justice* 20(2):51–68.

Felkenes, G. (1984) "Attitudes of police officers towards their professional ethics," *Journal of Criminal Justice* 12(3):211–220.

Ferdinand, T. (1980) "Police attitudes and police organization: Some interdepartmental and cross-cultural comparisons," *Police Studies* 3(3):46–60.

Fogelson, R. (1977) *Big-city police.* (Cambridge, MA: Harvard University Press).

Forst, B., J. Lucianovic, and S. Cox (1977) *What happens after arrest? A court perspective of police operations in the District of Columbia.* (Washington, DC: Institute for Law and Social Research).

Friedrich, R. (1977) *The impact of organizational, individual, and situational factors on police behavior.* Ph.D. dissertation, Department of Political Science, University of Michigan.

Fyfe, J. (1978) *Shots fired: An examination of New York City police firearms discharges.* Ph.D. dissertation, School of Criminal Justice, State University of New York at Albany.

Geller, R. and K. Karales (1981) *Split-second decisions: Shootings of and by the Chicago police.* (Chicago: Chicago Law Enforcement Study Group).

Golden, K. (1982) "Women in criminal justice: Occupational interests," *Journal of Criminal Justice* 10(2):147–152.

Haar, R. (1997) "Patterns of interaction in a police patrol bureau: Race and gender barriers to integration," *Justice Quarterly* 14(1):53–85.

Hale, C. and W. Wilson (1974) *Personal characteristics of assaulted and non-assaulted officers.* (Norman, OK: Bureau of Government Research, University of Oklahoma.

Halsted, A., M. Bromley, and J. Cochran (2000) "The effects of work orientation on job satisfaction among sheriffs' deputies practicing community-oriented policing," *Policing: An International Journal of Police Strategies and Management* 23(1):82–104.

Hickman, M. and B. Reaves (2001) *Local police departments, 1999.* (Washington, DC: Bureau of Justice Statistics).

Hochstedler, E. (1981) "Testing types: A review and test of police types," *Journal of Criminal Justice* 9(6):451–466.

Homant, R. and D. Kennedy (1985) "Police perceptions of spouse abuse: A comparison of male and female officers," *Journal of Criminal Justice* 13(1):29–47.

Hoover, L. (1995) "Education," in W. G. Bailey (ed.) *The encyclopedia of police science,* 2nd ed. (New York: Garland):245–248.

Horne, P. (1980) *Women in law enforcement.* (Springfield, IL: Charles C. Thomas).

Hou, C., A. Miracle, E. Poole, and R. Regoli (1983) "Assessing determinants of police cynicism in Taiwan," *Police Studies* 5(4):3–7.

Kennedy, D. and R. Homant (1983) "Attitudes of abused women toward male and female police officers," *Criminal Justice and Behavior* 10:391–405.

Kirkham, G. (1976) *Signal zero.* (New York: JB Lippincott).

Langworthy, R. (1987) "Comment: Have we measured the concept(s) of police cynicism using Neiderhoffer's cynicism index?" *Justice Quarterly* 4(2):277–280.

Lefkowitz, J. (1975) "Psychological attributes of policemen: A review of research and opinion," *Social Issues* 31(1):3–26.

Lord, L. (1989) "Policewomen," in W. G. Bailey (ed.) *The encyclopedia of police science.* (New York: Garland Publishing):491–502.

Lundman, R. (1980) *Police and policing: An introduction.* (New York: Holt, Rinehart & Winston).

Lynch, G. (1976) "Contributions of higher education to ethical behavior in law enforcement," *Journal of Criminal Justice* 4(4):285–290.

Manning, P. K. and J. Van Maanen (eds.) (1978) *Policing: A view from the street.* (Santa Monica, CA: Goodyear).

Martin, S. (1980) *Breaking and entering: Policewomen on patrol.* (Berkeley, CA: University of California Press).

McCampbell, M. (1986) *Field training for police officers: State of the art.* (Washington, DC: National Institute of Justice).

McNamara, J. (1967) "Uncertainties in police work: The relevance of police recruit's backgrounds and training," in D. Bordua (ed.) *The police: Six sociological essays.* (New York: John Wiley, 1967:163–252).

Milton, C. (1972) *Women in policing.* (Washington, DC: Police Foundation).

Morash, M. and R. Haar (1995) "Gender, workplace problems and stress in policing," *Justice Quarterly* 12(1):113–140.

Muir, W. (1977) *Police: Streetcorner politicians.* (Chicago: University of Chicago Press).

National Advisory Commission on Criminal Justice Standards and Goals (1973) *Police.* (Washington, DC: U.S. Government Printing Office).

Neiderhoffer, A. (1967) *Behind the shield: The police in urban society.* (Garden City, NJ: Doubleday).

O'Neill, M. (1974) "The role of the police: Normative role expectations in a metropolitan police department." Unpublished Ph.D. dissertation, State University of New York at Albany.

Paoline, E., S. Myers, and R. Worden (2000) "Police culture, individualism, and community policing: Evidence from two police departments," *Justice Quarterly* 17(3):575–605.

Piliavin, I. (1973) *Police-community alienation: Its structural roots and a proposed remedy.* (Andover, MA: Warner Modular Publications).

Polk, E. (1995) "The effects of ethnicity on career paths of advanced/specialized law enforcement officers," *Police Studies* 18(1):1–21.

Poole, E. and R. Regoli (1979) "Changes in the professional commitment of police recruits: A case study," *Journal of Criminal Justice* 7(3):243–247.

Reaves, B. and A. Goldberg (2000). *Local police departments, 1997.* (Washington, DC: Bureau of Justice Statistics).

Regoli, R. (1976) "The effects of college education on the maintenance of police cynicism," *Journal of Police Science and Administration* 4(3):340–345.

Regoli, R., J. Crank, R. Culbertson, and E. Poole (1987) "Police professionalism and cynicism reconsidered: An assessment of measurement issues," *Justice Quarterly* 4(2):281–286.

Riksheim, E. and S. Chermak (1993) "Causes of police behavior revisited," *Journal of Criminal Justice* 21(4):353–382.

Rossi, P., R. Berk, and B. Eidson (1974) *The roots of urban discontent: Public policy, municipal institutions, and the ghetto.* (New York: John Wiley & Sons).

Sanderson, B. (1977) "Police officers: The relationship of a college education to job performance," *Police Chief* 44(1):62.

Seltzer, R., Alone, S., and G. Howard (1996) "Police satisfaction with their jobs: Arresting officers in the District of Columbia," *Police Studies* 19(4):25–37.

Senna, J. and L. Siegel (1987) *Introduction to criminal justice,* 4th ed. (St. Paul, MN: West).

Sherman, L. (1978) *The quality of police education.* (San Francisco: Jossey-Bass).

Sherman, L. (1980) "The causes of police behavior: The current state of quantitative research," *Journal of Research in Crime and Delinquency*:69–100.

Skolnick, J. (1994) *Justice without trial: Law enforcement in a democratic society,* 3rd ed. (New York: John Wiley & Sons).

Smith, A., B. Locke, and W. Walker (1968) "Authoritarianism in police college students and non-police college students," *Journal of Criminal Law, Criminology & Police Science* 50:440–443.

Sparger, J. and D. Giacopassi (1983) "Copping out: Why police leave the force," in R. Bennett (ed.) *Police at work: Policy issues and analysis.* (Beverly Hills: Sage):107–124.

Stokes, L. and J. Scott (1996) "Affirmative action and selected minority groups in law enforcement," *Journal of Criminal Justice* 24(1):29–38.

Storch, J. and R. Panzarella (1996) "Police stress: State-trait anxiety in relation to occupational and personal stressors." *Journal of Criminal Justice* 24(2):99–107.

Sullivan, P. (1989) "Minority officers: Current issues," in R. Dunham and G. Alpert (eds.) *Critical issues in policing: Contemporary readings.* (Prospect Heights, IL: Waveland):331–345.

Swanson, C. (1977) "An uneasy look at college education and the police organization," *Journal of Criminal Justice* 5(4):311–320.

Thurman, Q., P. Bogen, and A. Giacomazzi (1993) "Program monitoring and community policing: A process evaluation of community policing in Spokane, Washington," *American Journal of Police* 12(3):89–114.

Townsey, R. (1982) "Female patrol officers: A review of the physical capability issue," in B. Price and N. Skoloff (eds.) *The criminal justice system and women.* (New York: Clark Boardman Company):413–426.

Travis, L. and R. Vukovich (1990) "Cynicism and job satisfaction in policing: Muddying the waters," *American Journal of Criminal Justice* 15(1):90–104.

Trojanowicz, R. and T. Nicholson (1976) "A comparison of behavioral styles of college graduate police officers v. non-college-going police officers," *Police Chief* 43(8).

Van Maanen, J. (1973) "Observations on the making of policemen," *Human Organization* 32(Winter):407–418.

Violanti, J. and F. Aron (1995) "Police stressors: Variations in perception among police personnel," *Journal of Criminal Justice* 23(3):287–294.

Westley, W. (1970) *Violence and the police.* (Cambridge, MA: MIT Press).

White, S. (1972) "A perspective on police professionalization," *Law & Society Review* 7(1):61–85.

Wilson, J. (1968) *Varieties of police behavior.* (Cambridge, MA: Harvard University Press).

Witte, R. (1969) "The dumb cop," *Police Chief* 36(1):38.

Worden, R. (1989) "Situational and attitudinal explanations of police behavior: A theoretical reappraisal and empirical assessment," *Law & Society Review* 23(4):667–711.

Worden, R. and S. Brandl (1990) "Protocol analysis of police decision-making: Toward a theory of police behavior," *American Journal of Criminal Justice* 14(2):297–318.

chapter **11**

POLICE AND COMMUNITY

CHAPTER OUTLINE

With the advent of what Klockars (1985) characterizes as full-time, paid vocational policing, the relation between police and community becomes a paramount concern. This should surprise no one, as the transition from avocational policing to vocational policing also signaled a shift from the community doing its own policing to hiring its police work. When a community pays select people for police services, it becomes a consumer. As a consumer, it has certain demands. When the service being purchased is community social control by community-sanctioned force, the relationship between consumer and service becomes intriguing, serious, and extremely volatile.

Although it should come as no surprise that the relation between the community and its paid police is of paramount concern, we should wonder why, after all these years, we haven't been able to specify the relationship. In this chapter we explore issues that limit our ability to do so. First, we explore definitions of community. It will become apparent that one of the most difficult aspects of this specification problem is our inability to consistently define what we mean by community. *If we lack the ability to define community, we will obviously lack the ability to specify its relationship to policing.*

The second issue addressed in this chapter is the means of social control. We will explore the theories of Ross (1926) and Black (1976) to gain insight into the relationship between community structure and means of social control. These theories suggest that the means a community uses to control itself vary from one community to the next. This variation has obvious implications for the work of the police, as indeed police are an instrument that the community uses to sustain social control.

The chapter concludes by balancing community social-control needs and police roles. Our focus is on community variety and on the implications of this variety for policing. We conclude that community variation results in differing social-control needs and that these differences demand different forms of paid vocational police.

COMMUNITY: WHAT IS IT?

Most discussions of the relationship between police and community assume that *community* has been defined or that its definition is inconsequential (see Greene and Taylor, 1988, p. 205). This assumption has profound consequences. If we assume that all communities are the same, or that manifest differences have no implications for policing, we are positioned to develop a model of policing that is universally applicable. It becomes possible to suggest that all communities should have foot patrol, or mini-stations, or that policing should be centralized. However, if we allow that community differences have implications for policing, then we might also consider that foot patrol is appropriate for some locales but not others, that some places are well served by mini-stations while others are not, and that only some places are suitable for centralized police.

This chapter assumes that community variety has implications for policing, and so we begin by examining different conceptualizations of community. We will discover that there are many different conceptions of community, and these have implications for the nature of policing.

Geography or Interest

Many theorists have attempted to define *community*. In 1955 Hillery, using what amounts to a "snowball" sampling technique[1] surveyed the social science literature for definitions of community. He isolated 94 definitions and by content analysis isolated areas of agreement. Hillery concluded that the important elements in the definition of community are "area, common ties, and social interaction" (1955:118). He also noted that social interaction and common ties are more important elements than area. Finally, he observed that area, as a conceptual element of the definition, tends to be more important when rural communities are being defined than when community more generally is being defined.

Trojanowicz and Moore (1988) also undertook the task of reviewing the evolution of definitions of community. They suggest two principal conceptions: community of interest and geographic community. *Geographic communities* are defined spatially. That is, a geographic community is an area in which people or members interact. *Communities of interest,* by contrast, are more focused on common interests. Examples of communities of interest would include the professional community, the African-American community, the police community, the student community, and so on. Each of these communities of interest shares a set of group interests that transcends space and shapes the nature of their interaction.

Trojanowicz and Moore note that earlier in human history, geographic communities and communities of interest overlapped. Before technological changes in mass transportation, mass communication, and mass media, the capacity to interact was constrained by proximity. Simply put, it was not possible to interact with people unless they were your neighbors. The inability of people to interact "out of their area" had the additional effect of causing people in the area to have a shared set of interests. Thus, geographic communities also tended to be communities of interests.

Trojanowicz and Moore (1988) further observe that with technological changes—particularly in transportation, communication, and media—came the ascendency of communities of interest over geographic communities. It was now possible for people to interact without regard for geography and so the interests that structured interaction became the principal feature of community. This conclusion has much in common with Hillery's, which predated it by some 30 years. Hillery also noted that interaction and common ties were more important to the definition of community than was area.

What are the implications of this discussion for policing? First, if we assume that communities of interest are the focus of policing, we ignore the

---

[1]Hillery's sample of definitions was developed by tracing references to definitions of community from two primary sources. He also examined the references cited in books with "promising titles" shelved adjacent to sources traced from the primary sources (1955:112).

Police officer interacting with a member of the business community, a community of interest. (Jerry Berndt/Stock Boston)

empirical fact that policing occurs in space. That is, police officers perform their tasks in areas. Although there are a host of patrol assignment methods, it remains the case that police officers are assigned to beats, or districts (areas), and it is within those beats, or districts, that they ply their trade.[2] If we are to be concerned with police and community, then it is important that we recognize how communities are defined geographically for the purposes of doing police work.

Moreover, these spaces are complex. It may be true that in the "good old days," spaces and interests tended to be congruent, but it is clear now that this is not necessarily true. For example, the same space can be both residential, commercial, and a major thoroughfare. This means that the same space will have different interest constituents (people who live there, people who own businesses there, people who work for the people who own businesses there, and people who simply travel through the space to get to somewhere else). The question now becomes whose interests are to be served by an activity (policing) that occurs in a space where there are competing interests?

Finally, the interest-versus-geography discussion affords us the opportunity to explore policy making. Policy regarding police operations is a prod-

---

[2]It is rare, though not unheard of, that police will be assigned to interest groups. For example, most large police departments have a juvenile unit and some have crisis-intervention units, but for the most part (even in those departments that have specialty units), the task of policing is assigned to space with the stipulation that interests be served.

uct of goal-setting exercises. Roberg (1979), referring to Cyert and March (1973), concludes that goal setting is primarily a political process in which the most powerful interest groups (or coalitions) prevail. If this is true, then the policing of places is done in accordance with policy developed to satisfy the community of dominant interests. Thus, it is quite conceivable that people who travel through an area, if they represent an interest group powerful enough, could cause policing policy to be established for the area that favors travelers. The travelers might be satisfied as long as traffic moves, whereas the residents might demand that police resources be devoted to making the area safe for pedestrians or to promote retail business within the area. Clearly, this discussion implies tension among interest groups within an area, and higher tension as the number of interest groups increases.

From the juxtaposition of interests and space, we see several issues emerge. It is apparent that communities can be defined either way. It is also apparent that the definition has implications for policing (policing is done in space; police policy is set by interests). And it is apparent that there is more tension in an area if the area must serve multiple interests.

A Structural Definition of Community

The foregoing discussion provides insight into traditional definitions of community and their theoretical implications. There is also a structural definition of *community* that denotes it as "the arrangement of groups, organizations, and larger systems that provide locality-relevant functions" (Duffee, 1990:149).[3] As implied in the definition, community is highly variable; that is, there are many different arrangements that can provide "locality-relevant functions." The thesis also provides us with a vehicle for exploring this variation and teasing out implications for policing.

Warren (1978) identifies five functions that have locality relevance: production-distribution-consumption, socialization, social control, social participation, and mutual support. Briefly, the production-distribution-consumption function is the process by which goods and services are produced, distributed, and consumed. The socialization function is the process by which community members learn the rules, norms, and values of the community. The social-control function is the process by which the rules of the community are enforced. The social-participation function is the process by which community members develop a sense of identity with the community. Finally, the mutual-support function is concerned with the giving of aid to community members during a time of crisis. Communities are those structures that see to the provision of these functions.

Duffee (1990) has expanded Warren's work by developing a typology of community structures (arrangements of groups, organizations, and larger

---

[3]Duffee's discussion was developed from R. Warren (1963).

systems) that endeavor to deliver the five functions. The typology emerges from the interaction of what he labels vertical relations and horizontal articulation. *Vertical relations* refer to whether the locality-relevant functions are provided to the locality from local resources and efforts or from the resources and efforts of the larger social system. If the locality is dependent on the larger social system (state or federal government, for example) for provision of relevant functions, it scores high on vertical relations. However, if the locality relies on its own tax and economic bases to meet its needs, it scores low on vertical relations. Essentially, vertical relations are concerned with local autonomy. Localities that are relatively independent of the larger society are autonomous and therefore have a low vertical relations score.

Horizontal Articulation

Horizontal articulation refers to the locality's ability to reach a consensus about what is proper or needed in a situation. Consensual communities score high on horizontal articulation, whereas communities that are less able to form consensus score low. Two separate processes determine the degree of horizontal articulation. If a locality is independent of the larger society (i.e., it is autonomous, or has a low vertical relations score), the degree of cultural homogeneity[4] determines a community's ability to reach consensus. In autonomous communities, horizontal articulation is high if cultural homogeneity is high. That is, consensus is a product of shared values and norms. In autonomous communities that are culturally heterogeneous, horizontal articulation will be low. In these independent localities, people will have diverse values and norms and will find it difficult to agree or form a consensus.

Localities that are dependent on the larger society for provision of locality-relevant functions achieve consensus (or not) by another means—coordination through formal mechanisms (e.g., coordinating councils). In many communities United Way serves as a coordinating body through which private resources from "outside" the locality are integrated to provide locality-relevant services.

Dependent localities that organize the provision of locality-relevant functions from the larger society achieve the effect of consensus through formal coordination mechanisms and score high on horizontal articulation. Dependent localities that do not coordinate the receipt and use of resources from the larger society score low on horizontal articulation. These communities receive input from many different sources, but there is no screen

---

[4]*Cultural homogeneity* refers to the degree to which a community has a single culture. A community is culturally *homogeneous* if all members of the community share the same culture. A community is culturally *heterogeneous* if members of the community come from different cultures.

Policing a homogeneous community. (James L. Shaffer/PhotoEdit)

through which these resources must pass to ensure that they support a single view of things.

Figure 11.1 summarizes the foregoing discussion. In Cell I are localities called *fragmented* communities. These are places that are dependent on the larger society for provision of locality-relevant functions but that lack the capacity to coordinate those external sources. In Cell II are localities called *interdependent* communities. These communities are similarly dependent on the larger society, but they have mechanisms to coordinate these external resources to provide locality-relevant functions. In Cell III are *disorganized* communities, which are independent and heterogeneous communities. Finally, in Cell IV are *solidary* communities, which are autonomous, homogeneous, consensual communities.

Duffee's (1990) model extends the geography-versus-interests discussion in a number of ways. It explicitly recognizes space as an important element in the definition of community. Communities are places that are the focus of locality-relevant functions. The real power of the model, however, lies in what it can suggest to us regarding the locus and nature of interests. In this regard we will explore two questions: what does the location on the vertical dimension tell us about whose interests will be served, and what

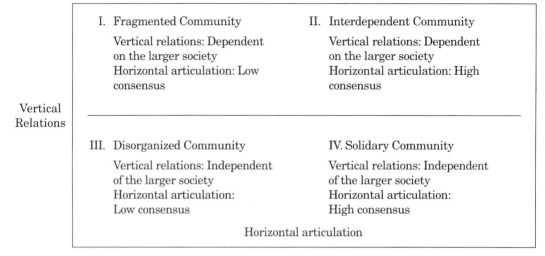

FIGURE 11.1 A Structural Taxonomy of Community. *Source:* Adapted from D. Duffee (1990) *Explaining Criminal Justice.* Prospect Heights, IL: Waveland, p. 155.

does the location on the horizontal dimension tell us about whose interests will be served?

Vertical Dimension. The *vertical dimension* of community refers to the degree to which a community is self-reliant. A self-reliant, autonomous community does not rely on the larger society for satisfaction of its local needs. It follows that the dominant interests in these disorganized and solidary communities would be local, as extralocal interests do not contribute to locality well-being. However, if the locality relies on the larger society for satisfaction of community needs, it follows that extralocal interests would predominate.

The locus of interests has considerable impact on local policing. As far back as Norman England, there was a distinction between the king's peace and local peace (Stennings, 1981). The king's peace, that of the larger society, took the form of law, whereas the local peace took the form of custom. Thus, police serving self-reliant communities (those independent of the king) were expected to enforce local custom in the manner dictated by local custom. By contrast, police serving dependent communities were expected to enforce the law in the manner dictated by the law.[5]

---

[5]It is essential that we recognize that no communities serve solely the law or local custom. Indeed, there are no modern communities that are completely independent or completely dependent on the larger society. The vertical dimension of community is a continuum that indicates more or less dependence rather than total dependence or independence.

Recent research by Burton et al. (1993) supports the idea that different police functions have their impetus from disparate locations. This study sought to determine the role of the police as established in state law. The researchers were surprised to find that in only one state was service mentioned in state law as a prescribed police role. They conclude that although the state defines law-enforcement and peacekeeping responsibilities for the police, it leaves specification of the public-service role to localities.

Horizontal Dimension. The horizontal dimension of communities also has an impact on local policing. Recall that horizontal articulation is the capacity of the entity to achieve consensus. In communities where consensus is likely, it is also likely that interests supported by that consensus will be more powerfully supported. Police in a solidary community, for example, would feel confident enforcing well-articulated local customs in customary manners. It is likely in solidary communities that police would intervene frequently to maintain the local peace. By contrast, police in disorganized communities would never be sure what the local custom was or how to enforce it. It is likely that police in disorganized communities would intervene only when absolutely necessary to maintain some ill-defined local peace.[6] Thus, communities high on horizontal articulation are likely to receive more enforcement than are communities low on horizontal articulation.

Variations in Community

The foregoing has demonstrated that there are a number of ways to define community because communities are complex and varied entities, and this variation has implications for policing. If, as seems reasonable, we accept that police policy is set by dominant interests (or coalitions) and plied on localities, then knowledge of interest groups is consequential (for elaboration, see Crank and Langworthy, 1992; Crank, 1994; Crank and Langworthy, 1996). Duffee (1990) suggests that the capacity to form consensus and autonomy are important community dimensions because they tell us who will be setting policy and how strongly the policy will be enacted.

If the community is dependent on the larger society but lacks the capacity to establish a consensus (i.e., it is fragmented), we would expect to see an emphasis on law enforcement with lesser regard for local custom. If the community is dependent on the larger society but has the capacity to develop a consensus (i.e., it is interdependent), we would expect to see an emphasis placed on the enforcement of both law and local custom. If the

---

[6]The distinctions between solidary and disorganized communities and associated police styles has much in common with J. Q. Wilson's (1968) service and watchman styles.

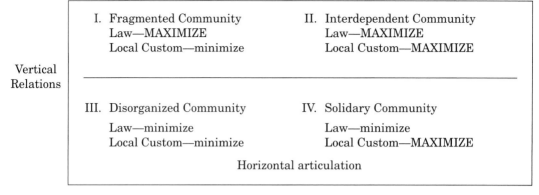

I. Fragmented Community
Law—MAXIMIZE
Local Custom—minimize

II. Interdependent Community
Law—MAXIMIZE
Local Custom—MAXIMIZE

Vertical Relations

III. Disorganized Community
Law—minimize
Local Custom—minimize

IV. Solidary Community
Law—minimize
Local Custom—MAXIMIZE

Horizontal articulation

FIGURE 11.2 Community Structure and Enforcement Emphasis

community is autonomous and heterogeneous (disorganized), we would expect to see a lesser concern with either law or local custom. Finally, if the community is autonomous and homogeneous (solidary), we would expect to see minimal concern for law enforcement but an emphasis on enforcement of local customs. Figure 11.2 summarizes these relations.

MEANS OF COMMUNITY SOCIAL CONTROL

We have established that there are varied definitions of communities, and that communities themselves are varied. We have demonstrated that variation in two dimensions has implications for enforcement efforts of the police. This section addresses variation in one locality-relevant function—social control. We focus on social control for two reasons. First, it is apparent that community social control is the locality-relevant function that permits the community to provide other locality-relevant functions (production-distribution-consumption, socialization, participation, and mutual support). Second, this text is concerned with police, whose role is defined by the capacity to employ nonnegotiable force to ensure conformity (Bittner, 1970; Klockars, 1985). Though many contend that police play a role in other locality-relevant functions, it can also be argued that the role they play is through forceful social control. Within this context, our discussion of community social control will explore theories of means of social control as they relate to community structure. We then relate this discussion to our structural theory of community, and develop implications for policing communities.

What Is Social Control?

Warren (1978:10) defines *social control* as "the process through which a group influences the behavior of its members toward conformity with its norms." Elaborating on what Warren means by "process" will help us place the role of police in social control in context. The focus on process is a concern with the means employed to cause conformity. A group, community, or society has innumerable means of encouraging conformity (Travis, 1995). It may endeavor to instill a set of values so that individuals will be self-controlled; it may promise rewards for conformity; it may threaten punishment if norms are violated; or it may simply force conformity. The fact that means are varied suggests that there are different agents of social control. Different individuals or groups may be assigned the tasks of instilling, rewarding, punishing, or enforcing behaviors.

The "process" of social control appears to be a multifaceted one designed to cause group members to conform their behavior to group expectations. Among the many groups and institutions that are assigned the tasks of social control, Warren lists local government, family, neighborhood, peer groups, church, school, television, and media as examples. Each of these agents of social control has the capacity to reward, punish, and enforce conformity and, through continuous social interaction, can reaffirm group norms.

Theories of Means of Social Control

From the foregoing discussion it is apparent that social control can take many forms. It follows, then, that the type and appropriateness of forms may vary. The following discussion will focus on two theories of means of social control, E. A. Ross's and D. Black's.

Ross's Thesis.   E. A. Ross, writing at the turn of the century (1926), was among the first sociologists to address the concept of social control. His book, *Social Control,* addressed not only the concept but the various means of achieving social control and the systems for maintaining it.

Ross identifies the following means of social control: public opinion, social suggestion, social religion, personal ideals, art, social valuations, law, education, belief, ceremony, and illusion. Having listed the means by which groups assure conformity to norms, Ross (1926:411) classifies each as either ethical or political. He believed that ethical means of social control gain their strength from "primal moral feelings" that shape individual conformity more from sentiment than from utility. Political means of social control tended to be those that are "deliberately chosen in order to reach certain ends" and are "likely to come under the control of the organized few." Ross's distinction between ethical and political means of social control thus turns on whether the means receives its controlling strength from individual acceptance of moral teachings (ethical) or is imposed by corporate actors to achieve corporate ends.

TABLE **11.1** Ross's Classification of Means of Social Control

ETHICAL MEANS	POLITICAL MEANS
Public opinion	Law
Social suggestion	Belief
Personal ideals	Ceremony
Social religion	Education
Art	Illusion
Social valuation	

Source: Adapted from E. A. Ross (1926) *Social Control: A Survey of the Foundations of Order.* New York: Macmillan, p. 411.

Table 11.1 summarizes Ross's classification of means of social control. Using this classification, Ross developed a theory of systems of social control. He saw the prominence of either political or ethical means as a function of the structure of the society. He suggests that "a political instrument operating through prejudice and fear" (1926:412) will predominate if

1. Elements of the population clash with one another.
2. Other means of social control too greatly restrict the will and welfare of individuals.
3. Underlying rules stereotype status differences.
4. Economic and opportunity inequalities are great and cumulative.
5. Parasitic relationships are maintained among races, classes, or sexes.

By contrast, Ross suggests, "ethical instruments, being more mild, enlightening, and suasive" (1926:412) will predominate if

1. The population is racially homogeneous.
2. The culture is uniform and diffused.
3. Social contacts within the society are frequent and amicable.
4. Requirements of individuals are few.
5. Social rules do not consecrate status differences or further parasitic relations.

For Ross it appears that the prominence of either the political or ethical systems of social control rests, first, on factors we could summarize as population heterogeneity and inequality and, second, on the degree to which individual options are restricted. Political control emerges where the society is heterogeneous, where status inequalities are great and "consecrated," and where the individual is overburdened by less formal, ethical controls. Ethical controls, on the other hand, emerge when the society is homogeneous, inequality is not great, and the ethical controls are not too restrictive of individual will.

Black's Thesis. Black (1976), writing 50 years later, echoes many of Ross's themes. He, too, suggests that the means of social control can be predicted from the nature and structure of the community. Black's focus was on governmental social control or law. Specifically, Black sought to define a set of propositions that would predict variation in the quantity of law in situations, among individuals, and across societies. Our focus here is on communities, so we restrict our discussions to Black's five society-level propositions.

First, Black contends that the quantity of law varies inversely with other types of social control (1976:107). This inverse function suggests that societies will rely more on law—governmental social control—if other forms of social control (most notably what Ross described as ethical means) fail to ensure order.[7]

Black also believed that the reliance on governmental social control varies directly with social stratification (1976:13). He defines *stratification* as inequality and, like Ross, suggests that as stratification increases, so will the reliance on law rather than on other social control. He saw that the relation between differentiation (population heterogeneity) and reliance on law as curvilinear (1976:39). In this case, Black's claim was that reliance on governmental social control increases with differentiation to a point where the differentiation creates interdependence (conceptually similar to symbiosis). At the point of interdependence, milder, enlightening, and more suasive means of social control begin to predominate.

Black's final two society-level propositions deal with the relationship between the quantity of law, the quantity of culture, and the quantity of organization. Black suggests that the reliance on law for social control varies directly with both the quantity of culture (1976:63) and the quantity of organization (1976:86). His argument regarding culture is that societies with great amounts of culture are more complex than other societies and must therefore rely more on governmental social control than on other social control. His argument regarding organization concerns the capacity for collective action. Black's postulate is that reliance on law—governmental social control—increases as a society's capacity for collective action increases.

Similarity of Ross's and Black's Theses. Black's thesis is quite similar to Ross's. Both view social structure as predicting the style of

---

[7]Black and Ross may not be in agreement on this point. Black argues the ascendancy of law when other social control fails. Ross argues the ascendancy of political means of social control (law) when ethical means are onerous.

social control. Both suggest that inequality and/or heterogeneity are directly related to formal means of social control (Ross's being political and Black's being governmental). Finally, both contend that their theories are simultaneously static and dynamic. That is, both Ross and Black contend that their theories predict the style of social control present in communities (static) and changes in styles of social control associated with changes in community structures (dynamic).

SOURCES OF POLICE AUTHORITY

Many police theorists have described styles of policing. For example, Wilson (1968) describes styles of departmental behavior. Others have described the working styles of individual police officers (e.g., Muir, 1977; Broderick, 1987; White, 1972). Although the following classification scheme has much in common with several of these, it is intended to be compatible with the preceding social-control and community-structure discussions. This scheme is based on sources of authority and, as we shall see, has implications for the nature of enforcement. Each of the following types may describe individual police officers as well as departments.

Officious law enforcers gain their authority from the larger society. The means of social control is the more formal "political," or "governmental," means rather than the less formal "ethical," or "other." Officious law-enforcement officers or departments use force (or threaten the use of force) to administer the law of the larger society. Officers and departments of this type focus on the rules of the larger society and assume responsibility for a very limited range of social-control activities—governmental social control through the use of force.

Personal authority as a distinguishing characteristic of police was first noted by Miller (1991:75). Miller observes that one of the features that differentiated early London police from New York police was the locus of their authority. The New York officers relied on personal authority (personal power) to do their work whereas the London officers relied on impersonal authority conveyed by the larger society. Police officers who rely on their personal authority use force to compel conformity to the officer's personal norms and values. The means of social control employed in these settings tends to be the less formal "ethical," or "other," rather than "political," or "governmental." Officers and departments of this type accord individuals wide latitude and are primarily concerned with "keeping the beat quiet."

Community-authority types of police and police departments gain their authority from the locality. In these settings officers use force to enact the will of the locality. Once again, police social-control efforts will be of the less formal type but as dictated by well-articulated community expectations. Officers and departments of this type focus on providing the services to the community that the community expects.

COMMUNITY, SOCIAL CONTROL, AND THE POLICE

In this section we integrate our discussions of community, social control, and styles of policing. First we address the relations among community, social control, and police styles. The focus is on the static view of the relationship among those three elements. Second, we address the dynamic view, focusing on changes in community and the implications of those changes for means of social control and for police style.

The Static View

Thus far in this chapter we have explored three sets of issues that shape the nature of policing in communities. Clearly, considerable variation exists in each of these elements. This discussion relates community types to social-control needs and police styles by addressing each type of community independently. It is important to remember that in these discussions we have been dealing with polar types, and that, in fact, communities will rarely if ever match the pure types portrayed here. Indeed, communities are more or less heterogeneous, more or less unequal, more or less autonomous, and more or less cohesive, never completely one thing or the other.

Disorganized Communities. Duffee (1990) describes disorganized communities as those that score low on vertical relations and horizontal articulation. Taken to the extreme, it is unlikely that a disorganized community would satisfy our definition of community, because extremely disorganized places could not minimally provide the required locality-relevant functions. In large measure, this failure would be a byproduct of the inability to provide sufficient social control.

To the extent that there is social control in a disorganized place, it is likely that it will be mostly of local origin. This is so because, by definition, disorganized places are autonomous,[8] which suggests that the larger society will not be imposing its will. The problem is that the locality does not have its own will to impose. Because the place is heterogeneous, the more suasive means of social control (other, ethical) will not have the capacity to ensure order. Because the place is independent of the larger society, there will be only minimal reliance on governmental social control. The lack of links to the larger society and the inability to form consensus result in a normative void. The direct consequence is a lesser total quantity of social control.

What style of policing are we likely to find in disorganized places? The lack of local social-control mechanisms, coupled with the lack of extralocal

---

[8]Even the word *autonomous* is a misnomer. Autonomy suggests a capacity to be independent, but disorganized places lack both a local capacity and the extralocal capacity.

social-control mechanisms, suggests that police will fill the normative void with their own definitions of order. It seems likely that police in disorganized places will rely on their personal authority to enforce their beliefs about what's "right and wrong." It is also likely that the frequency of police intervention will be low, focusing on serious disorder and major crimes. This style has much in common with Wilson's (1968) watchman style of policing; both rely on the individual officer to "keep the beat quiet" and not stir things up.

Interdependent Communities.   Interdependent communities are the direct opposite of the disorganized communities. They score high on both the vertical-relations and horizontal-articulation dimensions of community structure. These communities are dependent on the larger society for the provision of locality-relevant functions, but they are not at the mercy of the larger society. These communities can put their own "spin" on functions provided by the larger society because they have the capacity to form consensus and exert collective influence. However, the level of integration required to sustain such communities (local, extralocal, and between the locality and the larger society) is substantial. Indeed, Duffee suggests that this form of community is rare and perhaps unstable (1990:157).

The first feature of social control apparent in an interdependent community would be its quantity—there would be lots of it. The second feature would be the mix of governmental and other social control. In an interdependent community, the strong interests of the larger society would be served by an emphasis on law enforcement. But extralocal interests are not the only ones to be served in these communities. Strong local interests must also be served, resulting in an emphasis on local custom. Thus, social control in interdependent communities must simultaneously emphasize enforcement of law and local custom.

What kind of policing are we likely to find in interdependent communities (assuming we can find an interdependent community)? The dual emphasis on law enforcement and local-custom enforcement suggests that policing would take a mixed form that would simultaneously ensure that local and extralocal interests are served. Police in interdependent communities would simultaneously be "officious law enforcers" and rely on "community authority." While it is difficult to find examples of this kind of community receiving this mix of police service in the United States, authors describing policing in other countries have described this mix. Both Banton (1964) and Bayley (1991), describing policing in Scotland and Japan respectively, depict centralized police departments serving the interests of the larger society in a manner consistent with local custom.

Fragmented Communities.   Fragmented communities score high on vertical relations but low on horizontal articulation. These communities

Policing a disorganized community. (Ricky Flores/Impact Visuals)

are dependent on the larger society for the provision of locality-relevant functions, but, unlike interdependent communities, they lack the capacity to act consensually. The result is that fragmented communities receive functions and services from the larger society in a disjointed and uncoordinated manner.

Probably there would be less total social control in a fragmented community than in an interdependent community but considerably more than in a disorganized one. The quantity of social control would not be the only difference; it is also likely that the nature of social control would be different. Because fragmented communities rely on the larger society for satisfaction of their local needs, and lack the local capacity for consensus, the strong interests of the larger society will predominate. In this setting, governmental social control is likely to predominate in service of the law. Other, or ethical, forms of social control, serving local interests through enforcement of local customs, will not significantly shape social-control methodologies in fragmented communities.

Fragmented communities are likely to have officious law enforcers. This is so because the strong interests are extralocal, and the local concerns are subordinate. Police in fragmented communities will focus their enforcement energy on serving the interests of the larger society through enforcement of the law. In this setting, we would expect to see dispassionate law enforcers who go about their job impersonally, without reliance on local authority. The policing style associated with fragmented communities has much in common with Wilson's (1968) legalistic style.

Solidary Communities. Solidary communities are those we might see depicted in a Norman Rockwell illustration. These communities score low on vertical relations and high on horizontal articulation. They are self-contained, homogeneous communities that are independent of the larger society and normatively homogeneous. Since a solidary community is self-reliant and consensual, it will see to the provision of its own services in a manner it feels is appropriate.

The quantity of social control in a solidary community would be less than that in an interdependent community and more than in a disorganized one. It is also likely that solidary communities will have a larger quantity of social control than fragmented communities, because the former has the ability to rely on ethical means of social control.[9] Because solidary communities are self-reliant and consensual, local interests will predominate. Enforcement efforts will focus on sustaining local customs, with minimal regard for law. In this setting, ethical, or other, means of social control will predominate rather than political, or governmental, means.

Policing in solidary communities will probably be based on community authority. This is so because the strong interests are local, with extralocal concerns subordinate to them. Police in solidary communities will focus their energies on serving the interests of the locality through enforcement of local customs. In this setting, we should expect to see public servants tailoring their responses to situations to fit community expectations. This policing style has much in common with Wilson's (1968) service style.

Figure 11.3 summarizes the relations between community structure, interests served, predominant means of social control, and police styles.

The Dynamic View

Our concern here shifts from what we would expect to see if we looked at communities at a particular point in time to an examination of trends that affect communities. We will focus our attention on what Warren (1978) characterized as the "great change" and draw out the implications of this change for dominant means of social control and styles of policing.

---

[9]We should recall that Ross suggests that ethical means of social control could become so burdensome as to encourage a shift to presumably less burdensome political means of social control.

I. Fragmented Community	II. Interdependent Community
Interests: Larger society	Interests: Larger society and locality
Primary Means of Social Control:	Primary Means of Social Control:
Political-Governmental	Both Political-Governmental
	and Ethical-Other
Source of Police Authority:	Source of Police Authority:
Larger society	Both larger society and the
	community

Vertical Relations

III. Disorganized Community	IV. Solidary Community
Interests: None, idiosyncratic	Interests: Locality
Primary Means of Social Control:	Primary Means of Social Control:
Little social control—Political-	Ethical-Other
governmental	Source of Police Authority:
Source of Police Authority: Personal	Community

Horizontal articulation

FIGURE 11.3 Community Structure and Interests Served, Means of Social Control, and Police Authority

The "Great Change."   Warren (1978) was interested in defining the "great change" in order to develop a context in which to understand community change. He characterizes the *great change* as "a series of changes that have been taking place over a period of decades and even centuries" (1978:53). Following are the elements that together constitute the great change:

1. Division of labor
2. Differentiation of interests and associations
3. Increasing systemic relationships to the larger society
4. Bureaucratization and impersonalization
5. Transfer of functions to profit enterprises and government
6. Urbanization and suburbanization
7. Changing values

What are the consequences of these cressive changes[10] for community structure? Warren claims that the great change results in "increasing

---

[10]*Cressive changes* are unplanned changes to which systems must adapt. The "great change" is constituted of changes in the larger context to which systems must adapt.

orientation of local community units toward extracommunity systems of which they are a part, with a corresponding decline in community cohesion and autonomy" (1978:52). That is, division of labor, differentiation of interests and associations, bureaucratization and impersonalization, urbanization and suburbanization, and changing values contribute to the balkanization of places so that community cohesion is no longer likely. The loss of community cohesion diminishes the capacity to form consensus, resulting in a decline in horizontal articulation.

The great change also has implications for vertical relations. Increasing systemic relations to the larger society and the transference of functions to profit enterprises and government contribute to lost local autonomy and increased dependence on the larger society. Lost autonomy results in an increase in vertical relations.

Chart 11.1 summarizes the effects of the great change on community structure. The balkanization of localities results in diminished horizontal articulation. The increased dependence on the larger society results in increased vertical relations. Finally, the total effect of these cressive changes is a shift from essentially solidary communities to fragmented communities.

Implications for Social Control.   If the great change has implications for the structure of communities, it also has implications for the nature of social control in communities. Concerning social control, Warren (1978:61) observes a drift from primary-group controls (what Ross would characterize as "ethical" and Black as "other") toward secondary-group controls (political or governmental) associated with the great change.

Lost local autonomy, over time, results in a shift from the primacy of local interests to extralocal interests. This shift will probably also be associated with a shift in enforcement emphasis from local custom toward law. The decline in horizontal articulation exacerbates the drift toward law. Lo-

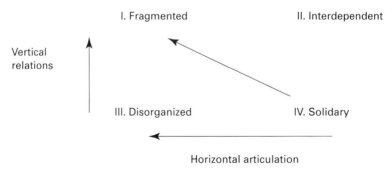

CHART 11.1 Effects of the "Great Change" on Community Structure

calities lacking the capacity to assert local interests find themselves less and less able to "insist" that governmental social control be constrained by local custom. The drift in means of social control associated with the great change is inexorably from an emphasis on local custom toward an emphasis on law.

It is also likely that the shift toward governmental social control is also a shift toward less total social control. As noted previously, in the contrast of solidary and fragmented communities, political means of social control are less efficient than ethical means. As communities drift from solidary toward fragmented communities, they lose the considerable moral suasive power of ethical means of social control and must rely instead on the less efficient force of political means.

Implications for Policing. Finally, the trends inherent in the great change also have implications for the nature of policing. Previously we have noted an association between the locus of police authority and the structure of communities. We noted that police in solidary communities perform as public servants enforcing local custom. By contrast, police in fragmented communities act as officious law enforcers enforcing law.

The great change occasions a drift along the fragmented community/solidary community axis toward fragmented communities. This suggests a shift in the locus of police authority from the locality to the larger society and a change in enforcement focus from local custom to law. Thus, Warren's (1978) great change is associated with a shift in policing from an emphasis on public servants enforcing local customs toward legalists enforcing the laws of the larger society.

CONCLUSION

We began this discussion with the observation that if communities vary, so might their social-control and policing needs. We also noted that the appropriateness of proposed police reforms might vary depending on the nature of the community, its social-control needs, and its expectations of police.

In the body of this chapter we demonstrated that communities vary, and that we have "theoretical" reason to believe social-control and policing needs do as well. We noted that some communities are well served using ethical means of social control whereas others must rely on political means.

We also noted that community variation is associated with sources of police authority. Police in fragmented communities get their authority from the larger society (law); police in solidary communities get their authority from the community (local custom); police in disorganized communities use personal authority; and police in interdependent communities gain authority from both the larger society and the community.

There is a growing body of research evidence that supports the notion that community characteristics are important correlates of police organization and activities. Kerley and Benson (2000) reviewed the impact of police involvement in community policing in two cities on improved community functioning. They conclude that such police efforts are likely to work better in more solidary or interdependent communities than in fragmented or disorganized communities. That is, police efforts to implement community policing programs are least likely to succeed in communities that have lower levels of horizontal relations. In terms of traditional law enforcement, Davenport (1999) found that community characteristics such as complexity (urbanism and population density) and turbulence (changes in population, urbanism, and density) are important factors limiting a police agency's ability to clear crimes. Allan Jiao (1998) has observed that police policy, or how the police organize and what functions they stress are linked to community desires. Other research has noted that public assessments of police effectiveness and general satisfaction with the police are at least partly dependent on neighborhood or community characteristics (Reisig and Parks, 2000; Priest and Carter, 1999).

Finally, we see that not only do communities differ from one to another, but also that they change over time. According to Warren's (1978) discussion of the "great change," there has been a drift from (1) solidary to fragmented communities, (2) enforcement of local custom to enforcement of law, and (3) police reliance on community to reliance on larger societal bases of authority.

Variation in community, social-control options, and differential bases of police authority suggest that police reforms will be differently received in communities. Police reformers, be they advocates of police professionalism, team policing, or community-oriented policing, tend to promote their particular reform without regard for community differences. It seems highly unlikely that any of these reforms will be universally applicable. Quite the contrary, the reforms appear differentially appropriate depending on the nature of the community.

CHAPTER CHECKUP

1. What is a *community?*
2. Communities can be described geographically or by interests. How are they different? Is it likely that they were ever compatible?

3. What is the *vertical* dimension of community?

4. What is the *horizontal* dimension of community?

5. What trends constitute the "great change?"

6. What is *social control?*

7. What effect does the great change have on community structure, social control needs, and styles of police behavior?

8. What are the sources of police authority, and how are they related to means of social control?

9. What community characteristics are associated with alternative means of social control?

REFERENCES

Banton, M. (1964) *The policeman in the community.* (New York: Basic Books).

Bayley, D. (1991) *Forces of order: Police behavior in Japan and the United States,* 2nd ed. (Berkeley, CA: University of California Press).

Bittner, E. (1970) *The functions of the police in modern society.* (Washington, DC: U.S. Government Printing Office).

Black, D. (1976) *The behavior of law.* (New York: Academic Press).

Broderick, J. (1987) *Police in a time of change.* (Prospect Heights, IL: Waveland).

Burton, V., J. Frank, R. Langworthy, and T. Barker (1993) "Research note: The prescribed roles of police in a free society: Analyzing state legal codes," *Justice Quarterly* 10(4):683–695.

Crank, J. (1994) "Watchman and community: Myth and institutionalization in policing," *Law and Society Review* 28(2):325–351.

Crank, J. and R. Langworthy (1992) "An institutional perspective of policing," *Journal of Criminal Law and Criminology* 83(2):901–926.

Crank, J. and R. Langworthy (1996) "Fragmented centralization and the organization of the police," *Police and Society* 6(2):213–229.

Cyert, R. and J. March (1973) *A behavioral theory of the firm.* (Englewood Cliffs, NJ: Prentice-Hall).

Davenport, D. (1999) "Environmental constraints and organizational outcomes: Modeling communities of municipal police departments," *Police Quarterly* 2(2):174–200.

Duffee, D. (1990) *Explaining criminal justice.* (Prospect Heights, IL: Waveland).

Greene, J. and R. Taylor (1988) "Community-based policing and foot patrol: Issues of theory and evaluation," in J. Greene and S. Mastrofski (eds.) *Community policing: Rhetoric or reality* (New York: Praeger): 195–223.

Hillery, G. (1955) "Definitions of community: Areas of agreement," *Rural Sociology* 20(2):111–123.

Jiao, A. (1998) "Matching police-community expectations: A method of determining policing models," *Journal of Criminal Justice* 26(4):291–306.

Kerley, K. and M. Benson (2000) "Does community-oriented policing help build stronger communities?" *Police Quarterly* 3(1):46–69.

Klockars, C. (1985) *The idea of police.* (Beverly Hills, CA: Sage).

Miller, W. (1991) "Cops and bobbies, 1830–1870," in C. Klockars and S. Mastrofski (eds.) *Thinking about police: Contemporary readings,* 2nd ed. (New York: McGraw-Hill):73–88.

Muir, W. (1977) *Streetcorner politicians.* (Chicago: University of Chicago Press).

Priest, T. and D. Carter (1999) "Evaluations of police performance in an African American sample," *Journal of Criminal Justice* 27(5):457–465.

Reisig, M. and R. Parks (2000) "Experience, quality of life, and neighborhood context: A hierarchical analysis of satisfaction with police," *Justice Quarterly* 17(3):607–630.

Roberg, R. (1979) *Police management and organizational behavior: A contingency approach.* (St. Paul, MN: West).

Ross, E. (1926) *Social control: A survey of the foundations of order.* (New York: Macmillan).

Stenning, P. (1981) *Legal status of the police.* (Ottawa, Canada: Law Reform Commission).

Travis, L. (1995) *Introduction to criminal justice.* (Cincinnati, OH: Anderson).

Trojanowicz, R. and M. Moore (1988) "The meaning of community in community policing," in *Community Policing Series No. 15* (East Lansing, MI: Michigan State University, National Neighborhood Foot Patrol Center).

Warren, R. (1978) *Community in America,* 3rd. ed. (Chicago: Rand McNally).

White, S. (1972) "A perspective on police professionalism," *Law and Society Review* 7(1):61–85.

Wilson, J. (1968) *Varieties of police behavior.* (Cambridge, MA: Harvard University Press).

PART FOUR

THE FUNCTIONS OF POLICING IN AMERICA

"To Serve and Protect" is a motto often found written on the sides of police cruisers. The motto summarizes the basic functions of the police in America but does not really tell us much about policing. For instance, whom do the police serve, how do they serve them, and when? Similar questions can be raised about the function of protection.

James Wilson (1968) identified the three basic functions of the police as law enforcement, service delivery, and order maintenance. Of the three, Wilson argued that the most important and most difficult function was that of order maintenance. Yet, when he analyzed citizen calls to the police, he discovered that the most common reason citizens call the police is to receive some sort of service, such as reporting traffic accidents.

The popular media and the police themselves portray their primary mission as that of law enforcement and crime control (Lundman, 1980). But study after study of police work has indicated that actual law-enforcement activities comprise a very small part of the police workload (Webster, 1970). More recently, Greene and Klockars (1991) report a study of police officer time in Wilmington, Delaware, where nearly half of officer time was devoted to crime-related matters. This figure, however, was obtained only after excluding time officers spent on administrative tasks and time when the officers were "clear" or "available." When all available officer time was used as the base, crime-related activity accounted for about one-quarter of officer time. Observations of police officers in three cities reveal similar results. Patrol officers in St. Petersburg, Florida, and Indianapolis were observed to spend relatively little time on crime-related activities, excluding general patrol (Parks et al., 1999). A similar study in Cincinnati revealed that patrol officers devoted less than 20 percent of their time to crime related activities (Frank, Brandl, and Watkins, 1997).

One result of this difference between what officers actually do and the emphasis on crime fighting by the police is role conflict. **Role conflict** occurs when our expectations about how a person or organization should behave do not match their behavior in reality. Thus, when we as citizens, or the police themselves, think that the primary purpose of the police is to catch crooks, we are disappointed and uneasy when we see that most police officers across the country make relatively few arrests and spend the majority of their on-duty time doing non-crime-related activities.

In Part IV we examine police work in reality, in terms of the three functions identified by Wilson (1968). We look at the types of services police provide to the community and the frequency with which such services are offered. Similarly, we examine the complicated nature of "order maintenance," and the police role in crime control and law enforcement. Correlates of these functions will be identified, and the relationships between law enforcement, service delivery, and order maintenance will be described. At the conclusion of this section we will have some appreciation of the variety of police work and of the factors that help explain what tasks are done, by whom, and how frequently.

chapter 12

LAW ENFORCEMENT
AND THE POLICE

CHAPTER OUTLINE

Although the police perform many functions in society, the prevailing public image of the police is that of crime fighters. Popular media, aided by the police themselves, have cultivated a definition of the police as law enforcers. Increasingly, police agencies are taking steps to consciously manage this public perception, including the assignment of personnel as "public information officers." These officers try to create a positive image of the agency in the media (Surette and Richard, 1995). The dominance of crime control in our understanding of the police role is partly a result of its actual importance, but it is also a product of the ambiguous nature of the police role. Of all the things police do, in many ways law enforcement is the easiest to accomplish and communicate about to the public.

THE POLICE AND CRIME

Mark Moore, Robert Trojanowicz, and George Kelling (1988:1) define the role of the police in America thus: "The core mission of the police is to control crime. No one disputes this." By this, of course, they mean that one goal of every police organization is crime control. No matter what else the police agency or officers might do, they must attend in some fashion to the problem of crime. Indeed, Peter Manning (1978) has observed that crime control is the impossible mandate of the police occupation.

The **occupational mandate** of the police is defined by the public and the police themselves as primarily a responsibility for crime control. They are viewed as the crime experts, who have both the right and the duty to define crime and devise responses to crime problems. Manning says that the mandate is impossible for a number of reasons, one of the most important being the practical constraint of individual rights in our society (1978:14–17). The police officer must balance the requirement to do something about crime with the requirements of due process. In the words of Jerome Skolnick (1994:9), the police must serve or achieve "order under law." In a democratic society such as the United States, the police must not only enforce the laws, they must obey them as well. In controlling crime this means that the police are not free to take whatever action might best reduce crime or reveal criminals. For example, random telephone taps might reveal a great deal of criminality to police, but in a democratic society the police are not generally allowed to invade the privacy of a citizen without showing some justification for their action. In the wake of the terrorist attacks of September 11, 2001, the United States is struggling to balance concerns for individual liberty with public safety interests. While Congress has expanded some police powers to conduct surveillance and federal agencies are more actively investigating terrorism, for the most part American police are constrained in their dealings with citizens.

Police efforts to control crime can be classed into two main categories: **proactive strategies and reactive strategies.** In *proactive strategies* the police take it upon themselves to discover crime and enforce the law. In *reactive strategies* the police respond (or react) to others' identifications of crime, as in answering citizen complaints. Further, crime control involves both an enforcement and prevention dimension. **Enforcement** is the application of the law, generally through arrest, on those who violate criminal

TABLE **12.1** A Typology of Police Crime-Control Efforts

Crime-Control Strategy	CRIME-CONTROL GOAL	
	Prevention	*Enforcement*
Proactive	I	II
Reactive	III	IV

statutes. **Prevention** covers attempts by the police to reduce the likelihood that crimes will occur. If prevention is successful, of course, there is no need for arrest.

The crime-control efforts of police in America can vary along these two distinct dimensions. They can emphasize either prevention of crime or enforcement of laws as their primary goal or function. They also can range along a continuum from proactive to reactive endeavors. Table 12.1 presents a theoretical typology of police crime-control efforts. We will return to this typology later, but first we must better define its terms.

POLICE CRIME-CONTROL STRATEGIES

Donald Black (1980:41–45) defines *law* as governmental social control and the police as a primary agency of the initiation, or mobilization, of that social control. His interest is in how governmental social control gets its caseload. For example, if someone steals your calculator, a variety of responses are available. You might confront the culprit; you and your friends might ostracize him, you might report the theft to parents, school officials, or other nonpolice authorities; or you might call the cops. Of course, you might also simply accept your fate as victim and resolve to be more careful with your calculator in the future. Alternatively, a passing police officer might observe the theft and make an arrest on her own initiative. The question Black faced was, given the variety of social-control options available, how do cases end up in the criminal system? He answers the question in this fashion (1980:43):

> A case can enter a legal system from two possible directions: A citizen may set the legal process in motion by bringing a complaint, or the state may initiate a complaint on its own authority, with no participation of a citizen complainant.

Black labeled the first direction, by citizen complaint, as *reactive* because the legal system (government) is reacting to the initiative of a citizen. The second direction, by government, is *proactive* because the government

TABLE **12.2** A Comparison of Means for the Mobilization of Law

REACTIVE MODEL	RELEVANT DIMENSION	PROACTIVE MODEL
Traditional street crimes	Legal intelligence	Regulatory and victimless crimes
Entrepreneurial, first-come, first-served model	Availability of law	Social welfare distributive model
Citizenry	Locus of discretion	Police agency

Source: Adapted from D. J. Black, (1973) "The mobilization of law." *Journal of Legal Studies* 2(1):125–49.

acts on its own. Black then compared the reactive and proactive methods of mobilizing the law on three dimensions: legal intelligence, availability of law, and location of discretion. Table 12.2 summarizes this discussion.

Legal intelligence refers to the knowledge a legal system has about law violations in its jurisdiction—what kinds of offenses are known to occur. Focusing on the police and criminal law (as distinct from civil law), the reactive method paints a picture of law violations that is dominated by the kinds of offenses about which citizens are most concerned and most likely to call the police. The reactive method of legal mobilization stresses street crimes. In contrast, a proactive method displays a relatively greater emphasis on regulatory and vice offenses. Both street crimes and the regulatory/vice violations occur in society. The difference is that the reactive model relies on complainant definitions of the crime problem and few speeders, patrons of gambling or prostitution, and violators of health or safety codes will report themselves to the police. In the proactive model, police officers detect and report regulatory and vice violations on their own initiative.

The *availability of law* is Black's (1980:52) phrase to describe citizen access to the legal system. Unlike the central question of a legal system's access to cases, this phrase refers to the means by which citizens can "receive" law. In the reactive case, an entrepreneurial model of availability exists. Law is there for anyone who wants it, on a first-come, first-served basis. If, as a citizen, you want law, you need only call the police. In the proactive situation, a social welfare, or distributive, model of law exists. All citizens, regardless of their personal desires, receive a share of law, which is more or less evenly distributed by the police. Thus, if the people in your neighborhood park their automobiles illegally, in the reactive situation you can call the police if you want parking law, while in the proactive case, the officer in the passing cruiser will distribute parking law (and parking citations) to the neighborhood.

The *location of discretion* is Black's (1980:56) term for describing the location of legal policy-making power. This power is the ability to decide when the law should be applied. Laws define behaviors and conditions that justify the intervention of the police, such as trespassing, burglary, theft, and the like. In practice, someone must decide if the facts of the current incident "fit" the law in such a way that the police are justified in intervening legally. This practical matching of the facts to the law is covered by Black under the term discretion. In the reactive situation, the citizen has the most discretionary power, principally because the citizen decides whether to even involve the police. In the proactive case, the majority of legal discretion rests with the police.

Black ends his comparison of reactive with proactive legal mobilization by linking the form of mobilization to the nature of the society in which it occurred. In short, he concludes that the more democratic a legal system, the more legal mobilization is likely to occur in a reactive fashion. The more restrained the power of government is, the more important the rights and interests of citizens and the more likely it is that citizens will determine which cases enter a legal system. In the United States, where we prize democratic values and individual rights, it is little surprise to note that most police crime-control actions lean toward the reactive model.

POLICE CRIME-CONTROL FUNCTIONS

The police are the gatekeepers of the criminal justice system in the United States. The **criminal justice system** is the formal social institution of our society that is charged with the prevention and control of crime (Travis, 1998). It is composed of the police, criminal courts, and criminal corrections. As a system, the criminal justice process consists of a series of interrelated decisions ranging from the detection of crime through discharge from custody. The police component of this system is responsible for the crime-control functions of detection, investigation, and arrest.

Detection

Detection is the discovery of crime. When police officers come to believe that a crime has probably occurred, that crime can be said to have been "detected," and the criminal justice process is initiated. The detection function illustrates the gatekeeper nature of the police role in criminal justice. If a citizen reports being the victim of a theft to the police, but the police do not, for whatever reason, believe the report, the justice process is not started. On the other hand, if the police do believe the complaint, the process has begun, because the police, as agents of the justice process, next decide what to do about the complaint.

Investigation

After detection, the police are responsible for **investigation,** the attempt to gather evidence about a possible crime and to identify a suspect. Investigation, as an information-gathering process, involves interviewing victims, witnesses, and suspects as well as collecting and analyzing hard evidence such as fingerprints, paint chips, documents, shell casings, and a host of other potential clues. The purpose of investigation is to confirm the fact of crime and to identify the lawbreaker(s).

Arrest

A successful investigation may lead to an arrest. The **arrest** is the taking of a criminal suspect into custody. Having established that a crime has occurred and that the suspect has, or probably has, committed the crime, the police place that person in custody, preventing his or her flight from prosecution. At this point, in terms of the criminal justice process, decision-making authority passes to judicial officers, who decide whether to charge the suspect, what charges to bring, whether to release the suspect on bail, and the like. In the overwhelming majority of criminal cases, it is the police who initiate criminal justice processing.

POLICE CRIME-CONTROL TACTICS

The primary crime-control tactic of American police is routine police patrol. Routine patrol evolved (or continued) from Sir Robert Peel's initial deployment of uniformed officers in London in 1829. Uniformed officers, whether in cars or on foot, are expected to patrol their assigned areas when not otherwise occupied. This patrol is designed to accomplish three main purposes: maintaining a police presence, enabling the police to respond quickly to emergencies, and detecting crime.

    The patrolling officers, by virtue of their uniforms and distinctively marked automobiles, maintain a police presence in the community. That is, people can see that the police are out and about. This presence is expected to both (1) deter potential offenders because the police are near and (2) reassure citizens that it is safe for them to go about their business.

    A second advantage to routine patrol is that the police officers are decentralized throughout the community. Thus, if there is a call for assistance, it is likely that a police officer will be near the location of the caller. The alternative of keeping all the officers at the station house would mean that any call for assistance would have to await the arrival of officers from the station. The quicker response of police on routine patrol is expected to enhance the ability of the police to "catch the crook in the act."

Finally, officers on routine patrol are expected to watch for signs of crime and intervene in suspicious circumstances. Thus, the patrol officer can detect crimes and prevent possible crimes through her or his own initiative as a result of being out in the community and observing the people on the patrol beat. In this regard, routine patrol has the potential for proactive policing.

A second major tactic of police in crime control is the use of specialists to investigate detected crimes. These specialists are usually called *detectives,* but it is actually the patrol officer who detects crimes, while the detective investigates them. On the report of a crime or probable crime by a patrol officer, the case is assigned to the detective unit of the agency (assuming the agency has one), and another officer conducts the followup investigation.

As Moore, Trojanowicz, and Kelling (1988:1) observe, "Professional crime-fighting now relies on three tactics: (1) motorized patrol; (2) rapid response to calls for service; and (3) retrospective investigation of crimes. . . . Although these tactics have scored their successes, they have been criticized within and outside policing for being reactive rather than proactive."

This criticism of traditional police crime-control tactics is based on the perceived failure of routine patrol and investigation to prevent crime and to attack the causes of crime. Rather than positioning themselves (through patrol) to respond to crime by quick reaction and followup investigations, recent critics of the police have suggested a reorientation toward prevention and preemptive efforts to control crime. In the past three decades police agencies across the country have tested ways to improve the ability of the police to control crime through changes in both patrol and investigation.

RESEARCH ON POLICE PATROL

For over a century the theory of preventive patrol by police that was popularized by Patrick Colquhoun and Robert Peel went untested. Police agencies routinely assigned the majority of their personnel to uniformed patrol duties and adopted technological advances (automobiles, telephone and radio communications, computer-aided dispatch, etc.) that would increase the responsiveness of patrol officers to calls for assistance.

Additionally, police agencies began to track crime reports and calls for service to assist in the scheduling of patrol officers and units. The size of a patrol beat and the numbers of cars assigned to a patrol district were linked to an analysis of the workload of the agency in that area. Thus, in a densely populated area that generated many calls for police service, more patrol units responsible for smaller areas might be assigned than in a less busy area. Similarly, more patrol officers might be assigned to the evening shift than to the day shift if evenings were peak times for the police.

Toward the end of the twentieth century, police patrol in most American jurisdictions, especially larger cities, involved two-officer teams in automobiles. These officers randomly cruised through an assigned area and answered calls for assistance that were dispatched to them over the police radio. Events during the 1960s—rising crime, civil unrest, deteriorating relations between the police and the community—set the stage for a questioning of traditional police practices.

Police and civic leaders were interested in finding ways of controlling the rising rates of crime. Additionally, there was a desire to improve police–community relations and encourage public cooperation with the police. Beginning in the 1950s, evidence on the limits of routine patrol began to accumulate, but it was the results of the Kansas City Preventive Patrol Experiment (Kelling et al., 1974) that spurred a re-evaluation of the importance of preventive routine patrol.

Patrol Effectiveness

In 1971 officers in the South Patrol Division Task Force of the Kansas City, Missouri, police department came to question the value of routine preventive patrol. In discussions with the Police Foundation, it was agreed that the patrol division would experiment with the effects of different levels of patrol. Fifteen patrol beats were classed into three categories of five beats each. The reactive beats were not routinely patrolled; police cars entered the area only in reaction to citizen calls for assistance. The proactive beats had two to three times the regular patrol by the addition of cars from the reactive beats. The control beats continued with the traditional level of routine patrol, one car per beat.

After a year of this experiment, researchers from the Police Foundation compared the three different levels of patrol in terms of public attitudes, levels of reported crime, levels of victimization, traffic accidents, response time, and similar factors. The researchers (Kelling, Pate, Dieckman and Brown, 1974:3–4) concluded,

> Given the large amount of data collected and the extremely diverse sources used, the overwhelming evidence is that decreasing or increasing routine preventive patrol within the range tested in this experiment had no effect on crime, citizen fear of crime, community attitudes toward the police on the delivery of police service, police response time, or traffic accidents.

As Klockars and Mastrofski (1991:131) more bluntly put it, "It makes about as much sense to have police patrol routinely in cars to fight crime as it does to have firemen patrol routinely in fire trucks to fight fire."

Earlier studies of police patrol impact in New York and England indicated that increasing from a situation of no routine patrol to one of some patrol had an effect on rates of crime and citizen perceptions, but given some patrol already in existence, significant changes in crime rates or attitudes required large increases in patrol officers. This finding gives credence to what is sometimes called the **mayonnaise theory of police patrol.** This term suggests that levels of police patrol are similar to levels of mayonnaise needed for making a sandwich. The fact is, while one may like a lot of mayonnaise on a sandwich, a little mayonnaise goes a long way. So, too, a little police patrol apparently goes a long way. Thus, starting patrol in an area that had no patrol affects rates of crime, but adding a little more patrol to an area that already has some coverage seems to make no difference in crime control.

One thing that emerged from the New York City studies of the impact of police patrol on crime was the phenomenon of **crime displacement.** Displacement is the effect of police patrol in one area of moving crime outside the patrol district (Green, 1995). Thus, if one starts or increases patrol in some section of the community, the number of crimes that occur in that section may decrease. However, the overall number of crimes in the community will remain unchanged because criminals simply shift their location out of the patrolled area. The effect of patrol, then, is not to reduce crime in general, but to displace it to another area.

A second finding of the New York studies on the impact of different levels of routine patrol is that police patrol seems to affect a relatively narrow range of public crimes. Crimes that typically occur on the streets and would be visible to patrolling officers, such as auto theft, muggings, vandalism, and the like are reduced when patrol resources are increased. More private crimes like rape, homicide, domestic violence, and other offenses that tend to occur inside homes and buildings away from public view are generally not affected by changes in levels of police patrol.

The impact of these studies of the effect of routine preventive patrol was to cause police leaders to reassess its use. If increases in levels of patrol were not likely to affect citizen feelings of satisfaction with police service or the amount of crime committed (in any significant degree), then patrol resources could be reallocated. Accordingly, the police in a number of U.S. jurisdictions began to experiment with different forms and structures of police patrol.

Patrol Staffing

Staffing of patrol units was one of the first issues to be tested. In 1975 the San Diego Police Department conducted a patrol-staffing experiment in which the basic question addressed was whether it made a difference if one or two officers were assigned to each police car. Kaplan (1978) summarizes the arguments by noting that one-officer patrol units are expected to reduce costs, increase patrol coverage, and reduce response time to calls for service, while two-officer units are expected to reduce the likelihood of officer injuries and increase the speed at which service calls are completed.

The San Diego study (Boydstun, Sherry, and Moelter, 1977) concluded that the one-officer patrol was a more efficient use of resources than the assignment of two officers to each car. Single officers in patrol cars were less likely to be assaulted, made more arrests, and wrote more crime reports than did patrols involving pairs of officers. Interestingly, one-officer patrols resulted in quicker response times, and when a backup car was assigned, both one-officer cars arrived at the scene more quickly than a single two-officer car. A replication in Kansas City showed the same results (Kessler, 1985). Kessler (1985:60) suggests that when one-officer cars are used, peer pressure on the officers to support each other when backup cars are dispatched results in faster response. Knowing that a fellow officer may need assistance motivates both to drive more rapidly and thus explains how two one-officer patrol cars can respond to calls more quickly than a single car.

To these findings that one-officer patrol meets the objectives of the police agency as well or better than two-officer patrols can be added the effect of this staffing on citizens. The data indicate that one-officer patrols result in more production (arrests, crime reports, etc.) and quicker response times, at less cost to the agency. Decker and Wagner (1982) assessed the effect of one-officer patrol on citizens as well as police. They found (1982:381) that single-officer patrols were less likely to result in injuries to the citizen or arrest of a citizen, and when a citizen was arrested, the arrest charge was less serious than when a two-officer patrol car responded. Thus, one-officer patrols in general appear to be more advantageous from the perspective of both the police and the community than do two-officer patrols.

A recent study of line-of-duty deaths among police officers by Lorie Fridell and Anthony Pate ("One-officer state police cars,", 1997) raised questions concerning officer safety. State police in one-officer cars had significantly higher chances of being killed. It may well be that officers for whom no backup is readily available, and/or officers who make frequent traffic stops, are at special risk. Like so much else in policing, it appears that patrol staffing and officer safety may be different under different conditions or in different agencies.

Response Time

One of the justifications for routine preventive patrol is that the officers will be better able to respond quickly to reports of crime. Police leaders have long thought this rapid response to calls would increase the chance that the patrol officers would catch the criminals at the scene. This belief in the value of rapid response was tested in Kansas City in 1977 (Van Kirk, 1978) and found not to be supported by the data. The research indicated that because citizens tend to delay reporting a crime to the police, a rapid response by the police once the crime is reported is not necessary.

Replication studies in four other jurisdictions confirmed this finding (Spelman and Brown, 1981). The implications of these findings are that there is little the police can achieve by more rapid response to crimes if the citizen/victim waits to report them. Cordner, Greene, and Bynum (1983) conducted a more detailed examination of response time and its various components in Pontiac, Michigan. They note (1983:150) that a major difference exists between crimes that involve the witness/victim directly (such as assault or robbery) and discovery crimes (like burglary or theft), where the victim does not detect the crime until sometime after it happens. In discovery crimes, they estimated that because of the time lag to discovery, police response time to most property crimes had to be measured in hours. Even personal crimes such as assault, they found, typically show a half-hour lag between occurrence and police response, largely because of a reporting delay by citizens. Importantly, Cordner, Greene, and Bynum (1983:150) caution against overgeneralizing from the negative findings of response-time studies: "It is fast becoming the conventional wisdom that 'response time doesn't matter' . . . Response time does matter."

For some crimes, notably burglary, quicker reaction by the police appears correlated with better investigations and increased likelihood of finding witnesses and physical evidence. With many assaults, however, the victim knows the offender, and a quicker response may not be that important. Although the relationships between response time and case outcome are complex, it seems that, in general, a quicker police response is preferred. But the causes of delay in police response, primarily reporting lag on the part of citizens, are beyond police control. If the citizen waits hours (for whatever reason) to report the crime, it makes little difference if the police take 2 minutes or 20 minutes to respond.

CHANGES IN PATROL

Police agencies have a tradition of dedicating extra patrol resources to identified problems. As Sherman (1991:189–190) puts it, "The very origins of the American police are tied to the permanent crackdown strategy." Faced with problems of crime or disorder, especially problems that attract widespread attention, the police have historically allocated extra resources in response (Kohfeld, 1983). What has changed in the past two decades, largely as an outgrowth of the Kansas City Preventive Patrol experiment and other research on police patrol, is that police agencies today are taking the initiative in mounting these crackdowns (Petersilia, 1987). They are also changing the ways in which they assign patrol officers and respond to citizen calls for aid.

Targeted Patrol

If random preventive patrol does not appear to be systematically related to either the control of crime or the attitudes of citizens, how can patrol efforts be better focused on crime control? In answer to this question, a number of jurisdictions changed the way in which their officers conducted patrols, or the ways in which the patrol force was deployed. In an attempt to increase the impact of patrol on crime reduction several police agencies replaced random patrol with specific goals for patrol officers.

According to William Bieck (1989:93), the impact of the Kansas City experiment has been widespread: "Following the collection and a cursory initial analysis of the data from the preventive patrol experiment, police administrators were at a loss to suggest an immediate alternative to preventive patrol." The Law Enforcement Assistance Administration (LEAA) supported a national initiative to study alternative strategies for police patrol, providing funding to almost 70 agencies for experiments with patrol.

The original Patrol Emphasis Project of the LEAA included support for developing a crime-analysis unit. This unit would gather information about crime problems to identify strategies by which the police agency could reduce the amount of crime. In 1976 the program was renamed the Integrated Criminal Apprehension Program, and nearly 70 police agencies developed or improved their crime-analysis capabilities. As Bieck (1989:93) notes, these projects developed a crime-analysis model still in use today. The model includes five steps: data collection, data collation, data analysis, product development, and feedback on data utilization to assess patrol effectiveness.

Using this basic model, police agencies created research units in which personnel reviewed the reports of patrol officers to determine patterns in criminal events and develop patrol strategies to intervene. For example, the unit might observe that robberies of convenience stores clustered between the hours of midnight and 2:00 A.M. Based on this information, patrol resources might be specifically directed to check these stores between these hours. The research unit would then monitor the effects of this effort to see if the rate of robberies was changed. With the advent of computerized crime mapping, this kind of analytic deployment of police patrol is even more common today (Rogers and Craig, 1994).

As a result of this initiative and their own actions, police agencies began to implement a variety of more proactive patrol strategies. In 1980 Thomas Repetto reported that over 70 percent of police agencies he surveyed had changed the organization of their patrol forces. Generally these changes involved the use of some patrol resources in a purely proactive way by releasing these officers from responsibility for responding to calls for as-

Aggressive patrol tactics have been used to control the level of weapons and drugs on the streets. (Lenore Davis/*New York Post*)

sistance. These released patrol units were then targeted at specific goals for patrol (Krajick, 1978).

In Wilmington, Delaware, the police department began what it called a **split-force patrol.** In this program approximately 25 percent of the patrol force was not committed to routine preventive patrol. Instead, this contingent of officers was assigned as needed to add increments of patrol to particular areas. Thus, if there was a problem with vandalism in some part of the city, this unit could be deployed as a saturation patrol of that area, especially during peak times of vandalism.

In Kansas City the police department experimented with alternative methods of patrol, including having patrol officers focus on specific suspected or known offenders and having them pay particular attention to specific locations. Known respectively as **perpetrator-oriented patrol** and **location-oriented patrol,** these efforts were meant to achieve a greater impact on crime from the planned allocation of patrol resources.

In New Haven, Connecticut, the police department instigated a program of **directed patrol.** At the start of each shift patrol officers received a set of instructions that directed them to engage in a particular set of patrol activities at some point in their shift. For example, an officer might be directed to leave the cruiser and engage in a foot patrol around an area school between 2:30 and 3:30 P.M. The instructions were based on an analysis of crime and problem data and designed to enhance the impact of patrol on these problems. The remainder of the officer's time was spent on routine patrol.

Other jurisdictions, such as Washington, D.C., and Minneapolis, Minnesota, targeted patrol resources at identified career criminals. In Minneapolis, the Target Eight program alerted all patrol officers to focus their attention on the surveillance and control of eight persons identified as serious offenders. In Washington, D.C., the Repeat Offender Project (ROP) assigned teams of two officers each to track and control identified repeat offenders. The thinking behind these and similar projects was that if the police could arrest the highest-rate offenders in the jurisdiction, these arrests would have an impact on the overall rate of crime. Thus, patrol officers were assigned to control the most "dangerous" offenders instead of protecting particular places or hoping to catch these offenders as part of random patrol activity.

Foot Patrol

One outcome of the finding that motorized preventive patrol levels are not associated with significant levels of citizen feelings of safety, satisfaction, or criminal victimization was a questioning of the need for the automobile. If discovery lag meant that rapid police response was not likely to result in greater rates of arrest, then patrol officers did not need automobiles to ensure quick responses to calls for assistance. Police and civic leaders increasingly showed interest in the use of foot patrol by police rather than automobile patrol.

George Kelling (n.d.:1) observes, "Foot patrol is regaining popularity as a police tactic." In the late 1970s two major experiments with foot patrol were conducted in Newark, New Jersey (Police Foundation, 1981), and Flint, Michigan (Trojanowicz, 1982). In both cities the researchers found that citizens liked foot patrol and felt safer as a result of having police offi-

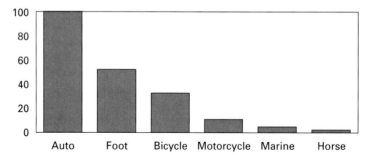

CHART **12.1** Percentage of Police Departments Routinely Using Types of Patrol, 1999. (*Source:* M. Hickman and B. Reaves (2001) *Local Police Departments, 1999.* Washington, DC: U.S. Bureau of Justice Statistics, p. 5.)

cers patrol their neighborhoods on foot. While the Newark data did not indicate any reduction in the levels of crime occurring in foot patrol areas, the Flint experiment showed a small reduction in crime. The Flint data also revealed a reduction in the number of calls for police service in the foot-patrol districts, which the researchers attributed to the availability of the neighborhood officer.

As a result of these studies and other efforts, foot patrol as a tactic of the police is being revived across the country. It has not, and probably cannot, replace motorized patrols, but where foot patrol has been instituted, it has generally been met with enthusiastic citizen support. Kelling summarizes the results of studies of foot patrol (n.d.:3) as reducing citizen fear of crime, increasing citizen satisfaction with the police, increasing officer understanding of citizens, and being associated with higher officer morale.

THE IMPACT OF PATROL

Throughout the 1980s and 1990s there arose what might be called a second generation of studies on the impact of police patrol efforts on levels of crime. This research has expanded our understanding of the impact of patrol on crime and disorder. Lawrence Sherman (Sherman et al., 1997:8-1) summarizes this understanding: "Police can prevent robbery, disorder, gun violence, drunk driving and domestic violence, but only by using certain methods under certain conditions."

After reviewing the research on police effectiveness with crime prevention, Sherman concludes (1997:8–33) that increased directed patrols in hot spots of crime could reduce levels of crime. So, too, proactive arrests of serious repeat offenders and drunk drivers, and arrests of employed domestic-violence offenders, have the effect of reducing the incidence of those crimes. **Hot spots of crime** are locations at which crime is significantly more common than other places (Sherman, Gartin, and Buerger, 1989). The notion of a hot spot is based on reviews of the reports of crime that show that certain places are more likely to be the scene of crimes. To use a traffic analogy, accidents are more likely at intersections than on straight stretches of road. Focusing police efforts in high-crime-rate spots is more likely to produce crime-control effects than spreading police resources across the entire jurisdiction.

Sherman and Weisburd (1995) report that focusing police patrol in hot spots was associated with modest reductions in crime and large reductions in disorder at those places. Koper (1995) investigated the link between the length of police presence at a location and the impact on crime. He concluded that police presence, even for short periods, had crime-reduction effects in hot spots, but that in other places, police needed to be on the scene for up to 15 minutes to produce similar reductions in crime. Sherman and Rogan (1995) report that raids on crack houses by uniformed officers had short-term (two weeks or less) effects of reducing calls for police service and reducing levels of reported crime in areas where the raids were made.

These findings and other research on the effect of differential application of police patrol in response to specific crime problems has expanded our understanding of patrol impact. The research indicates that patrol officers can influence rates and levels of crime as well as citizen feelings of safety and levels of disorder in communities. To achieve maximum effect, however, the allocation of patrol must be targeted. In addition, the amount of patrol applied to any particular problem is important. Too little attention fails to produce the desired result. Too much patrol means that a scarce resource is wasted. Research into the effects of foot patrol on crime and safety continues. This research is likely to not only improve the effectiveness of police patrol tactics, but also make us more realistic in our expectations of the impact of patrol.

THE INVESTIGATION PROCESS

The second major crime-control tactic of the police is investigation by detectives of reported crimes. Unlike preventive patrol, the detective specialist in policing is of a more recent vintage. Not until near the beginning of the twentieth century did municipal police agencies in America create the position of detective (Kuykendall, 1982). Fear of the impact of nonuniformed police detectives on individual liberty prevented the movement to detectives for several decades (Klockars, 1985).

Toward the end of the nineteenth century, however, a number of factors converged to create a demand for public police investigations of crimes and criminals. One factor was the inability or unwillingness of many citizens to hire private detectives to find and recover stolen property. Another was the public perception of a rising crime rate and an ineffectiveness of patrol officers in identifying and apprehending many offenders. Finally came the emergence of an acceptable model for the public detective. All these factors contributed to a growing acceptance of police detectives. In the United States, the success of the Pinkerton detective agency and various federal law-enforcement agencies provided models for the detective units of local police departments (Poland, 1989).

Jack Kuykendall (1982) developed a conceptual framework of the investigative process that includes an assessment of different types of investigation in terms of their "fit" with democratic values. Criminal investigations, he suggests (1982:137), could either be proactive (involving a predicted offense) or reactive (involving an accomplished crime), invisible (where police attention is concealed) or visible (where the public is aware of police action), and person-focused (targeted on a suspected individual) or event-focused (targeted on a specific offense or offense type). The combination of factors most threatening to democracy are proactive, invisible, person-focused investigations. For example, if the police suspect that you might commit some crime and thus assign a detective to surveillance of you and tap your telephone without your knowledge, your interests in privacy are compromised. These characteristics of expanded federal investigations of suspected terrorists after the World Trade Center and Pentagon attacks have fueled growing concerns that basic civil liberties are now at risk in the United States (Thomas and Isikoff, 2001). Fear of such police practices (and their possible abuses) account in part for the relatively late creation of detective divisions in police agencies.

The most democratic combination of factors, of course, is the reactive, visible, event-focused investigation. In response to a bank robbery, detectives interview witnesses in order to gather clues to identify the robber. The organizational scheme that most suits the democratic combination is the assignment of detectives to cases. The case typically represents a known criminal offense. The detective then investigates this offense in an effort to develop evidence that will link a particular person to the crime and allow for an arrest and conviction of the offender. This influence of concern for democratic values helps explain the traditional model of police investigations in the United States.

As we have already seen, uniformed patrol officers traditionally respond to citizen complaints of crime. These officers complete reports that are then assigned to detectives for followup investigations. In this model, the detective receives the case of the reported crime and is expected to interview witnesses, gather and assess physical evidence, and prepare the case for trial. The detective is reacting to a criminal event in which police interest has already been advertised by the response of a uniformed patrol officer.

In theory, the detective, by virtue of his or her expertise and lack of responsibility for responding to citizen calls for assistance, will be able to "crack the case." The patrol officer, lacking expertise and diverted by the demands of patrol, is incapable of devoting the time or attention needed to gather the evidence required to identify the offender and solve the case. Thus, the detective is the premier crime fighter among police, and for this and other reasons, a position as a detective is often a career aspiration for police officers (Poland, 1989).

RESEARCH ON INVESTIGATION

The Rand Institute undertook an evaluation of the criminal investigation process in 1973. The purpose of this study was to describe and assess the investigation process in hopes of identifying effective strategies and structures and assessing the contribution of various technological advances to successful investigation outcomes (Greenwood and Petersilia, 1975). The researchers surveyed 150 large police agencies and made site visits to 25 police departments to study the criminal investigation process.

As with the Kansas City Preventive Patrol experiment, the results of the RAND study of investigation questioned traditional assumptions about the importance and worth of detective units (Williams and Sumrall, 1982). Among other things, the Rand researchers found that the most important determinant of case clearance (the identification of the offender) was the information obtained by the patrol officer who initially responded to the complaint. That is, if the patrol officer identified witnesses or suspects, someone was likely to be arrested for the crime. If the patrol officer did not identify possible suspects, detectives were not likely to identify them either (Greenberg et al., 1977).

The Rand researchers found that detectives spent the majority of their time on tasks unrelated to the solving of crimes. Rather, detectives were critically important to the preparation of cases for trial—ensuring proper reports were completed, witnesses were interviewed, evidence was properly collected and stored, and similar administrative tasks. Also, the researchers suggested that certain investigative structures, such as task forces, especially those cooperating with the prosecutor's office, did appear to result in more cleared cases than did traditional detective work.

Following the results of this study, the National Institute of Justice funded efforts in several cities to implement changes in the investigative process suggested by the Rand team, under what was called the Managing Criminal Investigations program (Greenberg and Wasserman, 1979). The effects of changes in the investigation process were evaluated to determine how, if at all, they improved the process in those jurisdictions.

The results of the field tests of five sites using the Managing Criminal Investigations program were mixed. Some departments saved investigator resources, others did not. Some departments improved in investigative efficiency, others did not. Finally, no department found a significant increase in arrest, clearance, or conviction rates as a result of the program. Still, the response was favorable, as Greenberg and Wasserman (1979:7) report: "Despite what appear to be limited accomplishments, the MCI [Managing Criminal Investigations] staff in the five field test sites report that the program has been a success."

As a result of this research into the investigation process police administrators were required to reconsider traditional assumptions and practices regarding detectives. Through widespread dissemination efforts, including national conferences and seminars, the findings of the Rand project, the Managing Criminal Investigations program, and other research were quickly passed to police administrators across the country. As with the preventive patrol experiment, the research into criminal investigations led to some reforms in the detective function.

CHANGES IN INVESTIGATION

Two products of the research on the investigative process have become important fixtures in contemporary detective work. First, the importance of the responding patrol officer and her or his initial investigation to eventual case clearance has resulted in an expanded role for the patrol officer. Repetto (1980) notes that most police agencies responding to his survey reported giving patrol officers increasing responsibility for followup investigations. Similarly, Eck (1983) notes that most agencies now give patrol officers more responsibility in investigating crimes than was previously the case. Second, police agencies are increasingly using some sort of rating system to determine which cases will receive investigative attention.

One of the findings of the investigation research was that many cases are, simply, unsolvable. Certain factors present at the time the initial report is taken, such as the identification of a probable suspect, the location of witnesses, the collection of usable fingerprints, and the like, increase the chances that the offender will be identified and arrested (Greenberg et al., 1977; Greenwood and Petersilia, 1975). When these factors are missing, the case may well go unsolved. This finding has been replicated in other jurisdictions (Eck, 1983). As a result, police agencies have begun to assess the solvability of cases based on the presence or absence of those factors associated with higher clearance rates. Cases rating low in solvability are screened out early and do not become part of the detectives' active caseloads. As a result, investigative resources are dedicated to those cases most likely to be solved, with the apparent effect of increasing the clearance rates by focusing attention on the cases most likely to benefit from investigation (Williams and Sumrall, 1982; Petersilia, 1987).

If the traditional detective is not particularly effective in solving crimes, and if the number of cases requiring detective attention is greatly reduced, what should be the role of current investigative personnel? Faced with this question, police administrators have arrived at a variety of answers (Anderson, 1978). In some agencies, the response has been to downsize the detective bureau. In others, it has been to decentralize detectives and have them assist in, or direct, preliminary investigations by patrol officers. In still others, police administrators have opted to change the role of the detective to that of a more proactive crime fighter rather than a reactive crime solver.

Decoys and Stings

Gary Marx (1988) wrote about what he called "the new undercover police work," in which police officers aggressively and proactively seek out criminal offenders. **Undercover police officers** pose as criminals or victims in an effort to discover those citizens who have a criminal bent. The undercover operations create an opportunity for criminality, and the police then arrest those who avail themselves of that opportunity.

In **decoy units,** police officers place themselves in vulnerable positions. For example, an officer may act drunk, even feign being passed out, while conspicuously displaying a wallet or social security/welfare check envelope with cash in it. The decoy officer and several disguised backup officers then wait to see if someone will steal the money. If an offender takes the bait, the officers move in and make the arrest.

The New York City Police received national acclaim for such decoy practices when the Law Enforcement Assistance Administration picked its Street Crime Unit for recognition as a "model program" (Halper and Ku, 1976). Officers would pose as derelicts, tourists, or simply as imprudent residents and arrest those who sought to victimize them. In many cities similar units operate where undercover officers attempt to be the targets of criminals.

As Marx (1980) observes, officers using these tactics hope to attract the attention of criminals who would otherwise attack innocent civilians. The assumption underlying these efforts is that a certain level of criminality exists, and the use of decoy officers changes the target of the crime, not the likelihood of crime. The problem, as Marx sees it, is that the decoys may be too tempting as targets, so that otherwise law-abiding citizens succumb to the lure of an easy score.

An undercover officer feigns drunkenness to attract robbers. (AP/Wide World Photos)

Vice crime investigators have traditionally used deceptive, or **sting,** operations to apprehend those who sell drugs or provide other illegal services. The officers pose as customers of drug sellers or prostitutes, and when the offenders engage in the illegal activity—actually sell the drugs or agree to an act of prostitution—the officers identify themselves and make an arrest. In drug enforcement, this practice is known as the **buy/bust** (Moore, 1977). In recent years, the police have also begun to pose as prostitutes or drug sellers in hopes of catching the customers of vice operations.

A more recent innovation with the sting crime-control strategy of the new undercover police is the establishment of fencing operations. Here the police pose as buyers of stolen property and await thieves and burglars who will sell their stolen goods to the fence. After some period of time, the officers close down the shop and arrest all of those who sold stolen merchandise to the police. Weiner, Chelst, and Hart (1984) describe one such sting operation in Detroit.

Posing as a fencing operation, Detroit police ran a storefront sting in which they purchased stolen merchandise from citizens. Weiner and his colleagues (1984:300–301) report that the sting operation was successful in having many offenders convicted and sentenced to incarceration, but some 40 percent of those who sold goods to the sting only once were not arrested because of difficulties in establishing that the goods had been stolen. In contrast, Langworthy (1989) evaluated a sting operation in Birmingham in which most offenders were not sentenced to prison. Importantly, Langworthy's data indicated that the establishment of a fencing sting seemed to increase the level of crime in general, and the rate of crime in the area immediately around the sting location in particular.

While police patrol and patrol tactics and strategies have changed dramatically over the past three decades, investigations and the work of detectives have remained relatively stable. John Eck (1999) questioned why this would be the case. He suggests that detectives and investigation

Chart 12.2 Percentage of Local Police Departments with Primary Responsibility to Investigate Types of Crime. (*Source:* M. Hickman and B. Reaves (2001) *Local Police Departments, 1999.* Washington, DC: U.S. Bureau of Justice Statistics, p. 6.)

have not "kept up" with changes in patrol and other police activity for a variety of reasons. Much of investigation work is reactive, occurring after an offense, with a focus on identifying the person who committed a known crime, and building the legal case needed for successful prosecution. Detectives, as a group, represent a small part of total police personnel, and the detective tradition has not been focused on broad crime prevention and problem-solving efforts. Eck concludes (1999:4) that if crime prevention becomes the hallmark of policing in the future, "investigations are likely to become a support service of patrol at best, and at worst, a sideshow."

CHANGES IN AGENCY APPROACHES TO CRIME

Beginning in the 1960s, American police agencies experimented with different organizational structures. This experimentation has continued into the present and reflects, in large part, an attempt by the police to improve their ability to control and prevent crime. We have already reviewed some of the findings about the traditional police tactics of preventive patrol and criminal investigations. Although this research has affected police responses to the crime problem, another major influence has been the realization that citizen, or community, support is critical to successful crime control. We will return to this topic in Chapter 15. For now we need only set the stage for an assessment of changes in police agency approaches to crime.

The deteriorating relationship between the police and the public in the 1960s was evidenced by large-scale social unrest in the Civil Rights and anti–Vietnam War movements, the "due process revolution" of the U.S. Supreme Court, and the steady rise in crime rates that the police seemed powerless to stop. Dissidents engaged in both the Civil Rights and antiwar movements openly challenged the police and police authority. It was common for the more extreme members of both movements to label all police as "pigs," for example. The due-process revolution was also status-costly for the police, because the courts criticized then-current police practice, plus introduced penalties on the police. This action threatened both the status of the police in society and their group identity as the arbiters of the law (Pepinsky, 1970). The rising crime rate reflected police ineffectiveness and powerlessness.

Dissatisfaction with the police was cited as a cause of racial riots in the 1960s, and relations between the police and minority group members were strained in most cities. This situation has not changed substantially in many areas of the country, as the riot in Los Angeles following the acquittal of officers charged in the beating of Rodney King illustrated. During the 1980s, the city of Miami experienced race riots that also were sparked by conflicts between the police and the citizens. Two cases in which police officers killed minority citizens, the McDuffie and Johnson cases, resulted in large-scale rioting when officers were acquitted of criminal charges (Overtown Blue Ribbon Committee, 1984; Alpert, 1989). A similar pattern was ob-

During a violent demonstration after the first verdict in the criminal trial of officers accused of assaulting Rodney King, police arrest demonstrators. (Christopher Smith/Impact Visuals)

served in Cincinnati in the spring of 2001 when the fatal shooting of a young black man by Cincinnati police sparked three days of rioting.

In addition to the riots—a dramatic symptom of problems between the police and community—results of the first surveys of crime victims revealed that many citizens did not report crimes to the police, so the professional crime fighters in our society were unaware of how much crime was actually occurring. Further, many observers noted a tendency of citizens to refuse to cooperate with the police as witnesses or in other ways.

Faced with a crisis in community relations, police administrators sought ways to improve relations between the police and the public. One motivation for their eagerness to heal the rift was to improve their crime-control capability. Over the past two decades the combination of increasing knowledge about police crime-control tactics and about community relations has led to a number of innovative attempts at improving police law enforcement. Team policing, our next subject, was an initial effort at changing the way police business was done so that better crime control would result from a better relationship between the police and the community.

Team Policing

In the early 1970s several police departments changed the way in which police services were organized. A variety of programs and structures aimed at building a better rapport between the police and the citizenry were classed under the label **team policing.** As Sherman, Milton, and Kelly (1973:3) state, "Team policing is a term that has meant something different in every city in which it has been tried." They note that, even so, almost all such programs included the elements of geographic stability of patrol, maximum interaction among team members, and maximum communication with the community.

Sherman and his colleagues observed that the police agencies using team policing hoped that the program would improve their success in controlling crime (1973:100). They report, however, that none of the six programs they reviewed had been able to evaluate the impact of team policing on crime. Evaluations of other types of team-policing programs, such as the COMSEC Program in Cincinnati (Schwartz and Clarren, 1977) and the Los Angeles Basic Car Plan (Kerstetter, 1981), also failed to reveal a consistent impact on crime.

At its heart, team policing involves a decentralization of police operations. Typically, a neighborhood is defined by the police and a complement of officers is assigned to provide police services to that neighborhood. The officers are the primary or sole police in the area and often are responsible for patrol and investigation. The officers are expected to come to know the neighborhood and its residents, develop an understanding of the environment, and inspire the trust of citizens. The citizens are encouraged to talk with the police, and the police attempt to build rapport so that citizens will be more willing to cooperate with them. In the end, with community support and police understanding and knowledge of the community, it is expected that the team will be more effective at crime control.

Problem-Oriented Policing

Herman Goldstein (1979) suggested that police in America should revamp the way in which they approached their job by adopting what he called **problem-oriented policing.** Rather than reacting to specific incidents and the symptoms of problems in the communities they served, he urged, police administrators should seek the causes of citizen complaints to police and eradicate them. In some ways this call echoes an earlier assessment of the police function presented by Brown (1975). Brown advised that instead of responding to incidents, such as a burglary, police would find it more effective strategy to define burglaries as an issue and then devise strategies for responding to the issue of burglary. Further, agencies could identify issue aggregates, such as victimizing crimes, into which the issues of burglary, robbery, assault, and the like, could be collapsed. With these broader categories, the agency could begin to better understand its workload and needs.

Goldstein (1979) pointed out that police responses to citizen complaints or calls for assistance often represented treatment of the symptoms rather than the disease. The wise agency, he believed, would assess the kinds of events about which they received citizen complaints and seek to attack the causes of the problems. For example, if a relatively large portion of police business involved late-night robberies of convenience stores, the police would deal with the overall problem, rather than just focusing on the collection of robberies. They would do this by working with the stores to improve security, reduce temptation by limiting the amount of cash on hand in the stores, and provide additional police presence through patrol. The goal would be prevention of robberies through attacking the causes of the problem. Goldstein asks (1979:252), "How does a police agency make the shift to problem-oriented policing? Ideally, the initiative will come from police administrators."

Spelman and Eck (1987) describe problem-oriented policing in operation in Newport News, Virginia. They describe how the police initiated a problem-oriented approach to policing and report on its success in reducing robberies in the downtown area, burglaries in a housing project, and thefts on the waterfront. The process involved the police officers scanning their work for related incidents. This scanning revealed clusters of related types of police tasks which were then analyzed to determine causes and possible solutions. Next the officers worked with citizens to implement a program of action to deal with the problem. Finally, the police assessed the impact of their efforts to determine success. Spelman and Eck (1987:7) are supportive of problem-oriented policing and conclude,

> The problem-oriented police department thus will be able to take the initiative in working with other agencies on community problems when those problems touch on police responsibilities. . . . The result will be a more effective response to crime and other troubling conditions in our cities.

Community-Oriented Policing

In 1988 George Kelling (1988:1) wrote, "A quiet revolution is reshaping American policing." The revolution to which he referred is generally known as **community-oriented policing.** We will describe and discuss this movement later in Chapter 15, but for now we must place it into the context of changing police approaches to crime. Community-oriented policing is, in many ways, problem-oriented policing with a twist.

The twist here is that the police rely on the community to define problems and establish police policy. A number of writers have described and promoted community-oriented policing (Skolnick and Bayley, 1986; Manning, 1984; Trojanowicz and Bucqueroux, 1990; Wadman and Olson, 1990). Using slightly different terms, all have noted the increased role of the community in defining police problems and setting police priorities. More than the control or prevention of crime, the police are to be partners with the community in a general improvement of living conditions.

Of course, the hope is that this partnership will result in less crime by prevention efforts, less fear of crime by responding to citizen concerns, and more effective law enforcement through police–citizen cooperation. The essential difference between problem-oriented policing and community-oriented policing is the role of the public. In problem-oriented policing, the police define problems and design solutions. The public serve as an information source and a resource in responding to the problems identified by the police. In community-oriented policing, the police officers seek out citizen groups, or engage in community organizing to create such groups, and ask these people to identify the needs and problems of the community. The police then work with these community members to solve their problems. As Kelling (1988:1) writes,

> Organizing citizen's groups has become a priority in many departments. Increasingly, police departments are looking for means to evaluate themselves on their contribution to the quality of neighborhood life, not just crime statistics. Are such activities the business of policing? In a crescendo, police are answering yes.

THE POLICE AND CRIME: A TYPOLOGY

Table 12.3 presents a preliminary classification of the police efforts to respond to crime that we have examined in this chapter. You will recall that police crime-fighting efforts can emphasize a goal of crime prevention or of crime control. Problem- and community-oriented policing strategies both seek to prevent future crimes by dealing with the underlying causes of current problems. Team policing and traditional police practice, on the other hand, focus on detecting and apprehending criminal offenders. The traditional practices of routine patrol and detective casework, as well as the

TABLE 12.3 Classification of Police Crime-Control Efforts

Crime-Control Strategy	CRIME-CONTROL GOAL	
	Prevention	*Enforcement*
Proactive	Problem-oriented policing	Team policing
Reactive	Community-oriented policing	Traditional policing

newer community-oriented policing, are reactive strategies. In both cases, the impetus for police action comes from the community, and the police react to this community initiative. In team-policing and problem-oriented policing, however, the initiative and policy-making authority clearly rest with the police. Thus, these two strategies are more proactive.

The foregoing is, of course, a crude, preliminary classification. In many police agencies one can observe a team-policing program existing alongside traditional patrol, or a community-oriented policing program within a general framework of traditional practices. Also, while these approaches are related and in some sense follow each other in time, there is no uniform evolution from traditional practices through team policing and problem-oriented policing to community policing. Elements and examples of all four approaches can be found across America today. Finally, our description of these approaches ignores much of the nuance and overlap involved. Simply put, the differences between any given problem-oriented and community-oriented police program may not be as great as we have indicated here. Indeed, sometimes the only real difference is in the choice of name by the police agency running the program.

Quality of Life Policing

Perhaps nowhere in policing is this choice of name more apparent than in recent efforts to develop what has come to be called "quality of life" policing. Sometimes called "**zero tolerance**" policing, or "broken windows," "nuisance abatement," "order maintenance" or other titles, these police efforts seek to prevent crime by strengthening existing informal community social control. The core concept is that crime happens in places where social control is weak. Minor incivilities, like litter and graffiti, signal a lack of social control and contribute to a feeling among citizens that it is not their responsibility to maintain order in the neighborhood. Police efforts aimed at improving and strengthening community social control aggressively target nuisance offenses such as loitering, littering, public intoxication and the like. If even minor infractions are not tolerated, the thinking goes, then there will also be no serious crimes.

The research evidence on the effectiveness of quality of life policing is mixed (Miller, 2001). Some researchers conclude that a police focus on quality of life problems results in dramatic reductions in crime (Kelling and Coles, 1996; Silverman, 1999). The available evidence suggests that the relationship between minor crime and serious offenses is complex. Novak et al. (1999) report a test of aggressive policing of disorder on crime and conclude that police efforts at enforcing alcohol control laws was not associated with reductions in other crimes. Weiss and Freels (1996) report a similar lack of crime-prevention effects of increased traffic enforcement in Dayton, Ohio. In summarizing the available research, Sherman (Sherman et al.,

1997) suggests that a police focus on disorder and low-level offenses appears to be associated with reductions in other crimes. He worries, however, that such a focus undermines the legitimacy of the police in the view of the public, who may perceive the police as simply harassing them. He also voices concern about the longer-term effects of police actions, like arrests for minor offenses, on citizen support for the police, writing (1997:8–25), "The effects of an arrest experience over a minor offense may permanently lower police legitimacy, both for the arrested person and their social network of family and friends."

The purpose of the classification scheme in Table 12.3 is primarily to illustrate different orientations toward crime control that might characterize a police agency. Police agencies in any of the four categories can (and do) devote at least some effort towards "quality of life policing." All American police agencies are organized to respond to crime. To the degree that all such police agencies will respond to a burglar alarm, so, too, are all of them at least somewhat reactive in their dealings with crime. One difference is that in some agencies the organization, its subunits, and its officers exercise more initiative than in others. Another is that in some agencies the police are likely to intervene in anticipation of crime, whereas in others they will tend toward restraint until after a criminal event. The point is that these different approaches to crime control help explain police behavior in practice.

SOME CORRELATES OF POLICE CRIME CONTROL

The police are the gatekeepers of the criminal justice process and are responsible for the detection and investigation of crime and the arrest of suspected criminals. The major tactics of the police in responding to crime are patrol and investigation. Although the police are responsible for a variety of functions and services within society, crime control is a primary goal of all police organizations.

In seeking to respond to crime, a police agency can employ a variety of approaches, ranging from reactive to proactive strategies and emphasizing the prevention or control of crime. Police agencies in the United States can be classified according to which combination of goals and strategies dominates their approach to crime. This approach, in turn, helps us to understand differences in how police agencies do law enforcement and crime control.

In many ways this chapter is about the correlates of police approaches to crime control. For example, we have seen that although in a democratic society policing tends to be reactive, it is possible for American police to differ along a continuum from more reactive to more proactive. This difference is reflected both in the types of tactics the organization might use to respond to crime and in the types of crimes it detects and investigates. As with other aspects of policing in America, the correlates of police approaches to crime control are to be found at three levels; the community, the organization, and the people.

Community

How the police organize for crime control and how successful they are at responding to crime is partially a function of community characteristics. In smaller, more close-knit communities, the police are more familiar with the citizenry and with the crime problem in the jurisdiction, and the public is more familiar with the police. Cordner (1989:153) suggests that this difference accounts for generally higher clearance rates in rural communities than in urban communities. In a similar vein, crime and crime problems in general are likely to be greater in urban areas than in rural ones. McCarthy (1991) has also reported a relationship between community characteristics and arrest probabilities in that arrest rates correlate with urbanism, poverty, percentage of the population that is nonwhite, and unemployment rates as well as crime rates.

In short, the definition and composition of the crime problem, resources available to the police, and citizen cooperation with the police are all related to police crime-control efforts. These community characteristics set a context in which the police work to control crime. Community characteristics may also affect how the police organize to respond to crime by virtue of the likelihood that organized citizen groups exist that are interested in, and capable of, contributing to community-oriented policing programs, for example. Similarly, to the degree that community characteristics are associated with different types of police leadership, they extend an influence over organizational correlates as well. Pursley (1976) reports that communities with professional city managers and that had experienced population growth were more likely to hire professional police administrators from outside the local area.

Organization

The police organization itself exerts an influence on the crime control activities of its personnel. Indeed, our description of approaches to crime deals with organizational efforts to respond differently to crime problems in a police agency's jurisdiction. Observers have noted that one of the primary characteristics associated with a police agency's adoption of problem- or community-oriented policing is the agency's leadership (Skolnick and Bayley, 1986; and Kelling, 1988). These more professional leaders, they suggest, are willing to experiment with and change the structure and tactics of the police department.

Departmental size and type also appear to correlate with arrest decisions when officers have the opportunity to enforce the law. In two studies of drunken-driving law enforcement, departmental characteristics were associated with different levels of enforcement. In Maine, Myers, Heeren, and

Hingson (1989) reported that officers from large municipal police agencies were significantly less likely to stop suspected violators than were officers from smaller agencies or those from state police or sheriff's departments. Similarly, Mastrofski, Ritti, and Hoffmaster (1987) found that officers from larger departments were less likely to make arrests than were those from smaller agencies.

Although size of agency has been the variable correlated with enforcement decisions, the authors speculate that size is related to the ability of the organization, and particularly the supervisors, to control or influence the law-enforcement decisions of officers. Assuming that laws against drunken driving would or should be fully enforced, the researchers contend that characteristics of the larger organization—more anonymity among officers, heavier workloads, less direct supervision of officers, and the like—allow greater individual discretion in enforcement situations.

People

In addition to the effects of community and organizational factors, the actual enforcement of the law by police officers is also contingent in part on the characteristics of the officers themselves. In a study of officer effects on arrest decisions where community and organizational expectations are unclear, Juhnke and Bermann (1988:243) found that officer attitudes were significantly related to arrest decisions. These attitude differences, in turn, correlated with years of police experience (1988:249–250) in that more experienced officers were more likely to arrest than were less experienced officers.

William Walsh (1986) studied the arrest rates of patrol officers in one precinct of New York City. He studied four classes of officers ranging from those who made no felony arrests in one year through those who made 25 or more felony arrests. Walsh found that these differences in rates at which officers made arrests correlated with officer characteristics. Just as important, he found that officers were more likely to make arrests based not only on personal motivations but also on group dynamics within the patrol squad (1986:360).

Michael Brown (1981) observed that police officers decide whether to enforce the laws, especially in misdemeanor and traffic-violation situations. Their decisions, he suggested (1981:183), are founded on their assessments of the seriousness of the offense or their desires for activity more than on the law or agency policy. Therefore, part of the practice of police crime control is a product of the individual officers themselves.

Police crime-control practice—what happens on the streets—seems to be the product of a variety of forces. The seriousness of the offense, the structure and policies of the police organization, the expectations of the community, and the characteristics of the officer all combine to produce the outcomes. Although the crime-control decision in any given circumstance may be unpredictable, an understanding of the forces involved in police approaches to crime helps explain general patterns or tendencies in police decisions.

CHAPTER CHECKUP

1. What is the impossible mandate of the American police?
2. On what two dimensions can police crime-control efforts vary?
3. What does Donald Black (1980) mean by the term *mobilization of law?*
4. Why is patrol considered the backbone of policing?
5. How has police patrol changed as a result of research findings?
6. What is *investigation,* and how has it changed in recent years?
7. What are *problem-oriented* and *community-oriented* policing?
8. How are police crime-control efforts affected by people, organization, and community?

REFERENCES

Alpert, G. (1989) "Police use of deadly force: The Miami experience," in R. Dunham and G. Alpert (eds.) *Critical issues in policing: Contemporary readings.* (Prospect Heights, IL: Waveland):480–495.

Anderson, D. (1978) "Management moves in on the detective," *Police Magazine* (March):3–12.

Bieck, W. (1989) "Crime analysis," in W. G. Bailey (ed.) *The encyclopedia of police science.* (New York: Garland):89–100.

Black, D. (1980) *The manners and customs of the police.* (New York: Academic).

Boydstun, J., M. Sherry, and N. Moelter (1977) *Patrol staffing in San Diego: One- or two-officer units.* (Washington, DC: Police Foundation).

Brown, M. (1981) *Working the street: Police discretion and the dilemmas of reform.* (New York: Russell Sage Foundation).

Brown, W. (1975) "Local policing—A three dimensional task analysis," *Journal of Criminal Justice* 3(1):1–15.

Cordner, G. (1989) "Police agency size and investigative effectiveness," *Journal of Criminal Justice* 17(3):145–156.

Cordner, G., J. Greene, and T. Bynum (1983) "The sooner the better: Some effects of police response time," in R. Bennett (ed.) *Police at work: Policy issues and analysis.* (Beverly Hills, CA: Sage):145–164.

Decker, S. and A. Wagner (1982) "The impact of patrol staffing on police-citizen injuries and dispositions," *Journal of Criminal Justice* 10(5):375–382.

Eck, J. (1983) *Solving problems: The investigation of burglary and robbery.* (Washington, DC: National Institute of Justice).

Eck, J. (1999) *Problem-solving detectives: Some thoughts on their scarcity.* (Seattle, WA: Seattle Police Department).

Frank, J., S. Brandl, and R. Watkins (1997) "The content of community policing: A comparison of the daily activities of community and 'beat' officers," *Policing: An International Journal of Police Strategies and Management* 20(4):716–728.

Goldstein, H. (1979) "Improving policing: A problem-oriented approach," *Crime & Delinquency* 25(2):236–258.

Greene, J. and C. Klockars (1991) "What police do," in C. Klockars and S. Mastrofski (eds.) *Thinking about police: Contemporary readings.* (New York: McGraw-Hill):273–284.

Green, L. (1995) "Cleaning up drug hot spots in Oakland, California: The displacement and diffusion effects," *Justice Quarterly* 12(4):737–754.

Greenberg, B., C. Elliot, L. Kraft, and H. Procter (1977) *Felony investigation decision model: An analysis of investigative elements of information.* (Washington, DC: U.S. Government Printing Office).

Greenberg, I. and R. Wasserman (1979) *Managing criminal investigations.* (Washington, DC: U.S. Department of Justice).

Greenwood, P. and J. Petersilia (1975) *The criminal investigation process: Volume 1: Summary and policy implications.* (Washington, DC: U.S. Department of Justice).

Halper, A. and R. Ku (1976) *An exemplary project: New York police department street crime unit.* (Washington, DC: U.S. Government Printing Office).

Juhnke, R. and J. Bermann (1988) "Police discretion: Relationship of experiences to officer's beliefs and arrest decisions," *American Journal of Criminal Justice* 12(2):243–253.

Kaplan, E. (1978) "Evaluating the effectiveness of one-officer versus two-officer patrol units," *Journal of Criminal Justice* 7(4):325–355.

Kelling, G. (1988) *Police and communities: The quiet revolution.* (Washington, DC: U.S. Department of Justice).

Kelling, G. (n.d.) *Foot patrol.* (Washington, DC: U.S. Department of Justice).

Kelling, G. and K. Coles (1996) *Fixing broken windows: Restoring order and reducing crime in our communities.* (New York: Simon and Schuster).

Kelling, G., T. Pate, D. Dieckman, and C. Brown (1974) *The Kansas City preventive patrol experiment: A summary report.* (Washington, DC: The Police Foundation).

Kerstetter, W. (1981) "Patrol decentralization: An assessment," *Journal of Police Science and Administration* 9(1):48.

Kessler, D. (1985) "One- or two-officer cars? A perspective from Kansas City," *Journal of Criminal Justice* 13(1):49–64.

Klockars, C. (1985) *The idea of police.* (Beverly Hills, CA: Sage).

Klockars, C. and S. Mastrofski (eds.) (1991) *Thinking about police: Contemporary readings.* (New York: McGraw-Hill).

Kohfeld, C. (1983) "Rational cops, rational robbers, and information," *Journal of Criminal Justice* 11(5):459–466.

Koper, C. (1995) "Just enough police presence: Reducing crime and disorderly behavior by optimizing patrol time in crime hot spots," *Justice Quarterly* 12(4):649–672.

Krajick, K. (1978) "Does patrol prevent crime?" *Police Magazine* (September 1978):3–15.

Kuykendall, J. (1982) "The criminal investigative process: Toward a conceptual framework," *Journal of Criminal Justice* 10(2):131–145.

Langworthy, R. (1989) "Do stings control crime? An evaluation of a police fencing operation," *Justice Quarterly* 6(1):27–45.

Lundman, R. (1980) *Police and policing: An introduction.* (New York: Holt, Rinehart & Winston).

Manning, P. (1978) "The police: Mandate, strategies and appearances," in P. Manning and J. Van Maanen (eds.) *Policing: A view from the street.* (Santa Monica, CA: Goodyear):7–31.

Manning, P. (1984) "Community policing," *American Journal of Police* 3(2):205–227.

Marx, G. (1980) "The new police undercover work," in C. Klockars and S. Mastrofski (eds.) *Thinking about police: Contemporary readings,* 2nd ed. (New York: McGraw-Hill):240–258.

Marx, G. (1988) *Undercover: Police surveillance in America.* (Berkeley, CA: University of California Press).

Mastrofski, S., R. Ritti, and D. Hoffmaster (1987) "Organizational determinants of police discretion: The case of drinking-driving," *Journal of Criminal Justice* 15(5):387–402.

McCarthy, B. (1991) "Social structure, crime, and social control: An examination of factors influencing rates and probabilities of arrest." *Journal of Criminal Justice* 19(1):19–30.

Meyers, A., T. Heeren, and R. Hingson (1989) "Discretionary lenience in police enforcement of laws against drinking and driving: Two examples from the State of Maine, U.S.A.," *Journal of Criminal Justice* 17(3):179–186.

Miller, D. (2001) "Poking holes in the theory of 'Broken Windows,' " *Chronicle of Higher Education* (Feb. 9):A14–A16.

Moore, M. (1977) *Buy and bust.* (Lexington, MA: D.C. Heath).

Moore, M., R. Trojanowicz, and G. Kelling (1988) *Crime and policing.* (Washington, DC: U.S. Department of Justice).

Novak, K., J. Harman, A. Holsinger, and M. Turner (1999) "The effects of aggressive policing of disorder on serious crime," *Policing: An International Journal of Police Strategies and Management* 22(2):171–190.

"One-officer state police cars raise the risks to cops' lives," (1997) *Law Enforcement News* 23(461):1, 14.

Overtown Blue Ribbon Committee (1984) *Report on the Overtown disturbance.* (Miami, FL: City of Miami).

Parks, R., S. Mastrofski, C. DeJong, and M. Gray (1999) "How officers spend their time with the community," *Justice Quarterly* 16(3):483–518.

Pepinsky, H. (1970) "A theory of police reaction to Miranda v. Arizona," *Crime & Delinquency* 17(2):379–392.

Petersilia, J. (1987) *The influence of criminal justice research.* (Santa Monica, CA: Rand).

Poland, J. (1989) "Detectives," in W. G. Bailey (ed.) *The encyclopedia of police science.* (New York: Garland):142–145.

Police Foundation (1981) *The Newark foot patrol experiment.* (Washington, DC: The Police Foundation).

Pursley, R. (1976) "Community characteristics and policy implications: Some exploratory findings among municipal police chiefs," *Journal of Criminal Justice* 4(4):291–302.

Repetto, T. (1980) "Police organization and management," in R. Staufenberger (ed.) *Progress in policing: Essays on change.* (Cambridge, MA: Ballinger):65–84.

Rogers, R. and D. Craig (1994) "Geographic information systems in policing," *Police Studies* 17(2):67–78.

Schwartz, A. and S. Clarren (1977) *The Cincinnati team policing experiment: A summary report.* (Washington, DC: Police Foundation).

Sherman, L. (1991) "Police crackdowns: Initial and residual deterrence," in C. Klockars and S. Mastrofski (eds.) *Thinking about police: Contemporary readings.* (New York: McGraw-Hill):188–211.

Sherman, L., P. Gartin, and M. Buerger (1989) "Hot spots of predatory crime: Routine activities and the criminology of place," *Criminology* 27(1):27–55.

Sherman, L., D. Gottfredson, D. MacKenzie, J. Eck, P. Reuter, and S. Bushway (1997) *Preventing crime: What works, what doesn't, what's promising?* (Washington, DC: National Institute of Justice).

Sherman, L., C. Milton, and T. Kelly (1973) *Team policing: Seven case studies.* (Washington, DC: Police Foundation).

Sherman, L. and D. Rogan (1995) "Deterrent effects of police raids on crack houses: A randomized, controlled experiment," *Justice Quarterly* 12(4):755–781.

Sherman, L. and D. Weisburd (1995) "General deterrent effects of police patrol in crime 'hot spots': A randomized, controlled trial," *Justice Quarterly* 12(4):625–648.

Silverman, I. (1999) *NYPD battles crime: Innovative strategies in policing.* (Boston: Northeastern Univ. Press).

Skolnick, J. (1994) *Justice without trial,* 3rd ed. (New York: John Wiley).

Skolnick, J. and D. Bayley (1986) *The new blue line.* (New York: Free Press).

Spelman, W. and D. Brown (1981) *Calling the police: Citizen reporting of serious crime.* (Washington, DC: Police Executive Research Forum).

Spelman, W. and J. Eck (1987) *Problem-oriented policing.* (Washington, DC: U.S. Department of Justice).

Surette, R. and A. Richard (1995) "Public information officers: A descriptive study of crime-news gatekeepers," *Journal of Criminal Justice* 23(4):325–336.

Thomas, E. and M. Isikoff (2001) "Justice kept in the dark," *Newsweek,* (Dec. 10, 2001):37–43.

Travis, L. (1998) *Introduction to criminal justice,* 3rd ed. (Cincinnati, OH: Anderson).

Trojanowicz, R. (1982) *An evaluation of the Flint foot patrol project.* (East Lansing, MI: Michigan State University).

Trojanowicz, R. and B. Bucqueroux (1990) *Community policing: A contemporary perspective.* (Cincinnati, OH: Anderson).

Van Kirk, M. (1978) *Response-time analysis: Executive summary.* (Washington, DC: U.S. Department of Justice).

Wadman, R. and R. Olson (1990) *Community wellness: A new theory of policing.* (Washington, DC: Police Executive Research Forum).

Walsh, W. (1986) "Patrol officer arrest rates: A study in the social organization of police work," *Justice Quarterly* 3(3):271–290.

Webster, W. (1970) "Police task and time study," *Journal of Criminal Law, Criminology and Police Science* 61(1):94–100.

Weiner, K., K. Chelst, and W. Hart (1984) "Stinging the Detroit criminal: A total system perspective," *Journal of Criminal Justice* 12(3):289–302.

Weiss, A. and S. Freels (1996) "The effects of aggressive policing: The Dayton traffic enforcement experiment," *American Journal of Police* 15(1):45–64.

Williams, V. and R. Sumrall (1982) "Productivity measures in the criminal investigation function," *Journal of Criminal Justice* 10(2):111–122.

Wilson, J. (1968) *Varieties of police behavior.* (Cambridge, MA: Harvard University Press).

SERVICE AND THE POLICE

CHAPTER OUTLINE

The police in America are a primary source of aid and assistance to thousands of people every day. One of the more important functions of the police in our society is the provision of either direct service, or the referral to other service providers, for citizens in need. This service role of the police has evolved over time for a number of reasons.

SERVICE PROVISION BY POLICE

Because the police are available 24 hours each day, every day of the year, it is possible for a citizen to call the police and receive a response at any time. Faced with an emergency or other problem, citizens contact the police for assistance or guidance. James Wilson (1968:4–5) observed that the police provide a variety of services, from emergency medical aid to getting cats out of trees. Indeed, his classification of citizen complaints to the Syracuse police (1968:18) revealed that over one-third of the calls handled by police officers involved the provision of various services.

Wilson's findings agreed with an earlier estimate by Elaine and Ian Cumming with Laura Edell (1965), who estimated that more than half of the calls received from citizens involved the provision of assistance and service to citizens. They noted that police officers are called on to provide such services most often to the poor, because "poor, uneducated people appear to use the police in the way that middle-class people use family doctors and clergymen—that is, as the first port of call in time of trouble."

Other studies of the distribution of police work and citizen calls for police assistance have revealed a consistent pattern of frequent service activity by the police (Webster, 1970; Reiss, 1971; Bittner, 1974; Haller, 1976). The results of these studies have created a conventional wisdom that police devote most of their time and attention to "service" calls. As Greene and Klockars (1991:273) observe, "Virtually all introductory textbooks on policing now assert that 80 to 90 percent of police work is 'service-related,' and that less than 10 percent is 'crime-related.'"

The 1999 Law Enforcement Administrative Statistics (LEMAS) survey asked responding police agencies to identify special public safety functions they performed (Chart 13.1). The survey revealed that one-fifth of police agencies had primary responsibility for search-and-rescue operations, and nearly 60 percent had responsibility for animal control. About one-sixth of American police departments are responsible for the provision of emergency medical services, a fifth provide civil defense functions, and about one in ten also provide fire services (Hickman and Reaves, 2001:8).

Relying on data from the Wilmington, Delaware, police department, Greene and Klockars challenge the service conception of police work, as you may recall from the introduction to this section of the book. In their analy-

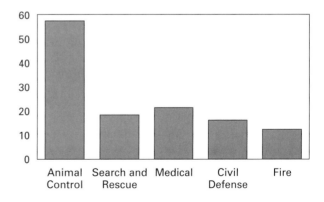

CHART 13.1
Public Safety Functions of Local Police. (*Source:* M. Hickman and B. Reaves (2001) *Local Police Departments, 1999.* Washington, DC: U.S. Bureau of Justice Statistics, p. 8.)

sis (1991:280–281), crime-related activities accounted for nearly half of police work. They suggest that controlling for uncommitted time (when officers are not responding to calls from citizens) changes the base and thus the percentage of police work attributed to service. Indeed, over 40 percent of officers' time was taken up doing things not involving public contact in their study. If the typical police officer spends three hours of every eight-hour shift doing tasks that don't involve public contact, there are only five hours of "free time" left. If the officer devotes four of those hours to crime-related activity, then 80 percent (four-fifths) of available time is spent on criminal matters. Thus, crime-related activity is a significant portion of patrol officer time available for interactions with the public.

This finding in regard to the Wilmington police department is actually consistent with an earlier study of the police conducted by Meyer and Taylor (1975). They noted that when only calls for service (or reactive police mobilization) were studied, about 75 percent of police work was service related. But if officer-initiated (or proactive mobilization) activity was the focus, crime-related work approached 50 percent of officer tasks. Meyer and Taylor (1975:143) observe, "Citizen mobilizations of the police are thought to reflect a generalized 'demand preference' for service activities." Thus, citizens call most often for service rather than for law enforcement. Officer-initiated activities, on the other hand, demonstrate a greater emphasis on law-enforcement or crime-control tasks. In large measure, this reflects the perception among officers that crime control efforts are real police work (Crank, 1998:117–119).

More recently, the Bureau of Justice Statistics has conducted surveys to ask people about their contacts with police (Langan et al., 2001). The 1999 survey revealed that about one-fifth of Americans over the age of 16 had some contact with police in that year. Over half of these contacts were the result of traffic stops by police. As Chart 13.2 shows, another one-fifth of contacts involved citizens reporting crimes, while 12 percent resulted from citizens asking for assistance and 13 percent centered around traffic accidents in which the citizen was involved either as a driver or as a witness. Only 7 percent involved citizens as suspects or witnesses to crime (Langan et al., 2001:2). Almost 62 percent of police-initiated contacts with citizens came as a result of motor vehicle stops. Only 12.5 percent of police initiated contacts with citizens were crime related, while almost 40 percent of citizen-initiated contacts involved criminal matters, mostly reporting crimes. Importantly, 60 percent of citizen-initiated contacts with the police did not involve criminal matters.

A debate of sorts has surrounded the notion of police social service provision for many years. Some observers, such as Bernard Garmire (1972:3) have argued that the community-service and law-enforcement roles of the police are incompatible. A single agency or single officer, Garmire believes, is incapable of successfully serving both functions. Thus, he argues (1972:5–8) for the creation of a two-tiered department composed of a law-enforcement

CHART 13.2
Citizen Contacts with Police by Reason for Contact. (*Source:* P. Langan, L. Greenfeld, S. Smith, M. Durose, and D. Levin (2001) *Contacts between Police and the Public.* Washington, DC: U.S. Bureau of Justice Statistics, p. 9.)

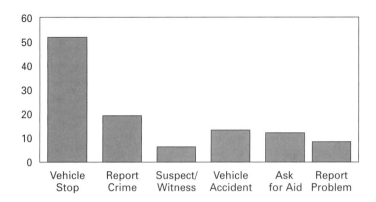

agency and a community-service agency. Others, including Egon Bittner (1975:40), claim that service is the most important function of the police.

Dilip Das (1987:8–11) describes what he terms a "service tradition" of the American police. He observes that early police and their forerunners (watchmen and constables) were responsible for a host of services, ranging from street repair and lighting through building inspections. Das (1987:10) suggests that to this formal assignment of a variety of service duties to the police must be added the reality of citizen demands for assistance:

> Apart from the formal development of the police as an urban service agency and their responsibilities for an endless range of miscellaneous work which was entrusted to them through urban administrative arrangement, there was yet another side to the police service role. This developed as a result of consumers' demands. As the cities became anonymous conglomerations of masses of people, the police were responding to more and more demands for intervention in private lives.

HUMAN SERVICES PROVIDED BY POLICE

Daniel Kennedy (1983) has suggested that the typical police department meets the requirements for being considered a human services agency. He (1983:42) identifies five themes that constitute a *human services ideology:* systemic integration of services, comprehensiveness and accessibility of services, client troubles defined as "problems in living," generic characteristics of helping activities, and accountability of service providers to clients. He concludes (1983:45) that "the police may properly be regarded as a human service agency."

Kennedy notes that the police serve to integrate services by their referrals of citizens to mental health, substance-abuse, medical, welfare, and other agencies. Because of their 24-hour availability jurisdictionwide, the police guarantee access to services. Kennedy notes the tendency of police officers to mediate conflicts rather than make arrests as evidence of their definition of client troubles as "problems in living." Similarly, the large amount of time police devote to rendering assistance to citizens, coupled with officer motivations to be helpful, evidence generic helping activities. Finally, especially given the reactive nature of most policing, it is apparent that the police are accountable to their clients.

This review of the police role in service delivery establishes that the police are as much a human service agency as is any government unit. Like welfare and health departments, the police are responsible for the general safety and security of the citizens. What sets the police apart, however, are their additional functions of law enforcement and order maintenance. The reality of coercive force and the resulting citizen suspicion of the police as helpers serves to confuse the human services classification of police work. Still, Kennedy (1983:44–45) concludes that, in general, the role of the police in America is consistent with a human services mission.

Egon Bittner (1974, 1975), in fact, suggests that it is the very ability to use force that places the police in the role of service provider. He writes (Bittner, 1975:44) that the capacity to use force is central to the police role, because

> in sum, the role of the police is to address all sorts of human problems when and insofar as their solutions do or may possibly require the use of force at the point of their occurrence. This lends homogeneity to such diverse procedures as catching a criminal, driving the mayor to the airport, evicting a drunken person from a bar, directing traffic, crowd control, taking care of lost children, administering medical first aid, and separating fighting relatives.

Especially in regard to service delivery, the police authority to use force means that efforts to resolve problems will not be opposed. The combination of constant and speedy availability with the capacity to enforce solutions makes the police an attractive service provider.

Indeed, recalling the history of the police in American cities, it is clear that public service was a primary responsibility of the early police. Monkonnen (1981) describes the role of early police in providing shelter, food, and employment assistance to the urban poor and homeless. This tradition of public service was disrupted with the reforms of policing that occurred in the earlier part of this century (Walker, 1977). One of these reforms, called *scientific policing* (Kelling and Stewart, 1991) focused police attention on law enforcement rather than service provision. Nevertheless, citizen demands for service from the police continued. In more recent times, police administrators see the value of service activities, particularly as a crime-prevention effort. Vern Folley (1989:558) notes that these leaders "suggest that law enforcement is a 'police service' " itself.

SERVICE FUNCTIONS OF POLICE

The reality of policing in America is that the law-enforcement and service functions of the police frequently overlap. We will explore this matter in more detail in Chapter 14, but it bears remembering that often the distinction between police crime control and police service is determined by the final actions of the officer. Thus, if the police invoke the criminal law, their actions are considered to be crime control. In contrast, if the police do not employ the criminal law, their actions can be considered as rendering service.

This use of the outcome of an encounter to define the nature of police reaction is perhaps most clearly visible in the case of dispute resolution. If the police are called to a domestic dispute, or an argument between neighbors, responding officers may or may not arrest the disputants. If an arrest is made, the police are engaged in crime control by reacting to an offense of some sort. If no arrest is made, the police will typically mediate the dispute or refer the parties to counseling. In the latter instance, the police have rendered service in an effort to assist the people involved with a problem in living. The normal tendency of police responding to these disputes is not to invoke the law by making an arrest (Buerger, Petrosino, and Petrosino, 1999).

The number and types of services police render to civilians vary by community and department. Some police agencies define themselves as service providers, whereas others prefer to emphasize crime control. Still, all local police agencies provide some non-crime-control services to their communities. These services may include emergency aid, licensing, provision of information, dispute resolution, lost-and-found services, and general safety functions.

Emergency Aid

Police officers are often the first to respond to life-threatening and less serious emergencies. Police officers are called to the scene of traffic accidents, heart attacks, fires and explosions, industrial accidents, chemical and other hazardous-material spills, airplane crashes, train wrecks, and all sorts of victimizing emergencies. In these cases officers are often required to provide emergency medical first aid and/or transport injured persons to medical facilities.

Natural and manmade disasters also elicit a police response. In the event of flood, hurricane, tornado, and other large-scale disasters, the police are expected to protect the safety of lives and property. Kemp (1989:171) observes that in such incidents the public expects the police to safely evacuate the area, assist with recovery services, and keep citizens adequately informed of the situation, among other things.

To these emergency-assistance activities can be added a host of other, perhaps less dramatic, forms of assistance expected of the police. Stranded travelers, motorists with car trouble, people who have locked themselves

Providing emergency medical aid is one of the service functions of the police. (Neil Schneider/*New York Post*)

out of their homes or automobiles and those experiencing similar difficulties frequently call on the police for assistance. Responding police officers may either provide direct assistance—for example, by breaking into the locked automobile—or refer the citizen to other service providers, such as a locksmith.

Licensing

Police agencies are often responsible for the issuance of a variety of licenses. Frequently citizens seeking permission to hold a parade or block party or carry a handgun must apply through the police. In addition to the issuance of licenses, the police are also responsible for any support services required of licensure. Thus, the citizen who secures a license for a parade typically also receives police assistance with traffic and crowd control at the event. Police agencies provide background investigations of applicants for many types of licenses including handgun permits, chauffer's driving licenses, and liquor licenses.

Passage of the Brady Handgun Violence Prevention Act of 1993 has expanded the role of police in firearms licensure (Gifford et al., 2000). Since 1994, over 22 million firearm transfer applications have been checked by law enforcement authorities. About half of these are checked directly by the FBI, but the other half are administered by state and local police agencies. Persons wishing to obtain handguns or other firearms from federally licensed firearms dealers must first pass a criminal history background check including the FBI's **National Instant Criminal Background Check System** (NICS). Twenty-four states do not have systems of background checks, and firearms dealers contact the FBI directly. Twenty-six states have identified "points of contact" where applicants go to secure needed background checks. In addition to the provisions of the Brady Bill, 38 states have state or local requirements for criminal background checks for firearms transfers. Of these, 19 assign this duty to state police agencies, 1 assigns it to courts, 5 to sheriff's departments alone, 5 to combinations of sheriffs and police departments, 3 to police departments alone, and the remainder to combinations of state and sheriff offices (2), state police and courts (2), or state and local police (1). Chart 13.3 shows the policing role in firearms transfer background checks in these 38 states.

As a result of their involvement in licensing, police agencies are naturally a source of information to citizens interested in applying for various permits. Further, the police agency typically maintains the files on licensees and applicants as well as providing any necessary inspection services required by the license. Police officers may be called on to periodically check safety at construction sites, investigate violations of liquor control laws, and otherwise ensure that citizens are complying with the requirements of various licenses. Ericson and Haggerty (1997) note how this involvement in licensure also contributes to the role of the police as a source of information.

Provision of Information

Much of the activity of patrol officers and police dispatchers consists of giving information to citizens about a host of issues. Travelers, for example, of-

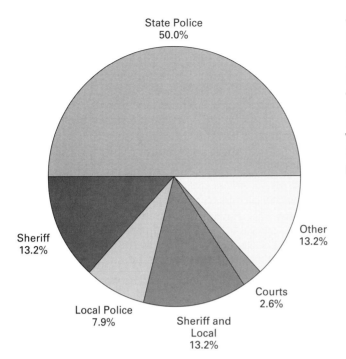

State Police
50.0%

Sheriff
13.2%

Other
13.2%

Courts
2.6%

Local Police
7.9%

Sheriff and
Local
13.2%

CHART 13.3
Firearms Transfer
Background Check
Responsibilities. (*Source:*
L. Gifford, D. Adams, and
G. Lauver (2000)
*Background Checks for
Firearm Transfers, 1999.*
Washington, DC: U.S.
Bureau of Justice Statistics,
p. 7.)

ten seek directions from police officers. Many times this information-giving activity is the police response to requests for assistance in areas that the agency has defined as beyond its scope.

Citizens might call the police to report a power outage or complain about needed street repairs. Often the police agency cannot do much about these types of problems. Rather than simply replying that the problem is not a police concern, the officers and dispatchers will refer callers to the appropriate office in which to lodge their complaint. On other occasions referring citizens to alternative services is the police strategy for solving the problem. Neighbors involved in a dispute over their property line, for example, might be referred to the civil courts.

More recently, police departments across the country have been given responsibility for notifying members of the public when sex offenders are released into the community (Farkas and Zevitz, 2000). Police agencies accomplish this notification in a variety of ways, including community meetings, door-to-door visits by police officers, posting fliers, mailing notices, and maintaining Internet sites (Adams, 1999). Several police agencies have also adopted what are called **reverse 911 systems.** These are automated telephone systems that allow the police department to send recorded messages to citizens. The systems are used to notify people about emergencies such as severe weather or simply to convey other information.

Dispute Resolution

One of the most common service functions of the police involves the resolution of disputes. Of course, one way to resolve a dispute (at least temporarily) is to arrest one or both parties to the argument. However, the police typically attempt to resolve disputes without recourse to the criminal sanction (Brown, 1981; Bayley and Garafalo, 1989; Kennedy, 1990). This resolution can be achieved through a number of strategies, including separating the parties, threatening arrest, direct counseling, and referral to appropriate agencies (Cooper, 1997).

The police are asked to settle disputes and disagreements between neighbors, spouses, retailers and customers, persons involved in automobile accidents, and almost any other circumstance where conflict between people causes a disturbance. All parties to a dispute think themselves to be in the right and typically expect the police officers to side with them. The officers, as neutral third parties, attempt to arbitrate or mediate the conflict.

Stephen Mastrofski et al. (2000) report a study of how officers decided to intervene in disputes. Based on observations of police officers in Indianapolis and St. Petersburg, they sought to explain police responses to citizen requests that the police control another person. The requests for control ranged from low (advise, persuade, warn, or threaten) to high (make the person leave or arrest the other person). In general, the police granted or partially granted the request by citizens for control of others. What the police did, however, was influenced by the amount of evidence available that the other person was at fault (or had broken the law), and the characteristics of the complaining citizen. The police were most likely to grant requests for control that came from citizens who appeared to have done nothing wrong, were credible (not intoxicated, mentally ill, or disrespectful), not closely related to the person they wanted controlled, and who asked the police to exert less control than legally possible. Given a credible complainant and sufficient evidence of crime, the police would arrest. Even here, however, the effect of the police was seen as limiting the complaining citizen's desire for more control. As might be expected in a dispute resolution situation, Mastrofski and his colleagues (2000) describe police officers as striking a compromise between the highest level of control requested by the complaining citizen and doing nothing to intervene.

Lost-and-Found Services

The local police agency serves as a community "lost and found" department. Persons who find money, wallets, or other property are expected to turn the found property in to the police, who will hold it in the event that the owner comes looking for it. So, too, with abandoned automobiles, the police are called to investigate the situation and remove or impound the car. The lost-and-found function extends to lost persons as well as property.

Missing persons are reported to the police, who then maintain a record of the identity of such people in the event that they are found or even actively seek out the missing person. Bishop and Schuessler (1989:332) report that over 10,000 missing-person cases are entered in the National Crime Information Center Missing Person file each month. In recent years there has been increasing concern about missing children in particular (Best, 1987; Hotaling and Finkelhor, 1990).

Perhaps the most dramatic missing-person cases occur when young children are reported as missing by their parents. Normally the typical police department will wait 24 hours before acting on the report of a missing adult, but when a young child is missing, response is quick. In the event that the child is not found in a relatively short period of time, the police department often mounts and organizes a search for the child. The human interest appeal of these searches guarantees widespread media attention.

Despite the great deal of concern about missing children today, what data are available indicate that the problem is not nearly as common as some have estimated. J. David Hirschel and Steven Lab (1988:43) studied the missing-person reports of one city and concluded that, "existing subjective estimates of

The police are frequently called to help locate and return missing children.

the number of missing persons may be somewhat high." Further, they noted that less than 3 percent of the over 800 cases they reviewed involved evidence that the missing person (adult or juvenile) had been taken by a stranger or was in danger at the time of disappearance (Hirschel and Lab, 1988:42–43).

A final category of the police lost-and-found function covers lost and stray animals. The position of animal-control officer is found in many police departments. This individual is responsible for catching and caging stray dogs and cats. In all local police agencies, general patrol officers may be called to the scene if a vicious or rabid animal is discovered. And, finally, when an exotic animal is loose, such as when a lion escapes from the local zoo, the police are called to assist in the search for and capture of the animal.

General Safety Services

As part of their overall mission to serve and protect, police officers become involved in a number of safety and preventive activities. Traffic control, for example, while incorporating a law-enforcement function in citing violators and arresting drunken drivers, is an essentially protective service. Officers directing traffic, investigating automobile accidents, and conducting safety inspections of vehicles are performing a safety service for the community (Gardiner, 1969). Hickman and Reaves (2001) report that 99 percent of police departments serve the functions of traffic law enforcement and accident investigation. Chart 13.4 shows the kinds of traffic and vehicle-related functions served by local law enforcement.

Many police agencies operate crime-prevention and child-safety programs designed to teach citizens how to protect themselves. Other programs are administered by the police in the hope of preventing future problems, including crime problems. Examples of such programs are DARE (National Institute of Justice, 1994) and a variety of youth after school programs (Chaiken, 1997) and other police–youth programs (Giacomazzi and

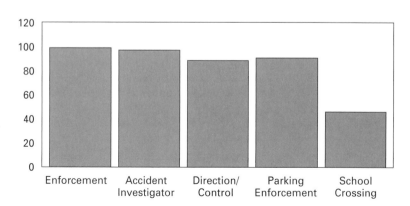

Chart 13.4
Traffic Functions of Local Police, 1999. (*Source:* M. Hickman and B. Reaves (2001) *Local Police Departments, 1999.* Washington, DC: U.S. Bureau of Justice Statistics, p. 8.)

Thurman, 1994). A survey of Texas residents reported that nearly 9 percent of those responding had attended some sort of community education function presented by the police (Hoover, Dowling, and Fenske, 1998). Similarly, the traditional responsibility of the watchman to observe dangerous conditions prevails. Police officers on patrol have discovered fires, dangerous road hazards, and other safety problems such as broken street lights or damaged buildings. Officers intervening in, or reporting, such circumstances are acting to enhance community safety.

RESEARCH ON POLICE SERVICES

Relatively little research deals directly with the service functions of the police. Most commonly, mention of police service activities has been limited to discussions of how little patrol officer time is spent on crime-control efforts (Webster, 1970; Boydstun, Sherry and Moelter, 1977; Tien, Simon, and Larson, 1978). As Whitaker et al. report (1982:61), "Despite their importance for performance measurement and planning, many police activities receive little attention and are not known in any systematic way by public officials . . . or the public at large."

The lack of available and detailed data on police service activities reflects the low priority given to such efforts by the police and the public. Particularly in the past few decades, the police role has increasingly been defined as that of crime control and law enforcement rather than service provision. Therefore, the importance of police service delivery has been found, in surface assessments at least, not in terms of how much of it the police accomplish, but in terms of how such services interfere with law-enforcement efforts. Yet, some data are available on the prevalence and types of police service delivery.

The Police Services Study

Between 1976 and 1978, the Police Services Study (PSS) intensively examined police patrol service in a sample of 60 neighborhoods within 24 jurisdictions in the Rochester, New York; St. Louis, Missouri; and Tampa–St. Petersburg, Florida, metropolitan areas (Whitaker et al., 1982; Ostrom, Parks, and Whitaker, 1978). This research involved observing the activities of patrol officers for 900 shifts of patrol work (15 shifts in each of the 60 neighborhoods). The results of this research indicated the types of services police patrol officers provided to the community.

For the 60 neighborhoods studied, 38 percent of police encounters with citizens involved criminal matters and another 22 percent were traffic related. The remaining 40 percent of police encounters were concerned with what the researchers called service (18 percent) and disorder (22 percent) calls. *Service encounters* included medical assistance, provision of information, dealing with dependent persons, and other general assistance. *Disorder assignments* included dealing with interpersonal violence and nuisance encounters such as noise disturbances, domestic arguments, and juvenile problems (Whitaker et al., 1982:168–169). In resolving these problems, officers made arrests in only 14 percent of cases. Aside from information-gathering activities such as interrogation or search, the most common officer action in encounters with citizens was to "lecture or threaten" the citizen, followed by "giving information."

The results of the PSS indicated that police officers intervene in as many noncriminal situations as criminal ones or possibly criminal ones. Even in criminal matters, the most common mode of officer intervention is to warn citizens or give them information rather than to apply legal sanctions. Much of what officers do, regardless of how they come to intervene with a citizen, is designed to resolve a problem without recourse to the law. In this regard, the PSS indicated that police officers render service to citizens a great deal of the time.

Whitaker and his colleagues conclude (1982:72),

> We have seen that in most neighborhoods, police patrols spend substantial portions of their time dealing with situations that do not involve crime. Often a majority of their time is spent on non-crime matters. Moreover, in most places police institute formal legal proceedings in only a small fraction of the encounters they have with citizens. Much of this activity concerns traffic violations or disorders rather than crime.

Earlier studies of police calls for service and patrol officer activities (Boydstun, Sherry, and Moelter, 1977; Tien, Simon, and Larson, 1978) reported similar results. The police devoted one-half or more of their time to interactions with citizens in matters not involving crimes in those studies. Further, the use of arrest or even citation was relatively rare. Instead, the police typically admonished, counseled, or advised citizens in an effort to resolve the problems that led to their intervention.

Identifying Service

One of the most troublesome problems in studying the police delivery of services is the diversity of situations in which they must act. For example, Tien and his associates (1978) did not have a category of *service* in their research. Whitaker and his associates (1982) used a *service* category in the classification of police officer activities, but this category did not include dealing with several types of nuisance calls, nor did it include police response to traffic accidents and road hazards.

Those studying police activities have not developed a standardized definition of what constitutes "service" activities. Even if such a definition were available, its application would be difficult because the motivation for service delivery may, in fact, be crime control. That is, it might be that police mount search-and-rescue efforts to find lost children so as to prevent the child from becoming a crime victim. If this is the case, the police effort, at least in the eyes of the police themselves, is one of crime control. Yet, as we have seen, study after study indicates that the police devote much attention to non-crime-related matters, and when criminal action possibly exists, they often still do not enforce the law.

Herman Goldstein (1979:242) argues that purpose of the police is problem solving: "In reality the police job is perhaps most accurately described as dealing with problems. Moreover, enforcing the criminal code is itself only a means to an end—one of several that the police employ in getting the job done." Sometimes an arrest is made for humanitarian rather than law-enforcement reasons. Thus, police officers may arrest children or homeless persons to ensure the provision of emergency medical care or lodging. Sometimes called *mercy bookings,* the use of arrest to ensure that a person receives some needed service is common in some places ("Forcing the mentally ill to get help," 1987).

The definition of police service activity is, then, very complicated. Although most of the available research distinguishes between service and law-enforcement activities based on either what the officer did (arrest/nonarrest) or what the citizen reported (crime/noncrime), the distinction is not so clear in practice. Instead, the police provide a problem-solving capability to the community that may take the form of a coercive (arrest) action. It is this capability that has led the police to deal with a number of recurrent problems that receive police attention.

Recurring Problems Requiring Police Service

Richard Lundman (1980: 104–108) has suggested that the typical police patrol officer responds to an overwhelming number and variety of situations. The officer is under departmental pressure to handle such situations efficiently (quickly), while the citizens involved view their problems as unique and complex. For a number of reasons, including the fact that the police officer can employ nonnegotiable force, the officer controls the encounter. *Nonnegotiable force* is Egon Bittner's term to describe the power of the police to coerce citizens, or to make them obey police commands.

To solve the conflict of needing to deal efficiently with situations complainants view as important and unique, the police develop a "shorthand" for categorizing problems and identifying solutions. Lundman (1980:105) describes it in these terms:

> Urban patrol officers therefore confront a serious problem: they are assigned too many calls to permit individualized policing of particular people and their troubles. That would interfere with efficiency. At the same time, they are constantly assigned calls requiring contact with

particular people who regard their difficulties as unique. Patrol officers construct and apply "typifications" and "recipes for action" as solutions to this problem.

The typification is a stereotype of the situation to which the officer is responding. For example, a *family beef* is a typification of domestic disputes based on the officer's prior work experiences and training. Though each domestic-violence situation may be unique, the officer prepares to respond to the typical "family beef." This classification process allows the officer to select a "recipe for action," or expected outcome. Again, based on experience, the officer knows what to do on arrival.

This process is not unique to police. All of us use typifications and their associated recipes for action in our lives every day. Physicians frequently deal with "the gall bladder" rather than with patient Smith. Professors often distinguish between "majors" and "nonmajors." College students select "introductory" or "advanced" courses. In each instance, a problem (medical diagnosis), person (student), or circumstance (class) is placed into a stereotype. This stereotype then guides behavior on the part of the person who uses it, at least until the specific problem, person, or circumstance violates the stereotype.

For police, Lundman (1980) suggests a basic distinction is drawn between those events likely to require "real policing" and other circumstances related to the job (see also Van Maanen, 1971). "Real policing" is the chas-

The police often deal with domestic violence and similar tragedies in the community. (David Handschuh/N.Y. Daily News Photo)

ing and catching of serious criminals, such as robbers, rapists, and murderers. These circumstances receive the full police response of rapid response with lights and siren, often involving multiple police units (Lundman, 1980:110–112). Other circumstances generally elicit a slow, deliberate response by a single patrol unit. It is these latter types of events that represent the bulk of police service activities, whether or not they ultimately result in an arrest. Two examples will permit us to examine police problem-solving services: special populations and traffic control.

Special Populations.   Social and economic changes in America over the past several years have increased the workload of police agencies in terms of their dealing with such categories of people as the homeless, mentally ill, and public inebriates. Whether from fear, annoyance, or humanitarian concern, citizens have traditionally called the police to handle problem persons in the community. The recent movement in the mental health field to deinstitutionalize the treatment of the mentally ill has resulted in larger numbers of mentally ill people in our communities (Panzarella and Alicea, 1997). Economic changes have likewise resulted in large numbers of homeless persons in the United States.

"The public repeatedly calls on law-enforcement officers for assistance with people who are mentally ill, drunk in public, and homeless," according to Peter Finn and Monique Sullivan (1989:1). Further, in recent years, requests for these services from police agencies have increased. At the same time, the options open to police officers in dealing with troubled and troublesome persons have been limited.

The President's Commission on Law Enforcement and Administration of Justice (1967) described the problem of public intoxication as leading to "2 million unnecessary arrests" each year. The commission noted that police were frequently called to deal with people who were drunk in public, and that although officers often used alternative dispositions, arrest of the public inebriate was a common occurrence. They recommended the decriminalization of public intoxication and the creation of detoxification centers throughout the nation. Rather than arresting offenders and using the criminal justice process, they urged, a medical solution should be applied to what they defined as a medical problem.

Many communities did indeed develop detoxification centers, but police use of these centers was limited for a variety of reasons. Force of habit, or familiarity with the use of arrest or informal orders to "move along," led to many police officers continuing their past practices. In many communities, adequate resources for the population of public inebriates did not exist. In still others, stringent admissions standards prevented the police from using these centers for those found drunk in public (Finn, 1988).

The mentally ill are another problem population with whom the police come into contact. The Police Service Study found that fewer than 1 percent of police encounters involved mentally ill or disturbed individuals, but other studies and conventional wisdom suggest that the problem, at least in urban areas, is more widespread. Linda Teplin (1986) reported that nearly 8 percent of police calls for service dealt with the mentally ill.

Beginning in the 1960s, mental health professionals initiated a policy of community treatment and deinstitutionalization of psychiatric patients (Scull, 1977). In the same time period, the rights of the mentally ill were strengthened as courts increasingly restricted the circumstances under which persons could be civilly committed against their will. The result was an increase in the number of persons with mental illness or other psychological/psychiatric troubles who were in the general population. A related effect was that it became more difficult for police to respond to these individuals, since the number of institutions was decreasing, and the procedures for commitment were more cumbersome (Finn and Sullivan, 1989). Linda Teplin (2000) has observed that persons with symptoms of mental illness are at higher risk of arrest than those not showing such symptoms. She suggests that this might be evidence of a "criminalization" of mental illness. That is, faced with limited options, the police might resolve problems involving the mentally ill by defining their behavior as criminal, and making arrests.

In contrast, Robin Engel and Eric Silver (2001) examined police handling of mentally disordered suspects and concluded that the police tended towards "informal" resolutions by ignoring, counseling, or otherwise calming and controlling the suspect without resort to physical force or arrest. Teplin (2000:11) has observed that experienced officers "often, . . . know just how to soothe the emotionally disturbed person, to act as a 'street-corner psychiatrist.' In this way, they help to maintain many mentally ill people within the community and make deinstitutionalization a more viable public policy."

The homeless are not, unfortunately, a new phenomenon. Rather, concern about the homeless and the numbers of homeless people who are estimated to exist in the population has grown tremendously in the past several years. Some experts estimate that as many as 3 million Americans are homeless. These homeless persons are a police concern for a number of reasons, including the fact that they might commit crimes. Perhaps as importantly, the homeless are also potential crime victims.

Although the contemporary concern about the homeless, and the problem this population poses for the police, have both increased recently, homeless persons have traditionally been a police problem. From the earliest days of the American police, when police departments provided lodging to the dispossessed, until the present, the homeless person has been a police target. Dilworth (1976) notes that concern about what to do with tramps and hobos was a key topic at meetings of American police leaders in the late 1890s and early 1900s. Similar discussions emerged during the Great Depression as thousands of unemployed people again took to wandering the country.

Today's homeless population appears to be different from those of earlier years. Many more women and children—and even intact families—find themselves within the homeless population than previously. The problems posed by this new population of the homeless are, however, essentially the same. These people are obviously suffering "problems in living" and are in need of basic services at a minimum. Further, they are potential criminals and potential criminal victims as a result of their precarious social and economic position. Yet, the ability of the police to intervene is limited.

Taken together, the mentally ill, public inebriates, and homeless (often there is overlap) constitute what are called **street people.** As Finn (1988:2) reports, "Even the most docile street people generate repulsion and fear among many residents, shoppers and commuters. The prospect of being accosted by a drunken, disoriented, or hostile panhandler can be as frightening for many people as the prospect of meeting an actual robber."

Avoiding places where one is likely to encounter street people, in turn, has negative consequences. Businesses suffer a loss of customers, and citizens suffer a loss of security and freedom. Usually there are too few shelters and treatment facilities to handle the full population of street people. Additionally, many of these persons do not seek admission to such facilities. Finally, the police are typically called to deal with troublesome street people—exactly the type of client that most shelters do not wish to admit. There is an unwillingness to accept police referrals in many of the existing shelters and treatment facilities (Finn, 1988).

In response to the growing problems of street people, police agencies across the country are developing responses to the needs of this population. Finn and Sullivan (1988) describe a dozen programs nationwide in which police agencies formed networks with social service organizations to respond to the problems of this population. The majority of these programs involved an agreement between the police and mental health agencies to deal with mentally ill persons coming to the attention of the police. One-third of the programs included arrangements for the handling of inebriates, and only one-sixth were designed to help with the homeless. The goals of these programs were to relieve police officials from the responsibility of dealing with persons whose problems were primarily psychiatric, medical, or economic. The programs also sought to prevent further criminal justice system attention being paid to the service population and to ensure that the police would refer the appropriate populations to the facilities involved. In this fashion, it was hoped that these networks would benefit all concerned—street people, service facilities, and the police themselves.

The goal of the police in dealing with street people or special populations is to resolve problems. Whether it is the problems of the street person or the problems of other citizens in dealing with street people, the police are called to resolve the issue. In some instances the problem is resolved by use of formal arrest, as when the police take homeless persons into custody to ensure they receive adequate housing and food for a while. In other instances the police refer and transport the person to some treatment or service facility.

Finn (1988) suggests two strategies employed by the police in dealing with street people. The first is strict enforcement, where the police routinely arrest and cite people and "maintain a strong pressure on them to keep moving." The second is benign neglect, where the police ignore street people as much as possible and intervene only in emergency or life-threatening situations. The development of the service networks described previously is an alternative to these traditional approaches. What all three approaches share is an emphasis on the *problem* of street people rather than on the *crimes* of street people.

Another special population includes those persons defined as "**vulnerable victims.**" Increasingly over the past few decades legislators have identified children and the elderly as being at especially high risk of abuse. One response to this recognition of the risks faced by children and the elderly has been the creation of special, mandatory reporting of child abuse or elder mistreatment. In most cases police officers are required to report instances of abuse, and in about one-third of the states, the police are responsible for investigating cases. The reporting and investigatory requirements are linked to crime control, the identification and arrest of abusers. In addition, however, these laws are designed to insure that child and elderly victims receive services. Often the police are expected to report suspected elder abuse to social-service agencies just as teachers must report suspected child abuse. A study of police reporting of elder abuse in Alabama by Daniels et al. (1999) reveals that police tend not to report elder abuse. Nonreporting occurs for a variety of reasons, including officer beliefs that the abuse was not serious, and the ability of officers to use criminal laws (domestic violence, assault, etc.) other than elder abuse. Daniels and his colleagues found that officers with more knowledge and experience with elder abuse reporting were less likely to report offenses. This, they conclude, is because these officers recognized that the system of investigation and service delivery was ineffective. They write (1999:223), "increased reporting of neglect is likely to increase dissatisfaction with the system [and] may partially explain the relatively low levels of reporting despite police beliefs that they are legally responsible to report." The central argument advanced by Daniels et al. is that states have placed an increased reporting and investigation burden on police who realize that the added effort does not translate into improved service for elder abuse victims. In response, the police often opt not to report suspected cases of elder mistreatment and either ignore less serious cases, or use other criminal laws to respond to more serious cases.

Traffic.   Noel Bufe (1995) notes that traffic-control responsibilities of the police date from the horse-and-buggy days. Since the early part of the twentieth century, however, growing concern over the safety of automobile traffic has prompted a wider role for the police. As Bufe (1995:776) puts it, police traffic services cover "everything the police do that relates to highway and traffic safety."

He breaks these efforts into six elements of police traffic services: traffic service administration, traffic engineering, ancillary services, accident investigation, traffic direction and control, and traffic law enforcement. *Traffic service administration* refers to the management and operation of traffic services within a police agency. *Traffic engineering* duties of the police are normally limited to identifying and reporting engineering-related hazards such as damaged bridges, road surfaces, dangerous bumps or curves, and the like. *Ancillary services* cover everything from traffic safety education through motorist assistance. *Accident investigation* is an information-collection process that yields data important to traffic engineering and useful to the courts. As the term implies, *traffic direction and control* covers police activities designed to constrain and direct the flow of traffic.

As with handling special populations, traffic law enforcement can be considered a police service function because of the goal of enforcement. According to Bufe and Thompson (1991:163), "The principal goal of police traffic services is to increase safety on the streets and highways." By extension, the goal of traffic law enforcement is to increase safety.

Bufe and Thompson (1991:165) discuss enforcement as the principal tool of the police in their efforts to reduce traffic accidents. Thus, the purpose of enforcing traffic laws is to encourage citizens to voluntarily comply

Traffic law enforcement includes not only invoking the law by issuing citations, but also maintaining the visibility of police units and issuing warnings.

with the traffic laws in order to enhance public safety. The objectives of traffic law enforcement do not require that citations be issued. "Maintaining the visibility of police units and issuing oral and written warnings are all considered parts of enforcement" (Bufe and Thompson, 1991:166). Moreover, "such things as information on violations as factors in accidents and general safe-driving practices are often as effective as the issuing of a citation in curbing unsafe driving activity" (Bufe, 1995:777).

The Police Services Study indicated that, on average, over 20 percent of police officer encounters with citizens involved traffic matters, yet in only about one-third of these encounters did police officers issue citations (Whitaker et al., 1982:70–71). Officers threatened or lectured citizens in nearly half of these cases. Studies of the enforcement of driving under the influence statutes also indicate that officers frequently decide not to arrest persons for violation of drunken-driving laws, a relatively serious traffic offense (Meyers et al., 1987; Mastrofski, Ritti, and Hoffmaster, 1987).

The provision of traffic services is both an important and unpopular function of the police (Bufe and Thompson, 1991). The enforcement of traffic laws tends not to be popular with either the citizen or the police officer (Kirkham and Wollan, 1980:72–73). Hoover, Dowling, and Fenske (1998) observed that citizen ratings of officer fairness in traffic stops decreased if the officer issued a citation rather than give a warning. It is not surprising that research indicates that officers are relatively lenient in their law-enforcement efforts in traffic, focusing instead on the service aspects of this function.

CHANGES IN POLICE SERVICE DELIVERY

At least since the 1960s (President's Commission on Law Enforcement and Administration of Justice, 1967), some have argued that police agencies should recognize their large service role and respond to it specifically. Since a majority of what police agencies do involves no law enforcement, these observers have suggested "civilianization" of police agencies. Gaines, Southerland, and Angell (1991:252–3) explain that "civilianization is the use of civilians instead of sworn police officers in positions not specifically requiring the authority of a sworn officer." Over the past two decades, most American police departments have increased the numbers of civilian employees in policing (Crank, 1989).

Non-sworn personnel are hired to perform a number of tasks within police agencies from research and planning through parking meter enforcement. Police operators or dispatchers are often civilian employees, and increasingly these personnel are being given service-delivery tasks as referral sources so that patrol officers need not be dispatched to some calls (Scott and Percy, 1983; Payne, 1991). Other police functions have been turned over to civilians as well, such as clerical and maintenance duties, acting as school crossing guards and parking meter attendants, and forensic analysis. In more recent years there has been renewed interest in using civilians for many of the traditional service duties of police officers.

The President's Commission (1967) urged the creation of three roles in the police service; a civilian *community service officer* (CSO), and the sworn positions of *police officer* and *police agent*. The police officer would operate much like a traditional police patrol officer, while the police agent would be a specialist in some aspect of investigation of crime. The CSO would be responsible for many of the non-law-enforcement functions of the police, such as emergency medical aid, assistance with lost and missing persons, traffic direction, and the like. The CSO was envisioned as an apprentice police officer.

A number of jurisdictions experimented with the CSO concept with generally positive results (Carter, Sapp, and Stephens, 1989). In another instance civilian employees were hired to serve as police service aides (Tien, Simon, and Larson, 1978). These aides assisted police officers or handled matters themselves in about one-third of calls for police service. Both programs were found to be useful tools for recruiting young people, especially minority group members, into careers in law enforcement.

The interest in civilian personnel, like the development of service networks and police–social worker teams described earlier, represents efforts to specialize police roles. Defining police service activities as "non-law-enforcement" leads administrators to define these positions as civilian. If civilians handle these service tasks, sworn personnel can be reserved for those police tasks where full police powers (arrest and use of force) are required, such as law enforcement. The problem remains, however, that often the civilian aide will not be able to resolve the problem. Recall that one of the main reasons the police provide so many services is because they can impose nonnegotiable solutions. The civilian employee lacking police powers is often no more able to impose solutions than are the citizens calling for aid.

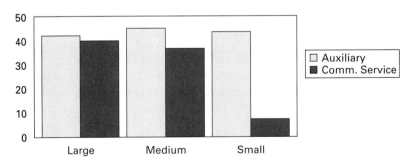

CHART 13.5
Percentage of Local Police Departments Using Auxiliary Police and Community Service Officers, 1999, by Size. (*Source:* M. Hickman and B. Reaves (2001) *Local Police Departments, 1999.* Washington, DC: U.S. Bureau of Justice Statistics, p. 4.)

CORRELATES OF POLICE SERVICE

All police agencies are required to provide some services to the citizens and communities they serve. Police service delivery, like police crime control, varies from place to place and from situation to situation. The types and styles of police service delivery correlate with factors at the three levels of community, organization, and people.

Community

The Police Service Study found that communities hold different expectations about the services police agencies should provide (Whitaker et al., 1982:32–33). Herman Goldstein has noted that the police are a multipurpose agency of municipal government. He wrote (1977:33), "Most of the noncriminal functions police now perform are not inappropriate tasks if a community concludes that the police agency is the logical administrative unit in which to house them." Thus, at least some of the differences in how police define their service obligations and the degree to which they engage in such activities is a reflection of community desires and needs.

Whitaker and his colleagues (1982:68) note, "There are differences in the kinds of problems officers confront in different places." They go on to say that not only do problems and police services differ from community to community, but they frequently also vary within jurisdictions by neighborhood (1982:72). In other words, the police appear to provide those services that the community requests. If this is true, then police service delivery is largely a reactive process.

It appears that police service functions are related to community characteristics as well. Police services are most often requested by the poor and in urban areas. Many police services are sought because no alternative service is available. Therefore, police agencies in poor urban neighborhoods are likely to face a heavier demand for a variety of services than agencies in more affluent, suburban neighborhoods. Paradoxically, Wilson (1968) observed that the service style of policing was most common in middle-class, suburban communities.

This apparent contradiction is really not that troublesome. Partly as a result of less pressing crime problems, suburban police agencies define their role as service, whereas urban police departments may be more likely to define their role as legalistic or watchman. Still, the frequency of calls for service, and the variety of situations for which police service is requested, may vary greatly between the two conditions. Consequently, the general-purpose urban police agency may deal with more criminal matters and may use the criminal law as a service strategy. The suburban departments, on the other hand, may deal with less crime and may use informal dispositions as a service strategy. Wilson relied on "arrest" as a key indicator of police department style, and we have seen that arrest is a strategy, or tool, of the officer.

More recently, LeBeau and Coulson (1996) investigated the link between neighborhood characteristics and types of police service requested. They found in one city that the timing of calls for service varied by neighborhood characteristics. Poor areas with high rates of unemployment experience more calls for service, and at different times, than do wealthier neighborhoods, in which most residents are employed. Both the demand for and delivery of police services seems to vary with time.

Organization

As with crime-control efforts, the activities of police officers are influenced by the constraints of the police organization. In regard to traffic enforcement, for example, both Wilson (1968:95) and Gardiner (1969) found that officers in different police departments issued traffic citations at widely different rates. In part, this difference is accounted for by the fact that the departments had varying norms or expectations of traffic citation "activity" among officers. Yet another difference among police agencies is the extent to which some service activities, such as mental health calls, traffic problems, or domestic disturbances, are the responsibility of specialist units.

Some police departments screen calls for service and refer citizens to other agencies over the telephone. Other departments routinely send a patrol officer to handle the call. In some agencies, especially the larger ones, certain service functions are specialized so that greater consistency (and arguably higher quality) in service delivery can be accomplished (Murphy, 1986). In other departments all officers are generalists and responsible for all types of service calls. Panzarella and Alicea (1997) report on a study of police handling of emotionally disturbed individuals. They found that while the agency they were studying had a special unit for these types of calls, the patrol officer was first on the scene, and the specialists were dispatched only when generalist patrol officers were unable to solve the problems.

Similarly, the amount of guidance or direction given to officers in terms of how to respond to service calls will vary among departments (Ruiz, 1993). Norms or quotas for traffic citations will generally ensure that officers issue a minimum number of tickets, though they will not necessarily ensure a stable pattern of traffic enforcement (Petersen, 1971). Likewise, the issuance of an arrest policy for domestic-violence cases will probably result in higher numbers of arrests in such circumstances but will not guarantee a uniform response of the department to the problem of domestic violence (Steinman, 1988). The types of services and ways in which they are delivered by police officers are partly a product of the structure and rules of the organization in which they work.

Belknap and McCall (1994) examined differences in referrals of domestic-violence victims to social services in one midwestern area. They found significant differences in regard to the likelihood of officers referring victims to services and in the types of services to which victims were referred. One of the strongest correlates of these differences was the agency for which the officer worked. Whether because of differences in training and resources or differences in agency philosophy, officer referrals varied depending on the organization for which the officer worked.

People

The delivery of police services to citizens is also partly a product of the people involved, both the citizens and the officers. We discussed earlier that the police officer selects between two primary strategies—arrest or informal disposition—in resolving citizen problems. This decision is often contingent on the demeanor and behavior of the citizen (Westley, 1970; Van Maanen, 1978; Reiss, 1971). The respectful, grateful citizen is treated informally, while the disrespectful or uncooperative citizen is handled formally. So, too, in many cases, the choice of a strategy in handling disputes depends in large part on the wishes of the citizen (Black and Reiss, 1970). If the complaining citizen asks for or wants an arrest, the police are more likely to make an arrest than if the citizen does not have such a preference. Recall that the complaining citizen is most likely to have his or her request granted by the police if he or she is "credible."

The other person involved in these service calls is the police officer herself or himself. Muir's (1977) study of police officers suggests that officers differ among themselves in terms of their desire to be service providers. Some officers see themselves as mediators and public servants. These officers are, theoretically at least, more likely to use informal strategies and to devote more time and effort to problem-solving activities than those officers who do not appreciate the service role. The way in which police deal with emotionally ill persons is also influenced by the attitudes of the officers involved (Panzarella and Alicea, 1997).

These and other attitudes of police officers appear to correlate with their actions in selecting between arrest and nonarrest solutions in traffic cases (Meyers et al., 1987). In the case of domestic violence, officer perspectives and attitudes affect their support for and compliance to departmental arrest policies (Steinman, 1988:2). Thus, the action taken by police in responding to a service call is a product, in part, of the characteristics of both the citizens and officers involved. Lynette Feder (1996) found that the presence of the domestic-violence offender at the scene was the most important predictor of an arrest being made by the police.

Again, policing in America represents a balance of forces at work in any given situation. Police service functions vary greatly in terms of the types of services and the frequency with which they are requested or provided. Although the outcome of any specific call for police service is difficult to predict, patterns of service calls and delivery strategies correlate with the characteristics of the communities, police organizations, citizens, and police officers involved. Knowledge of these characteristics improves our understanding of the police role in providing services to the community.

CHAPTER CHECKUP

1. How important is service delivery as a function of modern policing?
2. In what ways may the police be considered to represent a human services agency?
3. What types of services do the police routinely provide in American communities?
4. What was the Police Services Study?
5. How can service activities be distinguished from crime-control activities by police officers?
6. What are two recurring problems that require police services?
7. How have the police in America changed the ways in which they respond to service demands from the community?
8. How do the characteristics of communities, police organizations, and police officers affect the delivery of services by the police?

REFERENCES

Adams, D. (1999) *Summary of state sex offender registry dissemination procedures: Update 1999.* (Washington, DC: U.S. Bureau of Justice Statistics).

Bayley, D. and J. Garafalo (1989) "The management of violence by police patrol officers," *Criminology* 27(1):1–25.

Belknap, J. and K. McCall (1994) "Woman battering and police referrals," *Journal of Criminal Justice* 22(3):223–236.

Best, J. (1987) "Rhetoric in claims-making: Constructing the missing children problem," *Social Problems* 34:101–121.

Bishop, D. and T. Schuessler (1989) "Missing person file," in W. G. Bailey (ed.) *The encyclopedia of police science.* (New York: Garland):330–333.

Bittner, E. (1974) "Florence Nightingale in pursuit of Willie Sutton: A theory of police," in H. Jacob (ed.) *The potential for reform in criminal justice.* (Beverly Hills, CA: Sage):17–44.

Bittner, E. (1975) *The functions of the police in modern society.* (Washington, DC: National Institute of Mental Health).

Black, D. and A. Reiss (1970) "Police control of juveniles," *American Sociological Review* 35:63–77.

Boydstun, J., M. Sherry, and N. Moelter (1977) *Patrol staffing in San Diego.* (Washington, DC: Police Foundation).

Brown, M. (1981) *Working the street.* (New York: Russell Sage Foundation).

Buerger, M., A. Petrosino, and C. Petrosino (1999) "Extending the police role: Implications of police mediation as a problem-solving tool," *Police Quarterly* 2(2):125–149.

Bufe, N. (1995) "Traffic services," in W. G. Bailey (ed.) *The encyclopedia of police science,* 2nd ed. (New York: Garland):776–782.

Bufe, N. and L. Thompson (1991) "Traffic services," in W. Geller (ed.) *Local government police management,* 3rd ed. (Washington, DC: International City Management Association):159–184.

Carter, D., A. Sapp, and D. Stephens (1989) *The state of police education.* (Washington, DC: Police Executive Research Forum).

Chaiken, M. (1997) *Youth afterschool programs and law enforcement.* (Washington, DC: National Institute of Justice).

Cooper, C. (1997) "Patrol police officer conflict-resolution processes," *Journal of Criminal Justice* 25(2):87–101.

Crank, J. (1989) "Civilization in small and medium police departments in Illinois, 1973–1986," *Journal of Criminal Justice* 17(3):167–178.

Crank, J. (1998) *Understanding police culture.* (Cincinnati, OH: Anderson).

Cumming, E., I. Cumming, and L. Edell (1965) "Policeman as philosopher, guide and friend," *Social Problems* 12:276–286.

Daniels, R., L. Baumhover, W. Formby, and C. Clarke-Daniels (1999) "Police discretion and elder mistreatment: A nested model of observation, reporting, and satisfaction," *Journal of Criminal Justice* 27(3):227–238.

Das, D. (1987) *Understanding police human relations.* (Metuchen, NJ: Scarecrow).

Dilworth, D. (ed.) (1976) *The blue and the brass: American policing 1890–1910.* (Gaithersburg, MD: International Association of Chiefs of Police).

Engel, R. and E. Silver (2001) "Policing mentally disordered suspects: A reexamination of the criminalization hypothesis," *Criminology* 39(2):225–252.

Ericson, R. and K. Haggerty (1997) *Policing the risk society.* (Toronto: University of Toronto Press).

Farkas, M. and R. Zevitz (2000) "The law enforcement role in sex offender community notification: A research note," *Journal of Crime and Justice* 22(1):125–139.

Feder, L. (1996) "Police handling of domestic calls: The importance of offender's presence in the arrest decision," *Journal of Criminal Justice* 24(6):481–490.

Finn, P. (1988) *Street people.* (Washington, DC: U.S. Department of Justice).

Finn, P. and M. Sullivan (1988) *Police response to special populations: Handling the mentally ill, public inebriate, and the homeless.* (Washington, DC: U.S. Department of Justice).

Finn, P. and M. Sullivan (1989) "Police handling of the mentally ill: Sharing responsibility with the mental health system," *Journal of Criminal Justice* 17(1):1–14.

Folley, V. (1989) "Role of the police," in W. G. Bailey (ed.) *The encyclopedia of police science.* (New York: Garland):556–560.

"Forcing the mentally ill to get help" (1987) *Newsweek* (November 9):47–48.

Gardiner, J. (1969) *Traffic and the police: Variations in law enforcement policy.* (Cambridge, MA: Harvard University Press).

Gaines, L., M. Southerland, and J. Angell (1991) *Police administration.* (New York: McGraw-Hill).

Garmire, B. (1972) "The police role in an urban society," in R. Steadman (ed.) *The police and the community.* (Baltimore, MD: Johns Hopkins University Press):1–11.

Giacomazzi, A. and Q. Thurman (1994) "Cops and kids revisited: A second-year assessment of a community policing and delinquency prevention innovation," *Police Studies* 17(4):1–20.

Gifford, L., D. Adams, G. Lauver, and M. Bowling (2000) *Background checks for firearm transfers, 1999.* (Washington, DC: U.S. Bureau of Justice Statistics).

Goldstein, H. (1977) *Policing a free society.* (Cambridge, MA: Ballinger).

Goldstein, H. (1979) "Improving policing: A problem-oriented approach," *Crime & Delinquency* 25(2):236–258.

Greene, J. and C. Klockars (1991) "What police do," in C. Klockars and S. Mastrofski (eds.) *Thinking about police: Contemporary readings.* (New York: McGraw-Hill):273–284.

Haller, M. (1976) "Historical roots of police behavior: Chicago 1890–1925," *Law & Society Review* 10:303–333.

Hickman, M. and B. Reaves (2001) *Local police departments, 1999.* (Washington, DC: U.S. Bureau of Justice Statistics).

Hirschel, J. D. and S. Lab (1988) "Who is missing? The realities of the missing persons problem," *Journal of Criminal Justice* 16(1):35–45.

Hoover, L., J. Dowling, and J. Fenske (1998) "Extent of citizen contact with police," *Police Quarterly* 1(3):1–18.

Hotaling, G. and D. Finkelhor (1990) "Estimating the number of stranger-abduction homicides of children: A review of available evidence," *Journal of Criminal Justice* 18(5):385–399.

Kelling, G. and J. Stewart (1991) "The evolution of contemporary policing," in W. Geller (ed.) *Local government police management,* 3rd ed. (Washington, DC: International City Management Association):3–21.

Kemp, R. (1989) "Emergency management," in W. G. Bailey (ed.) *The encyclopedia of police science.* (New York: Garland):171–173.

Kennedy, D. (1983) "Toward a clarification of the police role as a human services agency," *Criminal Justice Review* 8(2):41–45.

Kennedy, L. (1990) *On the borders of crime: Conflict management and criminology.* (White Plains, NY: Longman).

Kirkham, G. and L. Wollan (1980) *Introduction to law enforcement.* (New York: Harper & Row).

Langan, P., L. Greenfeld, S. Smith, M. Durose, and D. Levin (2001) *Contacts between police and the public: Findings from the 1999 national survey.* (Washington, DC: U.S. Bureau of Justice Statistics).

LeBeau, J. and R. Coulson (1996) "Routine activities and the spatial-temporal variation of calls for police service: The experience of opposites on the quality of life spectrum," *Police Studies* 19(4):1–14.

Lundman, R. (1980) *Police and policing: An introduction.* (New York: Holt, Rinehart & Winston).

Mastrofski, S., J. Snipes, R. Parks, and C. Maxwell (2000) "The helping hand of the law: Police control of citizens on request," *Criminology* 38(2):307–342.

Mastrofski, S., R. Ritti, and D. Hoffmaster (1987) "Organizational determinants of police discretion: The case of drinking-driving," *Journal of Criminal Justice* 15(5):387–402.

Meyer, J. and W. Taylor (1975) "Analyzing the nature of police involvements: A research note concerning the effects of forms of police mobilization," *Journal of Criminal Justice* 3(2):141–145.

Meyers, A., T. Heeren, R. Hingson, and D. Kovenock (1987) "Cops and drivers: Police discretion and the enforcement of Maine's 1981 OUI law," *Journal of Criminal Justice* 15(5):361–368.

Monkonnen, E. (1981) *Police in urban America, 1860–1920.* (Cambridge, UK: Cambridge University Press).

Muir, W. (1977) *Police: Streetcorner politicians.* (Chicago: University of Chicago Press).

Murphy, G. (1986) *Special care: Improving the police response to the mentally disabled.* (Washington, DC: Police Executive Research Forum).

National Institute of Justice (1994) *The D.A.R.E. program: A review of prevalence, user satisfaction, and effectiveness.* (Washington, DC: National Institute of Justice).

Ostrom, E., R. Parks, and G. Whitaker (1978) *Patterns of metropolitan policing.* (Cambridge, MA: Ballinger).

Panzarella, R. and J. Alicea (1997) "Police tactics in incidents with mentally disturbed persons," *Policing: An International Journal of Police Strategies and Management* 20(2):326–338.

Payne, D. (1991) "Police dispatchers: Undertrained and underappreciated," *Footprints* 3(1–2):11–14.

Petersen, D. (1971) "Informal norms and police practices: The traffic ticket quota system," *Sociology and Social Research* 55(April):354–362.

President's Commission on Law Enforcement and Administration of Justice (1967) *The challenge of crime in a free society.* (Washington, DC: U.S. Government Printing Office).

Reiss, A. (1971) *The police and the public.* (New Haven, CT: Yale University Press).

Ruiz, J. (1993) "An interactive analysis between uniformed law enforcement officers and the mentally ill," *American Journal of Police* 12(4):149–177.

Scott, E. and S. Percy (1983) "Gatekeeping police services: Police operators and dispatchers," in R. Bennett (ed.) *Police at work: Policy issues and analysis.* (Beverly Hills, CA: Sage):127–144.

Scull, A. (1977) *Decarceration: Community treatment and the deviant: A radical view.* (Englewood Cliffs: Prentice-Hall).

Steinman, M. (1988) "Anticipating rank and file police reactions to arrest policies regarding spouse abuse," *Criminal Justice Research Bulletin* 4(3):1–5.

Teplin, L. (1986) *Keeping the peace: The parameters of police discretion in relation to the mentally disordered.* (Washington, DC: National Institute of Mental Health).

Teplin, L. (2000) "Keeping the peace: Police discretion and mentally ill persons," *NIJ Journal* (July):8–15.

Tien, J., J. Simon, and R. Larson (1978) "Police service aides: Paraprofessionals for the police," *Journal of Criminal Justice* 6(2):117–131.

Van Maanen, J. (1971) "Observations on the making of policemen," *Human Organization* 32:407–418.

Van Maanen, J. (1978) "The asshole," in P. Manning and J. Van Maanen (eds.) *Policing: A view from the street.* (Santa Monica, CA: Goodyear):221–238.

Walker, S. (1977) *A critical history of police reform.* (Lexington, MA: Lexington Books).

Webster, J. (1970) "Police task and time study," *Journal of Criminal Law, Criminology, and Police Science* 61:94–100.

Westley, W. (1970) *Violence and the police: A sociological study of law, custom, and morality.* (Cambridge, MA: MIT Press).

Whitaker, G., S. Mastrofski, E. Ostrom, R. Parks, and S. Percy (1982) *Basic issues in police performance.* (Washington, DC: U.S. Department of Justice).

Wilson, J. (1968) *Varieties of police behavior: The management of law and order in eight communities.* (Cambridge, MA: Harvard University Press).

ORDER AND THE POLICE

CHAPTER OUTLINE

The primary function of the police in America is to maintain order. We have seen that the police enforce the law and provide services. These two important, but sometimes conflicting, purposes are both ways to maintain order. In this chapter we will see how the police goal of order maintenance is achieved and how this task in turn influences law enforcement and service delivery.

THE POLICE AND ORDER MAINTENANCE

James Q. Wilson (1968:16) clearly asserts that order maintenance is the primary function of the police:

> The patrolman's role is defined more by his responsibility for maintaining order than by his responsibility for enforcing the law. By "order" is meant the absence of disorder, and by disorder is meant behavior that either disturbs or threatens to disturb the public peace or that involves face-to-face conflict between two or more persons.

Wilson distinguishes between law enforcement, where police more or less routinely apply the criminal law and begin the process of making a suspect liable to criminal penalties, and order maintenance, where officers use discretion in ambiguous circumstances to prevent or end disruptions. Unlike law-enforcement circumstances, where the actions of the officer are fairly predictable, the officer "approaches incidents that threaten order *not in terms of enforcing the law but in terms of 'handling the situation'*" (1968:31) [emphasis added].

Lorie Fridell et al. (2001:4) echo Wilson's position when they write, "While law enforcement is undeniably essential to maintaining good government, policing in a democratic society demands more. The police are essential to the fabric of society, not only as enforcers of first resort for federal, state and local laws, but also as moderators of behavior, keepers of the public peace and agents of prevention."

One factor that often complicates an analysis of police work is the almost infinite variety of means used to "handle" situations. Wilson (1968:31) notes that arrest is an option in order-maintenance activities. Goldstein (1977:28) observes that arrest is often used in order maintenance, whereas discretionary nonarrest decisions are common in law-enforcement situations. Thus, it is hard to distinguish when the police are enforcing laws from when they are maintaining order.

In fact, this distinction is relatively unimportant. What the police do, as Bittner (1974:30) notes, is intervene in situations where citizens see a need for someone to do something. In some cases, like serious crime (e.g., burglary, robbery, murder), the thing to be done is apparent—arrest the offender and thereby restore order. In other cases that people find threatening, uncomfortable, or disruptive (e.g., traffic accidents, disputes between neighbors, public intoxication), the thing to be done is not clear, as Wilson notes (1968:16). The task of the officer is to choose some intervention that will resolve this dispute or problem and preserve order, whether it is arrest, warning, counseling, or some other course of action.

The immediate task of the police officer, then, is the maintenance or restoration of order. The means by which the officer accomplishes this task differ from situation to situation and from one officer to another. Sometimes the criminal law and arrest are appropriate tools, sometimes less formal adjustments are better suited to preserving order. In traffic cases, for example, Bufe and Thompson (1991:167–168) suggest that officer decisions not to cite traffic violators, even when these decisions violate agency policies, are sometimes "the path of greatest public service." Should an officer cite a citizen for speeding if that citizen is rushing to the hospital during a medical emergency?

Order maintenance is, at the same time, the most important and the most difficult police function. It is important because, at base, the police were created to maintain order. We saw earlier, in Chapter 3, that creation of a formal police organization was the solution to growing fears about disorder in society. It is the most difficult because there is no clear definition of what constitutes order. In deciding how best to achieve or preserve order, police officers must exercise discretion.

POLICE DISCRETION AND THE REQUIREMENTS OF ORDER

Kenneth Culp Davis focuses attention on the exercise of **discretion** by criminal justice personnel. He writes (1969:4), "A public officer has discretion whenever the effective limits on his power leave him free to make a choice among possible courses of action or inaction." This definition says that a police officer has discretionary power whenever she or he can choose how to respond to a situation. In practice, that means that the patrol officer almost always has discretion. If we take the case of jaywalking as an example, we see the discretionary power of the typical police officer.

Assuming an ordinance against jaywalking, what happens if a police officer observes someone crossing the street improperly? The officer might cite the pedestrian for violation of the ordinance. Alternatively, the officer might stop the pedestrian and warn him or her not to jaywalk in the future. Or the officer may ignore the violation and do nothing at all. The officer may choose between courses of action (citation or warning) or inaction (ignoring the violation) within the limits of her or his power. *Discretion* means choice, or the power to choose.

Whether and how police officers intervene in all sorts of situations is a product of the exercise of discretion. In some circumstances the choice is relatively easy—arresting a bank robbery suspect. In others, the choice is more difficult—quelling a dispute between neighbors. In either event, the use of the criminal law is an option available to the officer.

Joseph Goldstein (1960) categorizes police discretion in terms of the application of the criminal law into two classes; invocation discretion and non-invocation discretion. As the term implies, **invocation discretion** refers to those situations in which the officer chooses to invoke or use the criminal law and thus issues a citation or makes an arrest. In contrast,

non-invocation discretion covers those circumstances where the officer could employ the law, but chooses not to do so.

Of the two classes of discretion, Goldstein is more concerned with non-invocation choices because these are, in his words, *low-visibility* choices. If the officer decides not to cite or arrest, no other authority will review that decision. If the officer arrests or cites, the officer's choice will be reviewed by a police supervisor, prosecutor, and/or court. Non-invocation decisions are troublesome because of their invisibility.

Suppose that a police officer observes two motorists speeding. The first is a male, whom the officer warns and does not cite. The second is a female, to whom the officer issues a citation for speeding. Suppose further, that this officer almost always cites female motorists and warns male motorists. Because the female motorists were, in fact, violating the law, their citations will be deemed justified by the officer's supervisor, the prosecutor, and the courts. Since the male motorists are unlikely to complain that they did not receive a deserved citation, the fact of discrimination will not be known, and the officer's behavior will not be checked or changed.

On the other hand, suppose that this officer routinely stops and cites female motorists whether or not they were breaking the law. Thus a woman traveling at the speed limit receives a citation for speeding. If the case goes to traffic court, the judge is likely to dismiss the charge, thereby checking, and controlling, the unfair discretionary decisions of the officer. The officer's decisions in both hypothetical scenarios are equally unfair and discriminatory, but the decision to invoke the law was reviewed and could therefore be corrected. This is why Goldstein (1960) is most concerned with the low-visibility, non-invocation decisions—they are not likely to be caught and corrected. Stephen Mastrofski (1999:7) observes, "Unfortunately, American police organizations remain virtually blind about what their police do in response to incidents that are not classified as a crime or traffic accident. . . . What the officer did remains a cipher unless an arrest was made, a citation was issued, or a crime report was filed."

In a later work, focusing specifically on police discretion, Kenneth Davis (1975) notes that by virtue of their ability to choose which laws to enforce, when, and against whom, the police effectively act as policy makers. It is the police, not the legislature, that decides which behaviors are to be controlled by the criminal law and which behaviors can be tolerated. Davis suggested that this power of the police be recognized by both the police and the legislature, and that the police administration devise rules and regulations to give guidance and structure to the discretionary decisions of officers.

Herman Goldstein (1977:124–126) echoes the sentiments of Davis, also calling for the structuring of police officer discretion through the development of rules and regulations. An example of such rule making can be seen in the common practice of police agencies developing tolerance limits for speed-limit enforcement. Many police agencies have policies to guide officers in their traffic enforcement efforts. Such a policy might, for example, instruct officers generally to ignore or at least not issue citations in situations where motorists are not exceeding the posted speed limit by at least five miles per hour (5 mph). Such an agency has a 5-mph tolerance limit, because officers are expected to tolerate "a little speeding," as long as it does not exceed a certain limit. This **tolerance** is based on both an expectation that motorists may bend the law a little and the fact that speedometers are neither always accurate nor standardized.

Obviously, this tolerance limit does not necessarily control officer actions. The police officer can still ignore those drivers exceeding even the tolerance limit and can cite those exceeding the speed limit but staying within the tolerance limit. What such a policy does, however, is instruct the officer that, other things being equal, it is permissible (even desirable) to ignore or warn those who marginally exceed the speed limit, and that it is expected that those traveling in excess of the limit by more than 5 mph will be cited.

Through such policies the police administrator is able to direct the discretionary behavior of officers. While not guaranteeing control over police officer discretion, these policies are expected to result in more consistent and uniform decisions by the police, regardless of which particular officer is involved. The policies are designed to promote "order" in the behavior of individual officers on patrol.

THE MAINTENANCE OF ORDER

Perhaps the most pressing problem facing the police in their efforts to maintain order is the ambiguous nature of the term **order** itself. James Wilson (1968:21–22) observes, "The difficulty, of course, is that public order is nowhere defined and can never be defined unambiguously because what constitutes order is a matter of opinion and convention, not a state of nature." The police must rely, at least in part, on public expectations and perceptions for their definition of order.

"Broken Windows" and Public Order

Fifteen years after identifying the importance of the order maintenance function for the police, Wilson collaborated with George Kelling (1982) to better describe the link between the police and order maintenance. Based on the experiences of the Newark Foot Patrol Project, they began by trying to reconcile the apparent incongruity that citizens felt safer with foot patrol in an area even though the actual rate of crime in patrolled areas remained unchanged or even increased. They attributed the increase in citizen perceptions of safety to the success of foot patrol officers in maintaining order. Wilson and Kelling (1982:30) summarize the impact of foot patrol officers in these words: "What foot-patrol officers did was to elevate, to the extent they could, the level of public order in these neighborhoods."

Wilson and Kelling titled their article "Broken Windows" as a way to describe the relationship between public order and public safety. Their thesis is that people who live in or travel through dilapidated or decaying neighborhoods fear for their safety. The source of this fear is not so much the actual likelihood that they will be victimized, but rather a sense that there is no concern and help available. If a window in a neighborhood building is broken and not repaired, it signals that no one is responsible for maintenance, and therefore that no one cares about what happens in that neighborhood. In short order, Wilson and Kelling claim, other windows will be broken, reinforcing the perception that the area is "out of control" and thus dangerous. A well-maintained area, on the other hand, signals that someone is in command.

The task of the police is to establish and maintain the rules in the neighborhood—to see to it that the broken windows are repaired. This effort at maintaining order pays off when residents and visitors see that behavior is controlled and thus feel safer. The feeling of safety, in theory at least, translates into a willingness to be out in public, which in turn makes it truly safer to be out (Skogan, 1986). The presence of larger numbers of law-abiding persons on the streets makes all of these people safer from attack. The conclusion of this argument is that the police should focus their attention on

The police provide crowd control at large gatherings to maintain public order.
(Tom Kelly)

neighborhoods with a few broken windows to prevent their further decay. By establishing or maintaining order in these areas, the police prevent the fear of crime and, in fact, improve the quality of life for residents.

Herman Goldstein (1977) anticipated the arguments of Wilson and Kelling. He urged police administrators to distinguish between combating the actual incidence of crime and fighting the fear of crime. Of the two tasks, he suggests, it is perhaps more important for the police to reduce citizen fear of crime. While not tying his argument to the order-maintenance function of police, Goldstein (1977:47–48) suggests that what police do to control fear might be different from what is required to control actual levels of crime:

> A police agency might put together an entirely different blend of services in dealing with fear, which would contain, in addition to a massive effort to educate the community, variations in the usual form of patrol, the increased use of technical surveillance equipment, and a campaign to acquaint citizens with methods for providing themselves with security at their own expense.

In a later book, Kelling and Coles (1996) more fully present the broken-windows theory of the link between order and crime. They compare a traditional criminal justice perspective with what they call *community-based prevention* (1996:240–241). Their conclusion is that police involvement in co-producing order with the community holds promise of bringing crime under control.

Lawrence Sherman (n.d.:2) summarizes contemporary thinking on the link between order and safety:

> Both physical and social signs of crime indicate disorder in the neighborhood and convey a sense that things are "out of control." Ultimately, disorder may attract such predatory violent crimes as robbery; a neighborhood that can't control minor incivilities may advertise itself to potential robbers as a neighborhood that can't control serious crime either.

That is, disorder as evidenced by physical decay and interpersonal incivility creates a climate conducive to crime. The maintenance of order by the police is a strategy to ultimately control the levels of crime that occur. But more than its possible effect on levels of crime, the maintenance of order is a goal of the police agency in its own right.

POLICE DISCRETION AND PUBLIC ORDER

Kenneth Davis (1975:73) writes,

> Criminal law has two sides—the formality and the reality. The formality is found in the statute books and opinions of appellate courts. The reality is found in the practices of enforcement officers. Drinking

in the park is a crime according to the ordinance, but quietly drinking at a family picnic without disturbing others is not a crime according to the reality of the law, because officers uniformly refuse to enforce the ordinance in such circumstances. When the formality and the reality differ, the reality is the one that prevails. When the officer says, "I won't interfere if you drink quietly," the words of the ordinance, the formality, are superseded by the enforcement policy, the reality.

In his study of police discretion, Davis uncovered numerous instances in which the Chicago police failed to enforce criminal statutes. This nonenforcement occurred despite the fact that state law and city ordinances directed police officers to enforce all laws at all times. It occurred despite the fact that the Chicago Police Department had a stated policy of full enforcement of the laws. In reality, patrol officers, not the department, decided enforcement policy. Davis writes (1975:2), "Chicago policemen have to decide what to do whenever they are confronted with a problem. Enforcement policy is made mainly by patrolmen."

The tendency of patrol officers is to underenforce the law in cases involving minor or less serious offenses (Wilson, 1968; Brown, 1981). Thus, the police exercise non-invocation discretion more often than they do invocation discretion. The patrol officer judges the appropriateness of a given response (to invoke the law or not) based on several factors. One of the first tasks facing an officer who responds to a disorderly situation is to gain control. As Pepinsky (1975:31) puts it, responding officers are motivated "to try to make manifest to themselves and to others that they are in control of police–citizen interactions, and that this control is legitimate and identifies the police as occupying a respectable status within society." In doing so, the police officer makes an important discretionary decision by defining the situation as criminal or noncriminal.

Pepinsky describes police responses to citizen complaints as being either universalistic or particularistic. A **universalistic response** is one where the officer bases the decision on characteristics of the situation—domestic assaults, neighbor disputes, vandalism. **Particularistic responses** are based on the characteristics of participants, and the police–citizen relationship—respectful or disrespectful, contrite or defiant, strangers or intimates.

Pepinsky (1975:32) suggests that universalistic approaches take time to develop and hence are more commonly used by veteran officers. These situation-based responses are designed to meet public expectations of the officer as the officer perceives them. For example, the officer may believe that the community expects or tolerates social drinking in parks by citizens attending picnics or softball games but is intolerant of public drunkeness in the park. Fistfights between teenagers might be normal and expected, but fights involving weapons are to be considered criminal.

Particularistic approaches are more closely tied to the immediate situation confronting the officer. Whereas universalistic tactics serve to reinforce community norms as perceived by the officer, particularistic ones support the officer's claim to legitimacy. Invoking the criminal law against a disrespectful citizen establishes the authority (or at least the power) of the officer. In conflicts between citizens of different status, police support for the higher-status citizen identifies the police with the higher, rather than lower, status group (Pepinsky, 1975:37–39).

Pepinsky's major point is that the officer's definition of the situation as being one that warrants formal law-enforcement activity is a discretionary decision. Particularly in what are order-maintenance situations, the officer is faced with ambiguous information from conflicting sources. Typically, both (or all) parties to a dispute feel that they are in the right and that their opponents are wrong. The responding officer seeks to resolve the dispute and restore order. Though the law may have been broken, law enforcement is not the officer's goal. The officer seeks to control the situation.

James Wilson (1968:32), borrowing from Egon Bittner (1967), gives the example of "a patrolman finding four men getting drunk in public; to control the situation—to prevent a disturbance that will bring a complaint, to break up the gathering, to prevent someone passing out and getting robbed—the patrolman may arrest one of the four and, having broken up the party, send the other three on their way. To him, which one gets arrested is not so important. To the judge, it is all-important."

In this example, the officer could, and some would argue should, arrest all four men. If the officer's purpose is law enforcement, arresting all four would be appropriate. On the other hand, if the officer seeks to control the situation and to disband the group, the arrest of any one of the four is likely to be a sufficient tactic. Of course, any nonarrest strategy that accomplished the officer's goals would also be acceptable.

The ambiguous nature of order-maintenance situations, coupled with the goal of maintaining order, gives rise to the exercise of discretion by police officers. Often the legal authority of the officer to intervene in a dispute is unclear, or the criminality of the case is not readily apparent. The path of least resistance is to avoid invoking the criminal law. Yet, the officer must take charge, and the criminal law is perhaps the most powerful tool available. Thus, the officer must choose whether and how to intervene.

Ambiguity

The nature of most order-maintenance calls to the police is unclear, at least to the responding officer. The dispatcher may instruct the officer to "see the woman" or simply report that there are "suspicious persons" at some address. Armed with that limited information, the officer must react to the circumstances he or she finds on arrival. Once at the scene, the officer seeks cues that alert him or her to the appropriate course of action.

In the words of James Wilson (1968:17), "Some . . . examples of disorderly behavior involve infractions of the law; any intervention of the police is at least under the color of the law and in fact might be viewed as an 'enforcement' of the law. A judge, examining the matter after the fact, is likely to see the issue wholly in these terms. The patrolman does not." What the officer does is dispense justice in the sense of attributing blame for the disorder.

When the officer arrests the husband in a case of domestic violence, the officer effectively says the husband was wrong and the victim was right. If an officer tells a neighbor to turn down the volume on her or his stereo, the officer tells the complaining citizen that he or she was right, the neighbor with the stereo is to blame for the disturbance. When the officer tells disputants to "knock it off," the message is that both (or all) parties are "to blame" for the disturbance.

Jerome Skolnick describes this peacekeeping responsibility of the police as *directing* the citizenry and *regulating* public morality (1994:54–55). Most often, the direction comes as a result of the officer's understanding of the public morality. If the public morality supports consideration of neighbors, the officer directs citizens to control the volume of their stereos, radios, parties, and so on. As public morality can be expected to differ among different subgroups of the public, it is almost always ambiguous. That there is a disturbance indicates that the participants, at least, hold differing views on what behaviors are tolerable or acceptable. The "losers" in the police intervention are likely to question the authority or right of the police officer to intervene, whether or not they comply with the officer's resolution.

The laws themselves, when they may have been broken in cases of public disturbance, are also ambiguous. What exactly is public intoxication or disorderly conduct, and how is the officer, arriving after the fact, to establish that someone is probably guilty of either offense? Davis believes that the **overreach of the law** causes police discretion because legislatures pass statutes that are too broadly worded, leaving it to the police to determine whether the law applies to a specific circumstance. The officer must assess the facts as she or he knows them in relation to the relevant statute. For example, does a prohibition on gambling apply to a friendly poker game among friends? Thus, the overreach of the law, covering behaviors that many people do not consider particularly wrong, along with the vague wording of the law, contribute to the need for officers to use discretion.

Furthermore, the facts are often ambiguous. Imagine an officer responding to a dispute between two neighbors over possession of a garden hose. Each neighbor claims the hose; neither has a receipt or other proof of ownership. One is accusing the other of theft. If you are that officer, what do you do?

Order-maintenance activities of the police are vitally important to a community but are the most difficult of police tasks. The officers involved frequently contend with ambiguous laws, unclear facts, and divergent citizen expectations of outcome. In picking their way through this tangle of uncertainties, officers exercise discretion to select what they think will be the best solution to the specific problem.

The Goal of Order

We have already noted that order is a condition not easily defined or identified. In this regard, the ambiguity that characterizes the order-maintenance efforts of police extends to their goal as well as their tactics. How does the officer know when order has been restored or maintained? As Wilson and Kelling (1982) note, order varies by neighborhood, so that what is tolerable in one area is intolerable in another. Further, Kelling and Stewart (1989:2) add the observation that order is likely to vary in the same area over time, so that what is acceptable or normal during the day may be unusual or disorderly at night. A number of observers have noted the problems in defining order and in establishing acceptable levels of order.

David Lynes (1996) has observed that police efforts to establish and maintain order are complicated by disagreements among citizens over what is appropriate and inappropriate behavior. He notes that the police are charged with the task of reproducing existing order. But in areas experiencing change, order itself is difficult to define, and police are ill prepared to adapt.

The primacy of order, however defined, also constrains the officer's choice of tactics. For example, during the 1960s, the initial response of police agencies to riots and other large-scale civil disturbances was to put on a show of force. Large numbers of police officers equipped with riot gear would be positioned in the area of the disturbance on the assumption that this display of power would deter demonstrators. Only after a few spectacular failures did police learn that such displays had the unfortunate tendency to provoke, rather than inhibit, violence (Ahern, 1972). The intervention of police, and particularly the use of arrest in some order-maintenance situations, can generate further conflict and controversy rather than establish police control.

A number of observers have remarked that the public will not countenance full enforcement of all the laws (Wilson, 1968; Pepinsky, 1975; Skolnick, 1994). Rather, the public and the police both expect that the laws will be selectively enforced—that the police will use their discretion to apply these broad laws in situations where order demands their application and ignore them when they are not necessary.

Full enforcement of all laws, even if possible, would most likely alienate police further from citizens, and in many cases might actually lead to disorder. Many cities have large public celebrations that attract thousands of spectators. Despite ordinances prohibiting public consumption of alcoholic beverages, for example, the police typically tolerate open displays of drinking. An attempt to control this behavior would probably overwhelm the resources of the police, but even if it did not, it would antagonize the crowd. Sufficiently provoked, the crowd could turn ugly, resulting in a riot. Thus, arrests for public drinking in these circumstances are reserved for those participants who have become disruptive after drinking. In these cases, the police action is viewed as justified by other citizens. In fact, they may be glad that the offending drunk has been removed.

As mentioned at the start of this chapter, order maintenance is the main goal of policing. The decision about whether to arrest and whom to arrest often depends on the officer's assessment of the effect of that arrest on order. Thus, the law-enforcement function is evaluated according to its contribution to the purpose, or goal, of order maintenance.

The same is true for many services rendered by the police. Traffic regulation in particular exemplifies the dominance of the order goal, as its primary purpose is to ensure order on the roads. Thus, when traffic is moving at a rate exceeding the posted speed limit, rather than detain all of the speeders, the police typically pursue only those traveling faster than the others.

In many ways, then, law enforcement and service delivery are two strategies by which officers can restore or maintain order in the community. The choice of strategy by police officers is based on an assessment of what action will best serve the interests of order maintenance. This judgment, in turn, reflects the officer's conception of order and his or her perception of what the citizens of the area define as tolerable.

Definitions of Order

Order cannot always be clearly defined. Some instances of disorder—riots, for example—are easily defined and observed, as are some instances of order, such as routine motorist behavior at a traffic light. But between these two extremes one finds a variety of behaviors. At some point behavior becomes too unpredictable, and order is threatened. Identifying this exact point where order becomes disorder is the problem.

Traditionally, the police have relied on citizens to define the point at which disorder has occurred through calling for police intervention. This is especially true of what can be called *private disorder.* Michael Brown (1981) distinguishes between those disputes that occur between strangers in public settings (*public disorder*) and those between intimates or in private settings. He suggests (1981:204), "The absence of any personal relationship between the antagonists and the ever present potential for a larger disturbance put a public disturbance in a different light from a private one." Although the police may rely on a citizen complaint in both instances, the officer feels freer to act on her or his own initiative in cases involving public disturbances than in those that occur in private.

In part, this reflects the officer's feeling that her or his perception of public expectations about order are appropriate in public settings. Private disputes, in contrast, involve the particular, personal definitions of order held by the disputants themselves. Conflicts between strangers are immediate. The sources of the conflict tend to be of recent origin. Disputes between intimates, however, may have a long history that will not be known to the officer. Ignorance of the origin of a dispute renders it difficult for the officer to assign blame in the order-maintenance endeavor. Here the officer tends to rely more heavily upon the desires of the complaining party to define what would be an appropriate response. As Brown (1981:303) puts it, "The reluctance of citizens to prosecute reinforces the predisposition to interpret the law in this manner (as not applying to dispute resolution)." Conversely, if the complainant desires a law-enforcement strategy, the police will generally comply.

A second source of definition for order comes from the police. Beyond the *public understanding* of what constitutes order, the responding police officer has an *occupational definition* of order in the immediate situation. That definition concerns (1) the officer's ability to take control and (2) the citizen's willingness to grant or recognize the authority of the officer. Brown (1981) and Van Maanen (1978) have described what is often known as the *attitude test.* Regardless of the nature of the dispute or who summoned the police, officers on the scene are interested in controlling the people involved during the time the officers are present.

Citizens who refuse to cooperate with officers, or who otherwise challenge the legitimacy of the officer's intervention, create disorder, or at least prevent the restoration of order, as far as the officer is concerned. To establish his or her authority so that the dispute can be controlled, the responding officer will deal harshly with the citizen who refuses to cooperate. Thus, the officer will focus first on making the immediate situation orderly and will later attend to the cause of the conflict, and possibly the long-term maintenance of order.

Order, as it is perceived by police officers, is a composite of several important dimensions. First it is a temporal dimension in that the officer emphasizes the immediate situation over any past or future disorderly events. Second, the officer tends to distinguish between public and private disorder, feeling more confident in dealing with public disorder on her or his own initiative. Finally, the officer seeks cues from the audience—the disputants, the complainant, and any other witnesses—about how best to handle the situation.

Developing Orderly Order Maintenance

Over the past few decades there has been an effort to control police officer behavior, especially in the gray areas of order maintenance. National campaigns to "criminalize" drunk driving and domestic violence had their genesis in dissatisfaction with police leniency. That police often ignored or trivialized drunk driving and spouse assault led some to call for

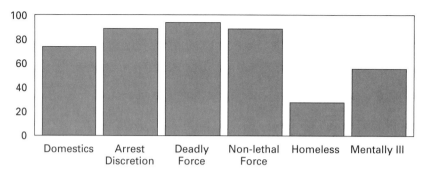

CHART 14.1 Percentage of Local Police Departments Having Written Policies, 1999. (*Source:* M. Hickman and B. Reaves (2001) *Local Police Departments, 1999.* Washington, DC: U.S. Bureau of Justice Statistics, p. 13.)

mandatory arrest and mandatory criminal penalties. Redefining these behaviors as serious crimes, and obligating the police to treat them formally were expected to change how officers reacted to these events. In addition to legislative changes, police departments have developed policies and procedures to guide officer actions in many ambiguous circumstances, as is seen in Chart 14.1.

At the same time, there has been a growing movement to expand the role of the police in dealing with "problems" so that police officers are expected to exercise broader discretion in responding to chronic problems such as domestic disputes (Buerger, Petrosino, and Petrosino, 1999). Recognizing that arrest is not always the most effective solution, police are increasingly expected to expand their reactions to crime and disorder problems (Mastrofski, 1999). The problem of domestic violence illustrates these recent developments.

Domestic Violence. The police response to domestic violence has been the focus of increasing attention and experimentation (Sherman and Berk, 1984; Johnson, 1990). The police are frequently called on to settle disputes among people, and disputes among family members are a substantial part of this workload. Because of the special and intimate relationship between the parties in family disputes, however, the police have been unwilling or unable to treat these incidents as law-enforcement matters. The domestic dispute has traditionally been seen as a social service matter, and an unpleasant one at that.

The traditional police response to domestic disputes has been to proceed slowly to the scene in hopes that the dispute will have ended before the officers arrive (Lundman, 1980). Once on the scene, the officers typically attempt to calm the parties and to mediate or arbitrate the dispute. Officers are, however, uncomfortable because of the nature of the dispute. As Potter (1978:41) describes it: "The officers separate the battling spouses, try to calm them down, and tell them to talk out, rather than fight out, their problems. They go on their way. For all the police know, the husband may start pounding the wife again as soon as they go out the door."

An almost legendary reluctance of victims of domestic assault to cooperate with prosecution was often given as the reason police did not attempt to invoke the criminal law. As James Wilson (1968:24) observes, in the middle 1960s: "A typical case, one which I witnessed many times, involves a wife with a black eye telling the patrolman she wants her husband, who she alleges hit her, 'thrown out of the house.' The officer knows he has no authority to throw husbands out of their homes and tells her so. She is dissatisfied. He suggests she file a complaint, but she does not want her husband arrested."

Police officers, like most of the public, are embarrassed and uncertain in incidents of domestic disputes. These things are private matters to most people, and not the kind of situations into which strangers, even police, wish to intervene. Further, in nearly one-half of this country's states, the officers must actually observe a misdemeanor assault or battery before they can make an arrest. The legal authority of the police to enter the home or make an arrest is generally cloudy. In short, the police face an ambiguous situation where there is an expectation that they should resolve the immediate dispute but little guidance about how that should be done (Brown, 1984; Sherman, 1985).

In the latter 1960s, consistent with a general trend to deemphasize the use of the criminal sanction, police agencies began to routinely prefer a policy of nonarrest and mediation in family disputes. In New York City the police experimented with the development of a specialized unit to respond to domestic disputes (Bard, 1970). The Family Crisis Intervention Unit was trained to deal with domestic problems and to provide both immediate counseling and referrals for followup services by community mental health agencies.

Similar programs were developed in other jurisdictions. Treger (1976) reported on the development of police–social work teams in which police officers worked cooperatively with social workers to solve the long-range problems of persons seeking police assistance, including those calling about domestic disputes. Potter (1978) reported on a number of referral programs throughout the nation in which police and social service agencies cooperated in assisting the victims of domestic violence. Higgins (1978) described similar police–social work teams operating in both the United States and Canada. Hanewicz and his colleagues (1982) described several similar projects but noted that most failed to ensure that referrals to social service agencies were followed by the disputants. They describe a program in Michigan (1982:495) in which sheriff's deputies cooperated with local social work agencies to encourage disputant contact with referral agencies. The effort of calling back people who had been the subjects of a domestic-dispute call resulted in a significant increase in their contacting community social service agencies.

Thus, into the latter 1970s, the emphasis in police practice in regard to domestic disputes was to refer parties to social or mental health service agencies and to avoid the use of the criminal law. Across the country a number of programs were implemented or planned that sought to improve the effectiveness of police intervention in such disputes by creating stronger links between the police and social service agencies. However, in this same period the noncriminal handling of domestic violence was the target of growing criticism.

In 1980 Sarah Berk and Donileen Loseke published an assessment of police responses to domestic violence. They (1980) conclude that the actions of the officers responding to such calls were based on the officers' perceptions of the situation. Arrest was used when the officer believed it was the only way to end the domestic conflict at that time. A replication by Worden and Pollitz (1984) essentially confirmed the findings of Berk and Loseke, concluding (1984:90), "Arrests are made when the circumstances indicate to the officer that the situation requires legal rather than less formal measures."

In 1977 the Police Foundation published a study on police work associated with domestic violence in Kansas City and Detroit (Breedlove et al., 1977). This study indicated that police were often called to the same address in response to domestic violence. In Kansas City in particular, the vast majority of homicides stemming from domestic violence occurred between people whom the police had earlier seen. In fact, half of these murders occurred in domestic situations where the police had responded five or more times in the previous two years.

Women's rights advocates took these and similar findings as evidence of a failure of the police to protect victims of domestic violence (Belknap, 1990). They wanted police officers to arrest and formally process offenders who committed domestic assault. At the same time, victims of domestic violence in cases where the police had been called earlier but had made no arrest began to press civil suits. Police agencies were being held liable in civil courts for failure to protect victims (Steinman, 1988).

Minneapolis Domestic-Violence Experiment. Responding to growing concerns about the police response to domestic violence, the Police Foundation conducted a domestic-violence experiment in Minneapolis (Sherman and Berk, 1984). The experiment sought to test the effects of three different strategies on rates of future domestic violence. Police officers were instructed, when responding to domestic violence calls, to either (1) arrest and transport the offender, (2) order the offender from the home for a while, or (3) attempt to mediate the dispute. The experiment assigned one of these three strategies to each domestic-violence call on a random basis.

After the police had responded to a domestic-disturbance call in Minneapolis, researchers tracked the outcome of the case. Every two weeks for the next six months, telephone calls were placed to the victim of the domestic violence to ask if any further violence had occurred. Additionally, the researchers checked official police records to see if any additional contacts between the offenders or victims and the police had occurred. At the conclusion of the experiment, Sherman and Berk concluded that the arrest strategy had the greatest effect on later violence. Of those offenders arrested, only half as many were involved in later domestic-violence calls as were those not arrested.

The result of these research findings and the other pressures coming to bear on police administrators about the handling of domestic disputes and domestic violence appears to have been a shift in police policy. In 1984 only 10 percent of American police departments had a policy to arrest domestic-violence perpetrators. By the middle 1980s such arrest policies were in place in nearly one-half of American police departments (Steinman, 1988). The emerging consensus of opinion was that arrest was the better method of responding to incidents of domestic violence.

Other researchers replicated the Minneapolis experiment in different cities to see if arrest of domestic violence offenders would have the same effects nationally. One replication in Omaha, Nebraska (Dunford, Huizinga, and Elliott, 1986), did not reveal significant differences in outcome related to different police interventions. The researchers conclude (1986:192), "Arresting suspects had no more effect on deterring future arrests or complaints (involving the same suspects and victims) than did separating or counseling them."

What is important about this examination of police responses to domestic violence is that it clearly illustrates how the police action is a means toward an end. Whether the weight of the evidence supports arrest or nonarrest policies as the best strategy for police in domestic-violence cases is not important to us. Rather, what we can see is that the police have a goal of either preventing a specific incidence of domestic violence or solving the problem of domestic violence. In attempting to solve this problem, police officers may use a variety of strategies, including arrest. If the question of domestic violence were one of law enforcement alone, the police would, theoretically at least, see the arrest as the end itself, not as a tactic.

The maintenance and restoration of order by police officers involves both law-enforcement and service activities and strategies. The choice of one tactic over the other, and the specific action the officer takes are the product of a balance of forces. George Kelling (1999) has examined this issue in depth and has called for the development of guidelines to help direct officers in their order-maintenance efforts. As with other aspects of policing we have examined thus far, these forces reflect the characteristics of the community, the police organization, and the people involved.

SOME CORRELATES OF ORDER MAINTENANCE

The exercise of discretion by police officers is a vital part of their efforts to maintain or restore order. Officers' decisions about whether and how to intervene in disorderly circumstances appears to correlate with three sets of factors we have used earlier to understand policing in the United States. The actions of officers correlate with the characteristics of the people involved in disorderly situations, those of the police organization, and those of the community they serve.

People

The people involved in disorderly situations fall into two main categories: (1) the police officers themselves and (2) the citizens with whom the police must intervene. Some of our earlier discussions have indicated how the characteristics of officers and citizens seem to affect the outcome of order-maintenance cases. Police officers have their own styles or approaches to disorderly circumstances, which seem to correlate with different patterns of intervention among officers. The demeanor, social status, and relationships between citizen participants in disorder also correlate with police action.

Police Officers. William Muir (1977) identifies four types of police officers, who differed as to their willingness to use coercive force and their empathy or understanding of citizens. These differences showed themselves in what Muir describes as distinctive styles of policing. Officers who had both empathy and a willingness to use force Muir calls the *professionals.* Officers who had empathy but were uncomfortable with force he labels the *reciprocators.* Those who were comfortable with force but lacked understanding are called *enforcers.* Finally, those who were uncomfortable with force and also lacked understanding he calls the *avoiders.* Muir based this typology of officers on his observations of the officers in order-maintenance situations. He describes distinctly different approaches to the maintenance of order, which he concludes are products of officer type.

Enforcers are likely to intervene powerfully with threats or use of physical force, or through the use of arrest. Avoiders in contrast, are loath to intervene and delay their arrival to the scene of order-maintenance calls in hopes that by the time they get there, the dispute will be settled. Reciprocators try to mediate and counsel the citizens involved in a dispute in an effort to resolve their problems. Professionals will intervene to take control of the situation, listen to all parties, and then encourage those parties to resolve the dispute themselves.

We have discussed the evidence on officer styles earlier (see Chapter 10). Despite some uncertainty about the validity of claims that specific types of police officers can be found, a rather large body of literature supports the conclusion that different officers react to situations differently. How an officer resolves a disorderly situation appears to be partly a product of that officer's attitudes, experiences, and perceptions.

   Citizens.   The second important category of people involved in order-maintenance activities of the police are the citizens themselves. These include the parties to the dispute, the complainant, and any witnesses. To be sure, it is often the case that the complainant is one of the parties to the dispute, but in others, a third-party citizen calls to complain. The responding police officer must deal effectively with these citizens if order is to be restored or maintained.

   The demeanor of the citizen has repeatedly been found to correlate with the probability of arrest by the officer. Citizens who challenge the police officer or who refuse to obey the commands or follow the directions of the officer are much more likely to be arrested than those citizens who are respectful and compliant (Lundman, 1998; Black, 1980; Westley, 1970; Black and Reiss, 1970). As we discussed, a citizen who interferes with the officer's ability to take control of the situation poses a particular threat to the officer. Although the tendency of the police is to underenforce the law in order-maintenance situations, the need to establish authority and control, to obtain "respect for the law," will often convince the officer to arrest in circumstances where arrest would otherwise not be likely. Crank (1998) notes that maintenance of respect for the police is a core value in the police culture.

   Black (1980) suggests that police decisions, and particularly arrest decisions, correlate with the status of the citizens. In particular, when disputes involve citizens of different social standings—a businessman versus a customer, a landlord versus a tenant, an adult versus a youth—the police will tend to support the higher-status person. This tendency is partly a function of rational risk assessment—the higher-status individual is more likely to make trouble for the officer than the lower-status person. More importantly, as Pepinsky (1975) notes, the officer is likely to see the view of the higher-status person as more legitimate and to desire to align himself or herself with the higher-status interest.

   The relational distance between disputants also correlates with police order-maintenance decisions (Black, 1980). The more closely related the disputants, the less likely the officer is to define the conflict as legal. Rather, the dispute will be defined as private or personal, and the officer will seek to avoid interjecting the formal, impersonal, criminal law into a personal arena. The natural tendency of the police to underenforce the law in order-maintenance situations is buttressed when those cases of disorder occur between intimates.

   Finally, the actions of the responding police officer correlate with the desires of the complainant. If the complainant demands or desires that the police make an arrest and grounds to do so exist, the likelihood of an arrest is increased. Similarly, despite the existence of sufficient grounds for arrest, if the complainant indicates a preference for nonarrest action, the officer is likely to honor the complainant's preference.

Order Maintenance as Interpersonal Interaction

While the literature suggests that the decisions of police officers in these cases correlate with the characteristics of the officers and citizens involved, there is much variation. For example, if the officer is an enforcer, the dispute is between strangers, and the complainant wants an arrest to be made, it is very likely that someone will be arrested. But a substantial amount of the time the officer will not make an arrest. Rather than being a simple addition of these correlates, there is reason to believe that the outcome of an order-maintenance call is partly a product of how these factors are combined. That is, at the level of people, order maintenance is a social interactive process.

Peter Manning and John Van Maanen (1978:216–218) suggest that the outcomes of police–citizen encounters are products of the interactions between the officers and the citizens. Each of these interactions contains five key elements that shape the behavior of the officer and the citizen and, in turn, help determine the outcome. These five dimensions are (1) questions of authority, (2) context of the encounter, (3) components of the interaction, (4) expected outcomes held by both parties and (5) the demeanor of both the citizens and the police.

Especially in order-maintenance situations, it is common for citizens to question the authority or legitimacy of police intervention. Domestic disputes, for example, are generally considered private matters and often occur within the home of the disputants. If the police are to intervene in these matters, they must establish their right or authority to do so. If the citizen grants or acknowledges police authority, the issue is resolved. If the citizen resists police intervention, the officer must establish control in some fashion (Sykes and Brent, 1980).

The context of the encounter refers to the time and place in which the intervention occurs. Police officers dealing with citizens, and the citizens themselves, may behave differently if they meet in public with an audience than if they meet in private. Similarly, police may be more apprehensive stopping citizens at night than during the day. In addition, what happens and how long the officer spends with the citizen may be determined by how busy she or he is or how close to the end of shift the encounter occurs.

The components of an interaction are the social "rules" that the participants feel should apply. Thus, the officer may feel that he or she should be cool in the face of tragedy or conflict. The citizen experiencing these things, naturally, is emotionally involved. To the extent that the officer and the citizen perceive their interaction as being guided by the same rules, it will go more smoothly.

Patrol officers intervene with citizens to prevent minor frictions from developing into serious cases of disorder or crime. (Erin N. Calmes)

Expected outcomes of the situation are a sort of special case of the components of the interaction. Not only do people have expectations about how they are to interact with each other, they also have expectations about the products of those interactions. If the citizen expects to be arrested or cited, and the officer fails to make an arrest, the citizen is pleasantly surprised and relieved. If the officer expects to make an arrest and the citizen expects a warning, the citizen is angry and disappointed. In conflict situations the picture becomes even more complex. If two parties to a conflict each expect the officer to support their side and arrest the other person, no matter what the officer does, someone will be angry and disappointed.

The final element of the interaction identified by Manning and Van Maanen (1978) is the demeanor of both the citizen and the officer. Simply put, the officer and the citizens involved in encounters take cues from each other about how to behave. An antagonistic officer may provoke rebellion and antagonism from the citizen. A contrite citizen may engender forgiveness and friendliness in the officer. How each party in an encounter behaves has implications for the behavior of the other parties.

Sykes and Brent (1980) report on the interactions between police and citizens in thousands of cases. A primary goal of the police in all circumstances is to gain control of the interaction. The researchers describe a continuum of officer behavior geared to taking control. In an effort to regulate the interaction, officers first ask questions. If the citizens answer the officers' questions, the officers have achieved definitional regulation. That is, answering the question is an implicit recognition of the officer's authority to ask the question and the citizen's obligation to respond or cooperate.

If the police are unable to regulate the situation by such definitional power, they move to issuing orders. Sykes and Brent call order-giving *imperative regulation.* If the officer commands a citizen to be quiet, and the citizen complies, the officer has established an ability (if not a right) to regulate the behavior of the citizen.

When neither definitional or imperative regulation succeed in allowing the officer to control the interaction, the officer will employ what Sykes and Brent (1980) call *coercive regulation.* As the term implies, the officer will take control of the situation by force if necessary. Threats of physical or legal harm or the actual use of physical force or arrest are means of coercive regulation. Sykes and Brent conclude that the less forceful tactics of definitional and imperative regulation are sufficient in the overwhelming majority of police–citizen encounters. The movement of the officer's behavior along this continuum from asking questions to making an arrest is contingent on the citizen's reactions to the officer. Thus, the behavior of the officer correlates with the behavior of the citizen.

Organization

We have already seen how organizational structure and regulations constrain the behavior of officers and the outcome of police–citizen encounters. Perhaps in the area of order maintenance, this constraint is most clear in the recent adoption of policies instructing officers to arrest in domestic-violence incidents (Steinman, 1989). Similar constraints can be found in departmental policies concerning tolerance limits and enforcement norms as they apply to traffic control.

The key issue in assessing the organizational correlates of police officer order-maintenance activities is the structuring and guidance of officer discretion. In his assessment of policing styles, James Wilson (1968:84–89) distinguishes among police interventions based on whether the action was police-invoked and whether it involved order maintenance or law enforcement. Law-enforcement situations, whether invoked by the police themselves or by citizen complaint are, in Wilson's view, subject to organizational control. Conversely, order-maintenance situations are less amenable to organizational constraint. Often police organizations have failed to adequately prepare officers for dealing with order maintenance (Lumb, 1995). This leaves the officer to her or his own devices in deciding how to achieve order.

Gary Sykes (1985) has argued that the nature of the order-maintenance or peacekeeping task facing police is such that it is not possible, and perhaps not desirable, to control officer discretion too strictly. As Sykes (1985:62) states it,

> When confronted with those who threaten a neighborhood's peace and security, there is an implicit community mandate for "somebody to do something" and to reinforce the image that "someone is in control." To deny the street officer the informal powers to act intuitively and spontaneously in the face of these problems cuts at the very heart of the police function as it has evolved in the neighborhood context.

The argument Sykes (1985) presents is the core of the current movement to community-oriented policing. There is evidence to suggest that police officers in large, centralized police organizations are isolated from, and act independently of, the community in which they work (Mastrofski, Ritti, and Hoffmaster, 1987). In their study of policing in Miami, Alpert and Dunham (1988:xv) found that officers did not alter their policing styles to suit the conditions of the neighborhoods in which they worked. They write, "Apparently, officers do not believe it is necessary to modify their styles of policing significantly when working in different districts. This indicates that police in the different neighborhoods do not have differing styles of policing to match the unique characteristics of the neighborhoods." Their recommendation (1988:xvii) is to decentralize police organization and control. Specifically, they recommend "policing with neighborhood administrative control." This is a solution to the order-maintenance dilemmas of policing that proponents of community-oriented policing espouse (Trojanowicz and Bucqueroux, 1990; Wadman and Olson, 1990).

Ignoring for now the implications of decentralization for equity in police protection across neighborhoods (Walker, 1984; Strecher, 1991; Travis, 1992), the current movement in policing supports neighborhood police organizations. Through the development of departmentwide policies and performance indicators, police organizations have constrained the ability of patrol officers to maintain order within neighborhoods that vary among themselves.

Department or citywide policies that guide officer actions in order-maintenance situations have constrained officer behavior, and these constraints exceed neighborhood boundaries. Because *order* is so difficult to define, and varies from neighborhood to neighborhood as well as over time,

jurisdictionwide policies and consistency in officer behavior may be coun-
terproductive to order maintenance within a neighborhood. For this reason,
the solution is proposed of structuring police organizations so that such
policies can be tailored to each neighborhood.

Whether and how an officer intervenes with citizens in an effort to
restore or maintain order depends at least partly on characteristics of the
organization. Departmental policies, officer training, and performance-
evaluation criteria used by the department can influence the officer's definition of order-maintenance situations. If the department emphasizes and
rewards law-enforcement activities, officers may prefer arrest as the way
to resolve disorderly circumstances. If the organization is large, officers
may be less inclined to tailor their actions to local conditions. While it is
only one of several correlates of police order-maintenance actions, the police organization can influence the outcome of police–citizen encounters.

Community

The contemporary interest in police order maintenance and its effect on
public safety is grounded in a renewed interest in the relationship between
the community and the police. Order, however defined, is linked to the community and the community characteristics. What is intolerable and disorderly in one location is acceptable, even normal, in another. Consequently,
the definition of order and the requirements of police intervention varies
with community characteristics.

Kelling and Stewart (1989:3) identify several characteristics of communities relevant to understanding order maintenance: the political culture of the city, the form of city government, the demographic composition
of the given neighborhood, and the extent to which neighbors feel threatened and have been able to mobilize. Order is more easily defined and maintained in homogeneous communities with shared values (Jiao, 1995).
Residents and police officers alike can tell which persons and behaviors are
normal or abnormal in the area. The more diverse the values and population of a neighborhood, the less clearly orderly behaviors can be distinguished from disorderly ones (Lynes, 1996).

As with their law-enforcement activities, much of the police workload
in response to disorder is reactive, resulting from citizen complaints. The
initial definition of a situation as requiring police intervention comes from
the residents of a community. In this way, the amount of order-maintenance
activity, its nature and scope, correlate with the willingness of citizens to
seek police assistance. The willingness to call the police, in turn, correlates
with both the level of tolerance for behaviors and the perception of the likelihood of police assistance.

Residents also differ in terms of their ability or willingness to work
with each other to ensure order and reduce crime in their neighborhoods.

Lorraine Mazerolle, Colleen Kadleck, and Jan Roehl (1998) reported a study of the role of place managers (landlords, building superintendants, etc.) on levels of disorder and crime. They found that the greatest effects on reducing disorder and crime occurred when place managers cooperated with others in the neighborhood. Having place managers simply call the police when crime or disorder occurred was not as effective in preventing or reducing these problems. John Eck and Julie Wartell (1999) report that place managers can have a large impact on crime in neighborhoods, but that many landlords have limited ability to manage and improve their property. In sum, when the owners and managers of neighborhood properties are motivated and organized to prevent crime and disorder, the need for police intervention is greatly reduced.

Researchers have long noted that police officers behave differently in different neighborhoods (Smith, 1986; Alpert and Dunham, 1988). Behavior that is tolerated in one neighborhood is combated in another based on officer perceptions of neighborhood values. Officer decisions to arrest or investigate differ from neighborhood to neighborhood. So, too, citizen perceptions of the police and attitudes toward the police correlate with the neighborhood in which they reside (Schuman and Gruenberg, 1972).

The definition of situations as threats to order, the police response to disorder, and the citizen's perceptions of the police are all associated with neighborhood characteristics. To the extent that order maintenance involves, in the words of Alpert and Dunham (1988), "reinforcing informal social controls," police efforts to maintain order must be understood within the context of the community or neighborhood where they occur. Poorer, less organized neighborhoods are places where the police are more likely to be called for assistance with a variety of issues, including order maintenance (Kurtz, Koons, and Taylor, 1998). The need for police intervention is also related to the existence of other social control mechanisms in the community. If the residents of a neighborhood are organized to ensure order without the police, the need for police order maintenance is lessened (Supancic and Willis, 1998).

CONCLUSION

Wadman and Olson (1990:52) observe, "The social well-being of a community also depends on order. Order maintenance, in a traditional sense, is the primary law-enforcement activity of most police departments." Beyond its role as a law-enforcement activity, we argue, order maintenance is the primary goal or purpose of policing.

In this fashion, law enforcement and service delivery can be viewed as tactics or tools for the maintenance of order. The decisions of officers about whether and how to intervene in the lives of citizens are strongly influenced by the officer's perceptions of public order. Hence, officers choose their interventions based on a desire to maintain or restore order. This choice of interventions on the part of officers makes police discretion a key component of order maintenance.

Because order is difficult to define and is ambiguous to the observer, order maintenance is the most difficult task of the police. The officer exercises discretion in evaluating the information available and choosing an intervention. The "acid test" of these choices is whether or not disputes, disruptions, and conflicts—disorder—are resolved. To resolve such conflicts, the first task of the officer is to gain control of the immediate situation.

The need to control the immediate situation renders each order-maintenance encounter an interpersonal interaction between the officer and the citizen(s), at least initially. The decisions and actions of police in attempting to control disorder are therefore products of a variety of forces. The characteristics of the officers and citizens involved, especially how they relate to each other, correlate with whether the officer uses force or the law as a tool to establish order. The decisions of officers, and the effectiveness of those decisions, are also associated with organizational constraints. Finally, the definition of order and the assessment of effectiveness are products of the characteristics of the community (or at least, the neighborhood) in which the officer is working.

Chapter Checkup

1. What is *order maintenance,* and how important is this function of the police?
2. What is *discretion?*
3. Why is non-invocation discretion considered to be low-visibility decision making?
4. What is the role of discretion in the maintenance of order by police?
5. What is the importance of the notion of "broken windows" for understanding police efforts at order maintenance?
6. In what ways does the lack of a clear definition of order hinder police efforts at maintaining the public peace?
7. How is *order* defined by the public and by police officers?
8. With reference to domestic violence, how have reformers tried to control police responses to order-maintenance problems?
9. In what ways do the characteristics of communities, police organizations, and police officers affect order-maintenance efforts of the police?
10. How do police officers attempt to establish order when they respond to citizen complaints?

REFERENCES

Ahern, J. (1972) *Police in trouble.* (New York: Hawthorn Books).

Alpert, R. and R. Dunham (1988) *Policing multi-ethnic neighborhoods.* (New York: Greenwood).

Bard, M. (1970) *Training police as specialists in family crisis intervention.* (Washington, DC: U.S. Government Printing Office).

Belknap, J. (1990) "Police training in domestic violence: Perceptions of training and knowledge of the law," *American Journal of Criminal Justice* 14(2):248–267.

Berk, S., and D. Loseke (1980) "'Handling family violence: Situational determinants of police arrest in domestic disturbances," *Law & Society Review* 15:317.

Bittner, E. (1967) "Police on skid row: A study of peace keeping," *American Sociological Review* 32(October):699–715.

Bittner, E. (1974) "Florence Nightengale in pursuit of Willie Sutton: A theory of police," in H. Jacob (ed.) *The potential for reform of criminal justice.* (Beverly Hills, CA: Sage):17–44.

Black, D. (1980) *The manners and customs of the police.* (New York: Academic).

Black, D. and A. Reiss (1970) "Police control of juveniles," *American Sociological Review* 35 (February):63–77.

Breedlove, R., et al. (1977) *Domestic violence and the police: Studies in Detroit and Kansas City.* (Washington, DC: The Police Foundation).

Brown, M. (1981) *Working the street.* (New York: Russell Sage Foundation).

Brown, S. (1984) "Police responses to wife beating: Neglect of a crime of violence," *Journal of Criminal Justice* 12(3):277–288.

Buerger, M., A. Petrosino, and C. Petrosino (1999). "Extending the police role: Implications of police mediation as a problem-solving tool," *Police Quarterly* 2(3):125–149.

Bufe, N. and L. Thompson (1991) "Traffic services," in W. Geller (ed.) *Local government police management,* 3rd ed. (Washington, DC: International City Management Association):159–184.

Crank, J. (1998) *Understanding police culture.* (Cincinnati, OH: Anderson).

Davis, K. (1969) *Discretionary justice.* (Baton Rouge: Louisiana State University Press).

Davis, K. (1975) *Police discretion.* (St. Paul, MN: West).

Dunford, F., D. Huizinga, and D. Elliott (1986) "The role of arrest in domestic assault: The Omaha experiment," *Criminology* 28(2):183–206.

Eck, J. and J. Wartell (1999) *Reducing crime and drug dealing by improving place management: A randomized experiment.* (Washington, DC: National Institute of Justice).

Fridell, L., R. Lunney, D. Diamond, and B. Kubu (2001) *Racially biased policing: A principled response.* (Washington, DC: PERF).

Goldstein, H. (1979) "Improving policing: A problem-oriented approach," *Crime & Delinquency* 25(2):236–258.

Goldstein, H. (1977) *Policing a free society.* (Cambridge, MA: Ballinger).

Goldstein, J. (1960) "Police discretion not to invoke the criminal justice process: Low-visibility decisions in the administration of justice," *Yale Law Journal* 69(March):543–594.

Hanewicz, W., C. Cassidy-Riske, L. Fransway, and M. O'Neill (1982) "Improving the linkages between domestic violence referral agencies and the police," *Journal of Criminal Justice* 10(6):493–503.

Higgins, J. (1978) "Social services for abused wives," *Social Casework* 59(May):266–271.

Jiao, A. (1995) "Community policing and community mutuality: A comparative analysis of American and Chinese police reforms," *Police Studies* 18(3–4):69–91.

Johnson, I. (1990) "A loglinear analysis of abused wives' decisions to call the police in domestic-violence disputes," *Journal of Criminal Justice* 18(2):147–160.

Kelling, G. (1999) *"Broken windows" and police discretion.* (Washington, DC: National Institute of Justice).

Kelling, G. and F. Coles (1996) *Fixing broken windows.* (New York: Free Press).

Kelling, G. and J. Stewart (1989) *Neighborhoods and police: The maintenance of civil authority.* (Washington, DC: U.S. Department of Justice).

Kurtz, E., B. Koons, and R. Taylor (1998) "Land use, physical deterioration, resident-based control, and calls for service on urban streetblocks," *Justice Quarterly* 15(1):121–149.

Lumb, R. (1995) "Policing culturally diverse groups: Continuing professional development programs for the police," *Police Studies* 18(1):23–43.

Lundman, R. (1980) Police and policing: An introduction. (New York: Holt, Rinehart & Winston).

Lundman, R. (1998) "City police and drunk driving: Baseline data," *Justice Quarterly* 15(3):527–546.

Lynes, D. (1996) "Cultural diversity and social order: Rethinking the role of community policing," *Journal of Criminal Justice* 24(6):491–502.

Manning, P. and J. Van Maanen (eds.) (1978) *Policing: A view from the street.* (Santa Monica, CA: Goodyear).

Mastrofski, S. (1999) *Policing for people.* (Washington, DC: Police Foundation).

Mastrofski, S., R. Ritti, and D. Hoffmaster (1987) "Organizational determinants of police discretion: The case of drinking-driving," *Journal of Criminal Justice* 15(5):387–402.

Mazerolle, L., C. Kadleck, and J. Roehl (1998) "Controlling drug and disorder problems: The role of place managers," *Criminology* 36(2):371–403.

Moore, M. and R. Trojanowicz (1988) *Policing and the fear of crime.* (Washington, DC: U.S. Department of Justice).

Muir, W. (1977) *Police: Streetcorner politicians.* (Chicago: University of Chicago Press).

Pepinsky, H. (1975) "Police decision-making," in D. Gottfredson (ed.) *Decision-making in the criminal justice system: Reviews and essays.* (Washington, DC: National Institute of Mental Health):21–52.

Potter, J. (1978) "Police and the battered wife: The search for understanding," *Police Magazine* (September):41–48.

Schuman, H. and B. Gruenberg (1972) "Dissatisfaction with city services: Is race an important factor?" in H. Huhn (ed.) *People and politics in urban society.* (Beverly Hills, CA: Sage):369–392.

Sherman, L. (1985) Domestic violence. (Washington, DC: U.S. Department of Justice).

Sherman, L. (n.d.) *Neighborhood safety.* (Washington, DC: U.S. Department of Justice).

Sherman, L., and R. Berk (1984) *The Minneapolis domestic violence experiment.* (Washington, DC: The Police Foundation).

Skogan, W. (1986) "Fear of crime and neighborhood change," in A. Reiss and M. Tonry (eds.) *Communities and crime.* (Chicago: University of Chicago Press):210.

Skolnick, J. (1994) *Justice without trial,* 3rd ed. (New York: John Wiley).

Smith, D. (1986) "The neighborhood context of police behavior," in A. Reiss and M. Tonry (eds.) *Communities and crime.* (Chicago: University of Chicago Press):313–341.

Snipes, J. and S. Mastrofski (1990) "An empirical test of Muir's typology of police officers," *American Journal of Criminal Justice* 14(2):268–296.

Steinman, M. (1988) "Anticipating rank-and-file police reactions to arrest policies regarding spouse abuse," *Criminal Justice Research Bulletin* 4(3):1–5.

Strecher, V. (1991) "Histories and futures of policing: Readings and misreadings of a pivotal present," *Police Forum* 1(1):1–9.

Supancic, M. and C. Willis (1998) "Extralegal justice and crime control," *Journal of Crime and Justice* 21(2):191–215.

Sykes, G. (1985) "The myth of reform: The functional limits of police accountability in a liberal society," *Justice Quarterly* 2(1):51–65.

Sykes, R. and E. Brent (1980) "The regulation of interaction by police: A systems view of taking charge," *Criminology* 18(2):182–197.

Travis, L. (1992) "Making history: Explaining the development of the police," *Police Forum* 2(2):6–10.

Treger, H. (1976) "Wheaton-Niles and Maywood police-social service projects: Comparative impressions," *Federal Probation* 40(3):33–39.

Trojanowicz, R. and B. Bucqueroux (1990) *Community policing: A contemporary perspective.* (Cincinnati, OH: Anderson).

Van Maanen, J. (1978) "The asshole," in P. Manning and J. Van Maanen (eds.) *Policing: A view from the street.* (Santa Monica, CA: Goodyear):221–238).

Wadman, R. and R. Olson (1990) *Community wellness: A new theory of policing.* (Washington, DC: Police Executive Research Forum).

Walker, S. (1984) " 'Broken windows' and fractured history: The use and misuse of history in recent police patrol analysis," *Justice Quarterly* 1(1):75–90.

Westley, W. (1970) *Violence and the police: A sociological study of law, custom, and morality.* (Cambridge, MA: MIT Press).

Wilson, J. (1968) *Varieties of police behavior.* (Cambridge, MA: Harvard University Press).

Wilson, J. and G. Kelling (1982) "Broken windows: Police and neighborhood safety," *Atlantic Monthly* 249(March):29–38.

Worden, R. and A. Pollitz (1984) "Police arrests in domestic disturbances: A further look." in G. Whitaker (ed.) *Understanding police agency performance.* (Washington, DC: U.S. Department of Justice):77–92.

PART FIVE

DILEMMAS IN POLICING

In this final part of the book we examine some persistent problems in policing in America and consider the future of policing. We have specifically chosen the word *dilemmas* for the title of Part V because it means questions or issues that lack a clear choice. The chapters that follow each address some aspect of policing in the United States where choices will be or have been made, but where different choices, perhaps better ones, are possible.

Chapter 15 examines the current movement toward community policing. We explore again the big-picture approach to understanding policing that was described in Chapter 1, applying it to the subject of community policing. We use the correlates of policing that we have described throughout this text—people, organization, and community—to understand variations in community policing. In doing so, we see that even this widespread reform in American policing is itself fragmented and complicated. This informs us that attempts to change the police will need to attend to the people, organizations, and communities involved. How to best alter policing practice constitutes a dilemma.

In Chapter 16 we investigate the problem of police misconduct. One dilemma faced by our society involves the question, Who polices the police? Related to this issue are a host of others, including the definition of police behaviors as being misconduct and the best means for controlling police behavior. As we shall see, the problems of corruption and other police wrongdoing have been with us for as long as we have had police. Clearly, we have not arrived at a solution to these dilemmas.

The final chapter attempts to preview the future of policing in America. While many questions remain unresolved, it would seem that a critical, beginning question concerns the relationship between the police and the citizenry. As we shall see, there are currently signs that policing may change in the future. Thus, we assess alternative scenarios. In one, the police will become more independent of the citizenry, more proactive, and less democratic. In another, the police will become even more closely tied to the citizenry and thus more reactive and democratic. Of course, a third alternative exists—that of no substantial change.

We began this book by discussing the fundamental tension between liberty and order that characterizes policing in America. In this section, and especially in Chapter 17, we return to this dilemma. The key to successfully solving the dilemmas of policing in America would appear to be finding an appropriate balance between liberty and order. Seeking this balance has implications for the relationship between the police and the communities they serve, the control of police misconduct, and, of course, the future of policing in America.

chapter **15**

COMMUNITY POLICING: TYING IT ALL TOGETHER

CHAPTER OUTLINE

We began this book by noting the variety in American policing and the fact that policing in practice is the product of a balance of forces. These forces have been identified as correlates of policing. For most of the book we have concentrated on three classes of correlates: people, organizations, and communities. Our position has been that policing as it happens is understandable in terms of how the people (citizens and officers), organization (police agency characteristics), and community (place and value/norm agreement) come together. Figure 15.1 presents this balance-of-forces model.

The current dominant model of policing in the United States, at least as measured by its central place in the literature and in discussions of policing, is community-oriented policing, or community policing (Dietz, 1997). A 1993 survey of 2,000 police departments revealed that almost one-fifth of responding departments had adopted community policing, and more than one-quarter were in the process of adopting the model (Wycoff, 1995). One

Figure 15.1
Policing as a Balance
of Forces

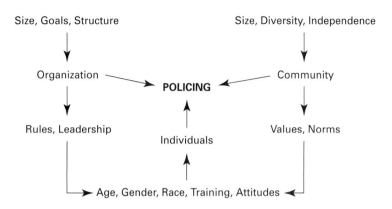

year later, respondents to a national survey indicated that 80 percent of police agencies and two-thirds of sheriffs had implemented community policing (McEwen, 1995:4). As Gary Cordner (1995:1) reports, "In little more than a decade, community policing has evolved from a few foot-patrol experiments to a comprehensive organizational strategy guiding modern police departments." In 1999, Matthew Hickman and Brian Reaves (2001:1) report, almost two-thirds of local police departments, serving over 85 percent of all residents had full-time community policing officers, almost double the number of departments that had full-time community officers in 1997. The number of community policing officers in state and local police agencies had grown from 21,000 in 1997 to nearly 113,000 in 1999.*

*Allan Jiao (1997) considers community policing to be one of four major models of policing. A **model** is a widely accepted concept that provides a way of understanding things. For example, an economic model helps us to understand how people make economic decisions. It assumes people do things that bring them the greatest return for their efforts, and that they wish to maximize benefits and minimize costs. We can thus apply an economic model to understand why someone might buy an automobile rather than lease one (or vice versa). We can use a similar psychiatric model to understand why someone might commit a crime.*

As Jiao (1997) notes, there are often competing models in existence at the same time. To continue our example, we can understand why someone might buy rather than lease a car, or commit a crime, using a psychiatric model. This model might suggest that the decision "means" something to the person. If that person needs to be in control, he or she may prefer to buy. Another person who who is afraid of commitment, might lease. Jiao identifies four contemporary models of policing as follows:

1. *Professional model*—based on the assumption that the police can effectively control crime and that the best means of doing so is through improvement of police personnel and strengthening the police organization.

2. *Security model*—based on a crime-prevention strategy of deploying officers as security guards to reduce the chances of crime, analyzing the security risks of different people and places, and reducing the opportunity for crime through changing the environment.

3. *Problem-oriented model*—based on an assumption that the police can best control crime by identifying the underlying causes of problems they are called upon to handle and attacking those causes. The emphasis of the organization and officers is placed on research, evaluation, and analysis.

4. *Community-policing model*—based on an assumption that the community exists and is the true source of power for social control. The emphasis of the organization and officers is on mobilizing and involving the community in efforts to prevent crime and disorder.

Most contemporary commentators on the police would suggest that the current movement to community policing represents a shift from the professional model to the community-policing model (Ponsaers, 2001). In this sort of shift, the definitions of the role of the police, the structure of the police organization, and the practice of police officers on the street are expected to change. Indeed, we cannot support the idea of different models existing unless those models produce differences in police operations.

Assuming that at least two models exist, the professional and community-policing models, and that we are undergoing a change from one to the other, how can this change be understood? Throughout this book we have argued that policing in America is the product of a balance of forces, especially community, police organization, and personal characteristics. The current spread of community policing allows us to apply this balance-of-forces perspective to American policing on a large scale.

The exercise that follows is necessarily limited. In one chapter we hope to describe and assess the development of a major change in police thinking and practice. The analysis included in this chapter skims the surface and highlights certain themes, including the correlates of people, organizations, and communities. This chapter is designed to explore community policing as a product of the balance of forces. Bear in mind that as an exploration, it is tentative and should be accepted only with great caution.

THE DEVELOPMENT OF COMMUNITY POLICING

Community policing can be understood as an outgrowth of two major forces in policing that began in the 1960s. Each of these contributed to the development of a model of policing that places the community in the center of crime-control efforts. Beginning in the 1960s, local policing became the focus of national attention. The President's Commission on Crime and

Administration of Justice (1967) exemplifies this attention. Responding to concerns about rising crime rates and civil disturbances related to the Civil Rights Movement, President Lyndon Johnson appointed a blue-ribbon committee to study crime and criminal justice in the United States.

At the start of this investigation, the dominant model of policing was the professional model (Kelling and Moore, 1988). The crime commission discovered that most crimes were not reported to the police, and that often the reason for nonreporting was citizen mistrust of the police. It was this crime commission that first used **victimization surveys** to assess the incidence of crime. These surveys revealed both that the police were unaware of much crime and that the fear of crime among citizens was a problem. Further, fear of crime was different from the actual level of crime. Many of those most afraid of crime had not been victims and were at low risk of becoming victims.

At the same time, the U.S. Supreme Court was in the midst of the due-process revolution, declaring many policing practices to be unconstitutional. The crime commission called for improved police–community relations, better educated police, increased numbers of minorities in policing, and controls on discrimination in police decision making (Radelet and Carter, 1994:19–21). Flowing from these recommendations, the U.S. Congress allocated money for the improvement of police education and for research into police practices. At the same time, private foundations began to invest in policing and criminal justice research, creating a knowledge base on which community policing was formed (Radelet and Carter, 1994:63–73).

This research produced the works we discussed earlier, including the Kansas City Preventive Patrol experiment and the Rand study of investigation. The research evidence continued to mount, showing that the police were limited in their ability to affect levels of crime. These findings led to developments along at least two separate fronts. First, there was increased interest in how best to involve citizens in efforts to control crime. Second, there was increased experimentation as to how the police could best reduce and prevent crime.

Perhaps the most direct forerunner to community policing was a number of studies on the impact of foot patrol on crime and citizen fear of crime. Robert Trojanowicz, in his study in Flint, Michigan, and the Police Foundation, in its foot-patrol studies in Newark, New Jersey, and Houston, Texas, observed similar outcomes. What citizens most wanted from the police were services and order maintenance. Observations of police and citizen interactions in the foot-patrol experiments led to the "broken-windows" theory of George Kelling and James Wilson (Kelling and Coles, 1996:19).

By the end of the 1970s there was already a growing consensus that (1) the police should be concerned about both the level of crime and the fear of crime (Goldstein, 1977) and (2) the police were reliant on citizens for information about crime. Police agencies mounted efforts to develop community crime-prevention programs, including neighborhood watch programs and efforts to improve police and community relations that included both team policing and the establishment of foot patrols (Toch, 1997). These developments opened lines of communication between citizens and the police, setting the stage for later efforts to develop the partnerships between the community and the police that are core to community policing.

Emergence of the Crime-Prevention Goal

Assessments of police impact on crime, as well as the effects of neighborhood watch programs, revealed the phenomenon of **crime displacement.** It seemed clear that changing police deployment could move crime around but not necessarily stop it from happening. Other research indicated that the traditional, reactive, incident-driven approach to crime control (where the police merely respond to specific citizen reports of crime without addressing the roots of the crime problems) was ineffective. Officers were frequently called back to the same locations, by the same people, for the same problems. Rather than treat the symptoms, some argued, we should address the causes and thus prevent later crimes (Goldstein, 1979).

Robert Trojanowicz and Bonnie Bucqueroux (1990:36) liken community policing to preventive medicine. The traditional police patrol officer served as an emergency room physician, treating any number of medical problems as they arose. The family physician, instead, works with the patient to anticipate and prevent later medical problems. So, too, the community police officer should work with the community to prevent later crime. As Wesley Skogan (1996:31) writes, "It is also widely assumed that crime prevention is probably more dependent on the community than on the police."

At the same time that research was indicating the limited ability of police to influence rates of crime, there was an increasing emphasis on physical crime prevention (Rosenbaum, 1986). Citizens, often with the encouragement of the police, were asked to take steps to reduce the chances of crime. These steps included marking valuables, improving lighting and locks, being cautious while out in public, and participating in neighborhood watch and similar programs. As community policing developed, the role of the police in crime prevention was highlighted. Physical crime prevention continues as a separate phenomenon (Smith, 1996; Taylor and Harrell, 1996), but crime prevention through environmental design has been firmly linked to community policing (Trojanowicz and Bucqueroux, 1990).

A growing emphasis on the prevention of crime, coupled with the recognition that the police are not well positioned to prevent crime without the support and cooperation of the community, defined the need for a partnership between the police and the community to control crime (Bureau of Justice Assistance, 1997). In addition to the control of crime, however, this partnership is expected to also improve the general quality of life in communities and work specifically to reduce the fear of crime. The police role has been redefined from that of crime control to a broader charge that includes community well-being. By working with the community to alter the causes of crime, the reasoning goes, the police are preventing crime from occurring in the future. Prevention of crime, rather than the apprehension of criminals is now defined as the primary goal of the police. As Dietz (1997:83) notes, with community policing, the police role involves not just the reduction in crime, but also a reduction in fear of crime and an improvement in the quality of life in the community.

With respect to the role of the police in crime control, the crime-prevention goal of community policing justifies police involvement in quality of life issues. In this regard, community policing represents the traditional purpose of the police—public safety. Indeed, Kelling and Moore (1988) claim that the movement to community policing is designed to achieve more effective crime reduction. Unlike apprehending offenders, where it is relatively easy to calculate a clearance rate to assess the percentage of crimes that have been solved by arrest, crime prevention is difficult to assess. How do we know if a crime has been prevented, and if prevented, that it was prevented by community-policing efforts (Palmiotto and Donahue, 1995)?

As we saw in Chapter 12, community policing has the ultimate goal of crime prevention through a reactive strategy. The police, working with the community, define problems of crime and order and develop solutions. The central role of the community in this model of policing is what sets it apart from problem-oriented policing approaches (Kratcoski, 1995). The rapid spread of community policing across the nation can be partially explained by the fact that community policing is elusive of definition. While the majority of police agencies say they have adopted or will be adopting a community-policing model, what those agencies actually do, and how they do it, varies greatly.

Varieties of Community Policing

Susan Sadd and Randolph Grinc (1996:1) observe, "Community policing could arguably be called the new orthodoxy of law enforcement in the United States." Despite the large numbers of police leaders who report they have adopted community policing, however, Sadd and Grinc note a lack of consistency in definition and practice. "Although the concept is defined in varying ways and its ability to meet its goals remains largely untested, community policing has gained widespread acceptance" (Sadd and Grinc, 1996:1). Wycoff (1995:1) notes, "Almost half of the police chiefs and sheriffs were unclear about the practical meaning of community policing."

Vincent Webb and Charles Katz (1997) observe that there is confusion not only among police officials, but also among members of the public. They suggest that little consensus exists in communities about what the police should do and what police duties are most appropriate and important. As might be expected, a variety of definitions helps to support a variety of practices. Gary Cordner (1995) offers one way of defining community policing, based on three important dimensions: philosophy, strategy, and program.

The **philosophical dimension** contains the central ideas on which community policing is founded. In general, Cordner (1995:2) writes that community policing entails a broadening of the police mission to include order maintenance and service delivery. It also includes an increased voice in policing matters for citizens. Community policing actively seeks citizen input to police policy decisions. Finally, community policing recognizes neighborhood variation. Consequently, it is expected that police practice and enforcement policy will need to be tailored to specific neighborhood conditions.

The **strategic dimension** refers to the ways in which the philosophical goals will be achieved. Here Cordner (1995:2–4) names three foci of community-policing strategy: geography, prevention, and substance. The *geographic focus* relates to the stability of officer assignment, so that the same officer(s) is responsible for a defined area. The *prevention focus* refers to the attempt by the police to look beyond specific incidents and solve underlying conditions that seem to lead to calls for service. The *substantive focus* covers the goal of producing real differences in the community as a result of police effort. It is a "bottom line" approach that keeps attention tuned to the problems of crime, fear, and disorder and requires officers to solve problems in these areas.

The **programmatic dimension** contains the programs, tactics, and behaviors of police officers and agencies that are designed to implement the strategies and achieve the goals of community policing. Cordner (1995:4–6) includes two components in this dimension; reoriented operations and problem solving. Community policing typically involves less reliance on random motor patrol, *reorienting* officer work toward foot patrol, directed patrol, and other operations. The *problem-solving* emphasis involves a rethinking of the police task away from responding to specific incidents toward an analytic approach to patterns of problems and a corresponding broadening of solutions to include actions beyond traditional investigation and arrest.

In his essay Cordner identifies eight factors in these three dimensions of community policing. Any one, or all eight of these factors may be altered by an agency's implementation of community policing. Further, each factor may be affected in a variety of ways. One agency might assign all officers to specific locations while another creates a separate unit of officers with stability of assignment. One agency might seek citizen input through formal public opinion surveys while another convenes town meetings. The point here is that the number of distinct components, and the range of alternative ways of changing each, produce an almost infinite number of possible community-policing models. Indeed, to date community policing in the United States, while perhaps a new orthodoxy, is characterized as much by its variety as it is by consistency. McLaughlin and Donahue (1997:47) remark, "Community policing is nebulously and discursively defined." The term, they contend, is so elastic that it is applied to almost any policing activity. They continue (1997:47), "Its indeterminate nature sometimes makes it hard to discern whether a police department advertising it is, in fact, doing it. It permits less scrupulous chiefs to appear progressive by repackaging traditional policing under a flashy, trendy label."

A brief description of policing practices that fall under the general heading of "community policing" illustrates the point. Given our approach to policing as the product of a balance of different forces, this variety is not surprising. Our task is to identify some of the ways that these forces combine to produce different outcomes in terms of community policing.

Community Policing in Practice

Sadd and Grinc (1996) describe neighborhood-oriented policing in eight cities. They observed eight distinct jurisdictions and programs. Although there were similarities across the study sites, each was unique in some regard. In some places community-policing efforts were designed to involve all officers and resulted in departmentwide decentralization of patrol. In others, specialist units were formed to work in defined areas. In still others, the police department identified problems and brought additional law-enforcement resources, such as tactical squads, to bear. In some places the police worked with residents to define problems, in others the police informed residents about their problems and proposed solutions. In certain cases, planning with the community preceded police efforts, in others the police efforts (increased enforcement, for example) preceded communication with citizens. Some used "neighborhood officers," others mobilized nonpolice community resources, and still others combined approaches.

A variety of programs and practices have come to be defined as "community policing." Despite the fact that proponents of community policing insist that it is a revolutionary model of policing that involves philosophical and organizational change (Trojanowicz and Bucqueroux, 1990), in practice what passes for community policing is often grafted onto existing policing practice. Some police administrators claim to be doing community policing because their agencies participate in DARE programming. In other agencies, the implementation of community policing involves the reassignment of motor patrol officers to such things as bicycle or foot patrol, without any major change in organizational goals or officer practice. Some police administrators do not pursue community policing because, they say, it is what they have always done.

Commenting on the spread of community policing in the United States, Herman Goldstein (Rosen, 1997:9) said, "I think the vast majority of claimed efforts to implement community policing are very superficial and very, very thin, to a great extent because of a lack of understanding of what it is we're trying to achieve in the larger context, and also because of a lack of real commitment." A confusion over the meaning of *community policing* helps to explain the wide variety of practices that are included under that label. Whatever the product of the balancing of forces in a given policing jurisdiction, it seems the outcome (or some part of it) in terms of police practice can be included under the label "community policing."

Suppose every police agency decentralized patrol responsibility, but none of the officers cooperated with residents? Suppose an agency employs fewer than five officers; how can one patrol officer for an entire jurisdiction be decentralized? What happens if, despite the best efforts of the police, members of the community resist working with the police, or if they disagree among themselves about what community problems are most in need of police attention? These questions indicate that even if community policing is the new orthodoxy, the specific characteristics of the communities, organizations, and people involved will influence how (and if) that orthodoxy is translated into police practice. We will now turn to an examination of these correlates of community policing.

COMMUNITY CORRELATES OF COMMUNITY POLICING

Horizontal Articulation

It seems redundant to speak of the community correlates of community policing. The core idea of community-oriented policing is to allow the community to define the problems of crime and disorder on which the police will concentrate. However, as Webb and Katz (1997:9) observe, "The research that has been done to date fails to find much community consensus on community policing activities." They cite a 1988 report by C. Murphy that found a variety of opinions and perceptions rather than the value consensus assumed by the community-policing model. As Robert Ford, chief of police in Port Orange, Florida, observed (quoted in Brady, 1996:7),

> The diversity within communities is phenomenal. Our belief now is that every time we go to solve a problem, we create another problem for ourselves from another group. So, for example, if the senior citizens are concerned about youth groups gathering, we solve that problem,

and that creates another problem with parents of youths who feel we are picking on the kids.

With reference to the community types described in Chapter 11, the theory underlying community policing assumes communities with relatively high levels of horizontal articulation. That is, there is an assumption of a shared understanding of what is proper and improper behavior, and a general consensus about what constitutes crime and disorder. Community policing will work best in the solidary or interdependent community, because police legitimacy rests on the authority of local custom. Community policing is unlikely to work very well in communities with low horizontal articulation—those lacking consensus.

One solution to this problem of differences in community consensus is to decentralize policing to a level where consensus can be achieved. Thus, in communities that have a mixture of interests, the police may use a neighborhood orientation. If the community has residential and business areas, each area gets its own neighborhood policing. These may be the same officers, but they deal with separate community clients. For instance, the community-policing officer may be the link between the local residents' association and the local business association. Here the officer may be "caught in the middle" between conflicting desires of these two organizations.

A second solution to this problem is for the police to sell the community on specific problems. For example, although senior citizens and the parents of youth may disagree about the seriousness of the problem of youth gathering on the streets, the police can identify drug sales as a problem and convince both groups that it is the most important problem facing the community. Absent a community consensus, the police can serve to compromise. In this case, however, it is the police and not the community who have identified the problem and selected a solution. While informing the public, the strategy is not truly a reactive one.

David Lynes (1996) discusses the problem of community diversity as it relates to the goals and practice of community policing. He writes (1996:501), "Police work places police officers along the fault lines of many if not most existing and developing social tensions." When there is disagreement and conflict among members of the community, when there is a lack of consensus about what is or is not orderly, the police officer (whether the community police officer or the professional police officer) must negotiate the boundaries. To the extent that even the community police officer is expected to maintain or establish a generalized notion of law and order, this task may be impossible. George Kelling and Catherine Coles (1996:138–141) describe this problem when assessing efforts to stop panhandling in New York City. Panhandlers were seen by most citizens and the police department as disorderly. An aggressive police campaign to stop panhandling was halted when a federal court ruled that the panhandlers had a constitutionally protected right to solicit money. That the case found its way into a court indicates the

existence of a difference of opinion about what was or was not disorderly behavior. Robert Bohm, Michael Reynolds, and Stephen Holmes (2000) found
similar problems in a study of community policing. They note that information about community problems came from only a minority of residents.
They also observe that "the community" must include not only residents in
the affected neighborhoods, but also all residents of the city. The police must
attend to the wishes and desires of the broader public (the city as a whole)
while attempting to respond to the identified problems that exist within any
given neighborhood.

Citizen Participation.   Yet another aspect of the community that
affects the implementation of community-oriented policing is the degree to
which citizens are involved in the partnership with the police. Sadd and
Grinc (1996) noted that one common implementation problem encountered
by all eight of the police departments they studied was a lack of citizen
participation. Members of the community may not wish to become involved
with the police for a variety of reasons, including a history of poor relations
with the police, a fear of retaliation by those persons targeted for police
attention, a lack of commitment to the community, and other similar
factors. One of the most challenging tasks facing community policing
officers is often that of mobilizing the community. Bohm, Reynolds and
Holmes (2000) report that the vast majority of citizens did not actively
participate in community policing.

As Webb and Katz (1997) note, community members often disagree
with the police about what types of activities are most important and most
beneficial. Community members tend to place higher values on traditional
enforcement activities by the police and lower values on the preventive activities. In this regard, the community shows a preference for the traditional role of the police as reacting to crime. They write (1997:21), "It may
be easy for the public to understand how gang investigations, drug sweeps
and bike patrols can directly impact on crime, but the processes whereby
neighborhood trash clean-ups and graffiti removal efforts reduce crime are
probably much more difficult to understand."

If the community can agree on the direct crime-control influence of
some community policing activities, it would seem that participation in
neighborhood watch programs would be something that community policing efforts would encourage. Wycoff (1995) found that police departments
reporting they had adopted community policing were more likely to be involved with neighborhood watch programs than those not having adopted
the new model. However, neighborhood watch programs vary across communities. Indeed, Smith, Novak, and Hurley (1997) report that the presence
of community organizations and neighborhood watch programs is related to
higher levels of crime. They suggest that one reason for this is that efforts
to develop neighborhood watch may be focused on high-crime communities.
In a study of the implementation of community policing in three small

cities, Kratcoski and Blair (1995:101) quote one officer's experiences with organizing block watch groups: "Only one [block watch program] remains active today. It's in a tight, cohesive area where there's not much turnover in neighborhoods."

There is increasing concern that limited citizen participation in community policing efforts is evidence of a failure of the idea of community policing. Several commentators have observed that when communities are characterized by conflict and dissensus (low horizontal articulation), community policing efforts may simply serve to increase the role of the police as instruments of elites used to control the disenfranchised (Oliver, 2000; Barlow and Barlow, 1999). One test of the effectiveness of community policing as a strategy to improve the quality of life is whether and how this effort improves communities.

Kent Kerley and Michael Benson (2000) observe that community policing efforts worked best in places with a high degree of cohesiveness and less well in places characterized by conflict and fragmentation. Nathan Pino (2001) also notes that community policing efforts have not been tested with regard to how they might contribute to the development of "social capital," or the ability of a neighborhood to organize itself for order maintenance and social control. To that end, David Duffee, Reginald Fluellen and Brian Renauer (1999) criticize current community policing efforts as being divorced from community theory. They suggest that community policing programs have focused too closely on symptoms (crime and disorder) and have not been designed to support and encourage community building. In sum, these observers suggest that the level of horizontal articulation is correlated both with the implementation of community policing initiatives and with their overall impact on crime and disorder.

Vertical Relations

Certain characteristics of communities, including size and urbanism, seem associated with the style of policing they receive. Recalling our earlier discussion of the big-city bias in American policing research, most of what we know about community policing comes from the experiences of very large, urban police departments. Weisheit, Falcone, and Wells (1994:12) note, "Crime is less frequent in rural areas, and that 'community policing' to which many urban departments now aspire has been a long-standing practice in rural police agencies." Their description of the rural community is congruent with the "solidary community." They suggest that residents of rural areas are generally distrustful of government, value independence, have knowledge of each other and of the police, and generally agree about what is proper or improper conduct. Much crime and disorder is avoided or sanctioned by informal social-control mechanisms. These communities are quite different from the large cities in which the national movement to community policing must be seen as revolutionary.

The implementation of community-oriented policing is linked to the vertical relations of communities. Communities that are more closely tied to external forces and developments are more likely to adopt community policing. Indeed, one of the effects of the federal initiative to put 100,000 more officers on the streets of America was the dissemination of community policing. Accepting federal support for additional personnel is conditioned on agreeing that these personnel will be assigned to community policing duties (Rosen, 1997). As one supervisor told Kratcoski and Blair (1995:103), "Grant monies help to sweeten the pot, since they function as inducements for getting officers to volunteer for innovative programs." The availability of additional, external funding has often been a decisive factor in the implementation of community policing initiatives.

Beyond funding, vertical relations have an effect in terms of the diffusion of an innovation. To the extent that a community's leaders, especially its police leaders, are attuned to national professional developments, that community's police agency is more likely to be aware of new ideas and more likely to implement them (Weiss, 1997). Skolnick and Bayley (1986) suggest that leaders in the early development of community policing were critical to the adoption of change in such places as Houston, Chicago, New York, and Newark. Professional police administrators who are linked to the policing field more than to a particular city or community are likely to be innovators and to keep "up to date" with developments in the profession. As Kratcoski and Blair (1995:103) observe, "Another dynamic of community-policing programs in small cities is that most of the programs are introduced through administrators who hear about it from professional contacts outside the local community."

Like the idea of formal policing itself, discussed in Chapter 4, the notion of community policing seems to be spreading from a few larger agencies to many smaller agencies. That the federal government has endorsed and supported community policing through the provision of funding support to police agencies and for research on community policing adds energy to the natural spread of the idea. The lack of precision in the definition of what community policing entails also contributes to its rapid adoption. Communities that are integrated with the larger society (high vertical relations)—which the "great change" (see Chapter 11) suggests is the case for most communities—are more likely to adopt community policing. Also, improved and increased communication among policing administrators seems to accelerate the adoption of community policing. Those communities that are less integrated or where police leaders have a local orientation are less likely to adopt community policing. Of course, these communities tend to have characteristics associated with being solidary, and are probably already policing on the basis of local custom.

Organizational Correlates of Community Policing

Community policing can be expected to produce (or be the product of) organizational change in the police agency. Recalling our discussion of police organizational designs, community policing, with its emphasis on the link between the officer and the community and the desire to be responsive to community concerns as they arise, would seem best served by an organic design. More traditional policing, with its emphasis on responding to crimes and apprehending offenders, would be best suited to a more mechanistic design. Since police organizations are responsible for the three main functions and the requirements of those functions vary in terms of complexity and discretion, no single organizational design is best. Even so, community policing would seem best served by an organizational structure that is more organic—that is, one having less hierarchy, less specialization, and more communication among members.

Proponents of community policing have often stated that to be successful, a police agency's adoption of community policing must entail fundamental organizational change. Vivian Lord (1996:504) observes, "A police department that commits to implementing community-oriented policing throughout its city is undergoing major organizational changes." Maguire (1997:554–555) concludes that commentaries on the organizational implications of adopting community policing indicate that "the organizational structures of community-policing departments are supposed to differ markedly from those of 'traditional' departments."

Embracing a wider role for the police organization—one that includes crime control, control of the fear of crime, and improvement of general community quality of life—expands the **domain** (area of responsibility) of the agency. Furthermore, accepting the definition of the police officer as problem solver and manager of her or his beat has implications for the hierarchical structure of the agency. Assigning officers to specific geographical beats implies increased spatial differentiation of the agency. Finally, granting the community policing officer general responsibility for quality of life issues implies less functional specialization. That is, the officer is expected to serve as general patrol officer, crime analyst, community liaison, and investigator.

Based on these expected effects of adopting community policing, Maguire (1997) assessed changes in the police organizational structures of departments that have adopted community policing and those that have not yet done so. He concludes that there are no significant structural differences between the two types of police agencies (1997:572). (His study was limited to large urban police departments.) Maguire attributes this lack of difference to four reasons. First, the length of the study, six years, might have been insufficient for the organizational changes to become evident. Second, police executives may not agree that structural changes are necessary to accommodate community policing. Third, other forces in the organizational environment, such as law-enforcement accreditation (Cordner and Williams, 1995) could constrain the organization and limit structural change. Finally, he speculates that large departments might not be implementing community policing in "wholesale" fashion.

Assuming that police organizations desire to implement community policing, and assuming that the practice of community policing involves a fundamental shift in how the police agency is operated, we can expect that different organizations will implement community policing in different ways. Wholesale adoption of community policing would involve an almost total decentralization of the police organization. Each community or neighborhood would have its own police officer or organization. Rather than one police department responsible for a dozen neighborhoods, there would be 12 police departments (at a minimum). Alternatively, large police departments are generally comfortable with specialized units. The creation of a community policing unit within a larger agency makes sense. If a large police agency has a specialist unit for traffic enforcement, vice crimes, and criminal investigation, it is not much of a stretch for that organization to create a community policing office to which specialists are assigned. Thus, large organizations having specialist units are likely to implement community policing through the creation of specialist units consistent with the existing organizational structure. In agencies lacking such specialist units, it is likely that community policing will be implemented agencywide, with general patrol officers and the police management broadening their job descriptions and accepting the new organizational goal of promoting community quality of life.

Evidence suggests that this is what happens. Most of the published literature on community policing has come from studies of large police agencies. These agencies tend to create specialist community policing units that work in conjunction with traditional patrol officers, detective units, and so on. The specialist officers are often freed from responsibility for taking radio-dispatched calls for service and are required to work with community members and groups to define problems and coordinate attempted solutions. Edward Maguire and Stephen Mastrofski (2000) report that, from an organizational perspective, community policing may be multidimensional. If there is only one definition/style of community policing, then we would expect all police departments to report similar structures, policies, and procedures. On the other hand, if community policing is multidimensional, involving a range of practices and structures, it is likely that police agencies claiming to have implemented community policing will report different operational and structural adaptations. Maguire and Mastrofski (2000) found evidence to suggest that community policing involves at least five separate dimensions. They observe (2000:14) that "it is possible that community policing might exist in many different shapes and forms throughout the United States."

One of the most well-known community policing initiatives was implemented in Chicago. Remember that Chicago was the home of O. W. Wilson, perhaps the greatest advocate of professional policing and specialized, hierarchical police organizations. The Chicago initiative, called CAPS (Chicago Alternative Policing Strategy) was started in 1993 (National Institute of Justice, 1995). The program divided the patrol division into two basic groups: beat officers assigned to the city's 279 distinct beats and rapid-response officers assigned to handle 911 system calls for service. The Chicago Police Department retained its detective, youth, gang, and other specialist divisions. The CAPS initiative has gone citywide and thus far seems to have been both well implemented and effective (Chicago Community Policing Evaluation Consortium, 1997). Yet, specialist assignments and units have been retained, and a managerial hierarchy exists, ranging from the lowest level (beat teams) to the area management team that oversees several districts. A comment on how citywide units such as the tactical division can be used by beat officers is instructive. The police department recently developed a form for use by beat team leaders to request assistance from the citywide units or other districts. As described in the Chicago Community Policing Evaluation Consortium, "It allows beat team leaders to seek help (after getting approval from the sector management team leader, district commander, and area deputy chief) from other divisions of units for documented priority problems on their beat. Beat team leaders are supposed to have a response to their request within 10 days" (1997:81).

This example, taken from a police agency that is on the cutting edge of community policing and organizational change, illustrates the enduring qualities of organizations. The implementation of community policing in large police organizations is likely to be consistent with the established structure and tradition of the organization. So, too, in smaller organizations which already have low levels of hierarchy, functional specialization and cover relatively small geographic areas, the organizational structure of the police agency already fits what community policing proponents often say is necessary.

It may be that over time the police organizations in large cities and small towns that implement community policing will come to be indistinguishable, or there may always be structural variation based on jurisdiction size and the organizational environment. What seems clear at this point is that organizational differences are associated with different strategies of implementation of the same reform. William King (2000) has explored the problems in understanding how police organizations change. He suggests that the sheer complexity of police organizational change hinders the development of understanding. Not only do the organizations vary, but so too do the types of changes and the environments (communities) in which the organizations exist. Jihong Zhao (1996) specifically studied the development and spread of community policing as an organizational change. His conclusion is that, in most cases, larger American police agencies adopted community policing in response to growing environmental pressure. That is, as the clients and owners (public and political leaders) voiced dissatisfaction with police operations, the agencies implemented community policing in response to criticisms.

Community policing is thought to entail not only changes in organizational structure but also in managerial style. Proponents of community policing suggest that this reform requires a participatory style of management and is incompatible with a "telling style" (see Chapter 9). Kratcoski and Blair (1995:103) report that efforts to implement community policing through administrative directives generate line-officer resistance. Sadd and Grinc (1996:8) likewise note that community policing initiatives need "to be actively 'sold' to patrol officers." Existing distrust of management by officers, a history of labor–management hostility, and the development of specialized community policing units have all been found to be areas of friction between police managers and police officers (Cordner, 1995; Sadd and Grinc, 1996; Lord, 1996; Travis and Winston, 1997).

A final question related to organizational correlates of community policing concerns what is the best organizational structure for the achievement of community policing goals. In a study of police officer task performance, Travis and Sanders (1997) found that performance of "community-policing tasks" varied by officer assignment. In organizations that assigned specialists to community policing, those officers performed community policing tasks (attending public meetings, working with neighborhood watch groups, etc.) significantly more often than those not so assigned. Further, officers in those agencies with specialists who were not assigned as community police officers appeared to deal with the public less frequently. Similar differences in the tasks performed by regular patrol officers as compared to community policing officers have been reported by other observers (Parks et al., 1999; Frank, Brandl, and Watkins, 1997). Thus, using specialist units might mean that some officers very actively promote community policing, but that the rest of the membership of the organization withdraws from interacting with the community. The net effect may be less overall police and community cooperation.

Given the relatively recent appearance of community policing and the current lack of research about the effects of different organizational structures on community policing practice, any solid conclusions must await further study (Zhao, Lovrich, and Thurman, 1999). Although we may not be able to identify any causal links with confidence at this point, the growing body of research in this area supports our contention that the police organization (and management) are important correlates of both the adoption of community policing and of the variations in how community policing is implemented.

PEOPLE AS CORRELATES OF COMMUNITY POLICING

We have already looked briefly at the characteristics of individuals as they relate to community policing in terms of citizen perceptions of the police role and how those perceptions differ across communities. This section focuses primarily on the characteristics of police officers and examines what the research literature tells us about individual officers and the movement to community policing. In the previous section we learned that officers with a community policing assignment engage in different activities at different frequencies than those not having such assignments. Is there such a thing as the best type of officer for community policing? How do the traits of individual officers affect efforts to implement community policing?

Sadd and Grinc (1996:10) note, "Whether they were assigned to INOP [community policing] projects or not, officers believed that certain individual 'styles' of policing were more suitable than others to community policing. The importance of the characteristics of individual police officers in the success of the INOP projects was a theme in all sites, and it manifested itself in several ways." In some places, the work of a few very dedicated officers was instrumental in success. In others, the inability of some officers to perform well as community policing officers was a hindrance to the program. The anecdotal evidence suggests that officer characteristics are important to project success.

A number of observers have noted that individual officers differ in their assessments of the appropriateness and effectiveness of community policing initiatives (Lord, 1996; LeClair and Sullivan, 1997; Kratcoski and Blair, 1995; Bromley and Cochran, 1999; Lewis, Rosenberg, and Sigler, 1999). Proponents of community policing recognize the importance of the individual officer to its success. At base, community policing involves *empowering* the police officer (Gaines, 1997) to identify problems and attempt solutions. However, this grant of latitude to the individual officer is based on the assumptions that the officer can perform these tasks and that the officer wants to do them. As Gaines (1997:4) observes, these assumptions may be erroneous.

To the extent that community policing creates independent officers seeking to solve problems in the community, the type and quality of police service will be increasingly dependent on the skills and characteristics of the individual officer. As described by Trojanowicz and Bucqueroux (1990:16), "The CPO [community policing officer] operates as a mini-chief in the assigned beat area, with the autonomy to do what it takes to solve the problems people care about most." Delegating this level of independence to the police officer poses important questions for the organization and for society. Will officers act on their own biases and prejudices? Will policing be consistent across communities and groups of people? How much variation in officer activity is tolerable? Russell and MacLachlan (1999) report that increasing officer and community participation in policy making may have the effect of confusing policy, undermining accountability, and increased officer frustration. In a sense, this is a case where "too many cooks spoil the broth." As Mastrofski, Worden, and Snipes (1996:1) put it, "As police discretion increases, so too does the risk that officers will be swayed by 'extralegal' considerations—factors outside the law, such as the suspect's race, sex, age, and demeanor." Based on observations of community policing officers in Richmond, Virginia, Mastrofski and his colleagues (1996:2) found that officers were guided primarily by legal considerations, but that officers in favor of community policing made fewer arrests overall. These officers used a variety of nonarrest dispositions, such as referrals and warnings.

One solution to the potential problem of too much individual discretion being exercised by community policing officers is to establish some level of consistency through selection and training (Palmiotto, Birzer, and Unnithan, 2000). There is evidence to suggest that higher educational levels of officers are sought by agencies implementing community policing (Carter, Sapp, and Stephens, 1988; Trojanowicz and Bucqueroux, 1990). Better-educated officers are less likely to generate citizen complaints, and they ought to be better prepared to perform the analytic functions required of the problem-solving community policing officer. Similarly, the desire to establish a closer relationship with the community has underscored continuing efforts at affirmative-action practices in American police departments. Sanders, Hughes, and Langworthy (1995) report little change in recruitment and selection practices between 1990 and 1994.

In addition to selecting officers better suited to the tasks of community policing, there is a growing emphasis on training for community policing. Dantzker et al. (1995) assessed the training process for community policing in Chicago. They note the importance of training in preparing officers for the new challenges they will face as community police officers. As always, one anticipated goal of officer training is more consistency in behavior and decisions across officers. A number of researchers have examined training programs, especially cultural-diversity training in hopes of improving the preparation of police officers for dealing with diverse communities (Blakemore, Barlow, and Padgett, 1995; Lumb, 1995). A relatively new innovation associated with community policing is the provision of training to citizens as well as officers.

Some police agencies currently operate **Citizen Police Academies** (Cohn, 1996). These academies are similar to police recruit academies, but the classes are composed of citizens. The hope is that by exposing citizens to police training, citizen understanding of police officer behavior, and citizen support for the police, will be enhanced. Beyond this sort of citizen cultural-diversity training to improve understanding of the police, other agencies, such as the one in Chicago, have embarked on programs to specifically train citizens for their role in community policing (Chicago Community Policing Evaluation Consortium, 1997).

Evidence suggests that the characteristics of individual officers are associated with the implementation of community policing efforts and with their success. This is not surprising, given that officer characteristics have traditionally been a correlate of policing in practice and that community policing enhances the status of the officer as decision maker. Concern over the exercise of discretion by officers is a hindrance to the implementation of community policing as it was theoretically developed. Police administrators struggle with ways to empower officers while controlling the chances that officers will engage in corrupt or wrongful behavior. Geller and Swanger (1995) have suggested that middle managers of police agencies are well situated to influence the exercise of discretion by officers in ways that are more collaborative and constructive than the traditional enforcement of departmental rules.

CONCLUSION

This chapter examined the current movement to community policing as an opportunity to apply the balance-of-forces approach to understanding policing. The adoption and implementation of community policing in American police departments varies from agency to agency. What research is available indicates that the characteristics of communities, especially vertical and horizontal articulation, are related to the adoption and successful implementation of community policing. Similarly, the characteristics of existing police organizations and leaders are related to the likelihood of adopting community policing, as well as the ways in which community policing initiatives will be implemented. Finally, police officer abilities and attitudes towards community policing have been found to be important factors in understanding the adoption and success of community policing. With a history of just over 10 years, it is too early to be able to definitively test for the effects of these correlates on community policing. Still, the evidence that is available suggests that the variety in community policing across the country is related to differences in communities, organizations, and people.

CHAPTER CHECKUP

1. What is *community policing?*
2. How can community policing be considered a new orthodoxy?
3. What is the link between crime prevention and community policing?
4. What characteristics of communities are linked to variations in community policing?
5. What characteristics of police organizations are linked to variations in community policing?
6. How have police departments tried to influence officers and citizens to prepare them for community policing?

REFERENCES

Barlow, D. and M. Barlow (1999) "A political economy of community policing," *Policing: An International Journal of Police Strategies and Management* 22(4):646–674.

Blakemore, J., D. Barlow, and D. Padgett (1995) "From the classroom to the community: Introducing process in police diversity training," *Police Studies* 18(1):71–83.

Bohm, R., K. M. Reynolds, and S. Holmes (2000) "Perceptions of neighborhood problems and their solutions: Implications for community policing," *Policing: An International Journal of Police Strategies and Management* 23(4):439–465.

Brady, T. (1996) *Measuring what matters: Part one: Measures of crime, fear, and disorder.* (Washington, DC: National Institute of Justice).

Bromley, M. and J. Cochran (1999) "A case-study of community policing in a southern sheriff's office," *Police Quarterly* 2(1):36–56.

Bureau of Justice Assistance (1997) *Crime prevention and community policing: A vital partnership.* (Washington, DC: Bureau of Justice Assistance).

Carter, D., A. Sapp, and D. Stephens (1988) *The state of police education: Policy direction for the 21st century.* (Washington, DC: Police Executive Research Forum).

Chicago Community Policing Evaluation Consortium (1997) *Community policing in Chicago, year four: An interim report.* (Chicago: Illinois Criminal Justice Information Authority).

Cohn, E. (1996) "The citizen police academy: A recipe for improving police–community relations," *Journal of Criminal Justice* 24(3):265–271.

Cordner, G. (1995) "Community policing: Elements and effects," *Police Forum* 5(3):1–8.

Cordner, G. and G. Williams (1995) "Community policing and accreditation: A content analysis of CALEA standards," *Police Forum* 5(1):1–9.

Dantzker, G., A. Lurigio, S. Hartnett, S. Houmes, S. Davidsdottir, and K. Donovan (1995) "Preparing police officers for community policing: An evaluation of training for Chicago's alternative policing strategy," *Police Studies* 18(1):45–69.

Dietz, A. (1997) "Evaluating community policing: Quality police service and fear of crime," *Policing: An International Journal of Police Strategies and Management* 20(1):83–100.

Duffee, D., R. Fluellen, and B. Renauer (1999) "Community variables in community policing," *Police Quarterly* 2(1):5–35.

Frank, J., S. Brandl, and R. Watkins (1997) "The content of community policing: A comparison of the daily activities of community and 'beat' officers," *Policing: An International Journal of Police Strategies and Management* 20(4):715–728.

Gaines, L. (1997) "Empowering police officers: A tarnished silver bullet?" *Police Forum* 7(4):1–7.

Geller, W. and G. Swanger (1995) *Managing innovation in policing: The untapped potential of the middle manager.* (Washington, DC: Police Executive Research Forum).

Goldstein, H. (1977) *Policing a free society.* (Cambridge, MA: Ballinger).

Goldstein, H. (1979) "Improving policing: A problem-oriented approach," *Crime & Delinquency* 25(2):236–258.

Hickman, M. and B. Reaves (2001) *Community policing in local police departments, 1997 and 1999.* (Washington, DC: Bureau of Justice Statistics).

Jiao, A. (1997) "Factoring policing models," *Policing: An International Journal of Police Strategies and Management* 20(3):454–472.

Kelling, G. and C. Coles (1996) *Fixing broken windows.* (New York: Free Press).

Kelling, G. and M. Moore (1988) *The evolving strategy of policing.* (Washington, DC: U.S. Department of Justice).

Kerley, K. and M. Benson (2000) "Does community-oriented policing build stronger communities?" *Police Quarterly* 3(1):46–69.

King. W. (2000) "Measuring police innovation: Issues and measurement," *Policing: An International Journal of Police Strategies and Management* 23(3):303–317.

Kratcoski, P. (1995) "Community-oriented policing: COP and POP," in W. G. Bailey (ed.) *The encyclopedia of police science,* 2nd ed. (New York: Garland):94–98.

Kratcoski, P. and R. Blair (1995) "Dynamics of community policing in small communities," in P. Kratcoski and D. Dukes (eds.) *Issues in community policing.* (Cincinnati, OH: Anderson):85–104.

LeClair, E. and A. Sullivan (1997) "An exploratory analysis: Assessing the attitudes of supervisors and rank-and-file officers toward community policing." Paper presented at the annual meeting of the Academy of Criminal Justice Sciences, Louisville, KY.

Lewis, S., H. Rosenberg, and R. Sigler (1999) "Acceptance of community policing among police officers and police administrators," *Policing: An International Journal of Police Strategies and Management* 22(4):567–588.

Lord, V. (1996) "An impact of community policing: Reported stressors, social support, and strain among police offices in a changing police department," *Journal of Criminal Justice* 24(6):503–517.

Lumb, R. (1995) "Policing culturally diverse groups: Continuing professional development programs for police," *Police Studies* 18(1):23–43.

Lynes, D. (1996) "Cultural diversity and social order: Rethinking the role of community policing," *Journal of Criminal Justice* 24(6):491–502.

Maguire, E. (1997) "Structural change in large municipal police organizations during the community-policing era," *Justice Quarterly* 14(3):547–576.

Maguire, E. and S. Mastrofski (2000) "Patterns of community policing in the United States," *Police Quarterly* 3(1):4–45.

Mastrofski, S., R. Worden, and J. Snipes (1996) *Law enforcement in a time of community policing.* (Washington, DC: National Institute of Justice).

McEwen, T. (1995) *National assessment program: 1994 survey results.* (Washington, DC: National Institute of Justice).

McLaughlin, C. and M. Donahue (1997) "Problem-oriented policing: Assessing the process," *Justice Professional* 10(1):47–59.

National Institute of Justice (1995) *Community policing in Chicago: Year two.* (Washington, DC: National Institute of Justice).

Oliver, M. (2000) "Book review essay," *Justice Quarterly* 17(1):231–234.

Palmiotto, M., M. Birzer, and P. Unnithan (2000) "Training in community policing: A suggested curriculum," *Policing: An International Journal of Police Strategies and Management* 23(1):8–21.

Palmiotto, M. and M. Donahue (1995) "Evaluating community policing: Problems and prospects," *Police Studies* 18(2):33–53.

Parks, R., S. Mastrofski, C. DeJong, and M. Gray (1999) "How officers spend their time with the community," *Justice Quarterly* 16(3):483–518.

Pino, N. (2001) "Community policing and social capital," *Policing: An International Journal of Police Strategies and Management* 24(2):200–215.

Ponsaers, P. (2001) "Reading about 'community (oriented) policing' and police models," *Policing: An International Journal of Police Strategies and Management* 24(4):470–496.

President's Commission on Crime and Administration of Justice (1967) *Task force report: Police.* (Washington, DC: U.S. Government Printing Office).

Radelet, L. and D. Carter (1994) *The police and the community,* 5th ed. (New York: Macmillan).

Rosen, M. (1997) "A LEN interview with Professor Herman Goldstein," *Law Enforcement News* 23(461):8–11.

Rosenbaum, D. (ed.) (1986) *Community crime prevention: Does it work?* (Beverly Hills, CA: Sage).

Russell, G. and S. MacLachlan (1999) "Community policing, decentralized decision making and employee satisfaction," *Journal of Crime and Justice* (22)2:31–54.

Sadd, S. and R. Grinc (1996) *Implementation challenges in community policing.* (Washington, DC: National Institute of Justice).

Sanders, B., T. Hughes, and R. Langworthy (1995) "Police officer recruitment and selection: A survey of major police departments in the U.S.," *Police Forum* 5(4):1–4.

Skogan, W. (1996) "The community's role in community policing," *National Institute of Justice Journal* (August):31–34.

Skolnick, J. and D. Bayley (1986) *The new blue line: Police innovation in six American cities.* (New York: Free Press).

Smith, B., K. Novak, and D. Hurley (1997) "Neighborhood crime prevention: The influences of community-based organizations and neighborhood watch," *Journal of Crime and Justice* 20(2):69–86.

Smith, M. (1996) *Crime prevention through environmental design in parking facilities.* (Washington, DC: National Institute of Justice).

Taylor, R. and A. Harrell (1996) *Physical environment and crime.* (Washington, DC: National Institute of Justice).

Toch, H. (1997) "The democratization of policing in the United States: 1895–1973," *Police Forum* 7(2):1–8.

Travis, L. and B. Sanders (1997) *Community-policing activities: The Ohio task analysis project, final report.* (Cincinnati, OH: University of Cincinnati).

Travis L. and C. Winston (1997) "Dissension in the ranks: Officer resistance to community policing, cynicism, and support for the organization." Paper presented at the annual meeting of the Midwestern Criminal Justice Association, Cincinnati, OH.

Trojanowicz, R. and B. Bucqueroux (1990) *Community policing: A contemporary perspective.* (Cincinnati, OH: Anderson).

Webb, V. and C. Katz (1997) "Citizen ratings of the importance of community policing activities," *Policing: An International Journal of Police Strategies and Management* 20(1):7–23.

Weisheit, R., D. Falcone, and L. Wells (1994) *Rural crime and rural policing.* (Washington, DC: National Institute of Justice).

Weiss, A. (1997) "The communication of innovation in American policing," *Policing: An International Journal of Police Strategies and Management* 20(1):292–310.

Wycoff, M. (1995) *Community policing strategies.* (Washington, DC: National Institute of Justice).

Zhao, J. (1996) *Why police organizations change: A study of community-oriented policing.* (Washington, DC: Police Executive Research Forum).

Zhao, J., N. Lovrich, and Q. Thurman (1999) "The status of community policing in American cities: Facilitators and impediments revisited," *Policing: An International Journal of Police Strategies and Management* 22(1):74–92.

CONTROLLING THE POLICE

CHAPTER OUTLINE

Since the beginning of formal, paid policing, there has been a concern about the possibility of abuse of police powers. As Lundman (1980) phrases it, a "dynamic tension" exists between the desire for personal liberty and the desire for public order. Consequently, initial efforts to establish police in both England and the United States faced opposition from those who distrusted the grant of police powers to government agents. The question they raised was, Who will watch the watchman?

We have seen that much of what police officers do is characterized by discretion—the ability to choose between different courses of action or inaction. How does society guide those choices to ensure that the police make the right decisions? This chapter examines types of police misconduct or controversial police decisions. We then describe and assess different methods of controlling police discretion that have been proposed or attempted in various jurisdictions.

POLICE MISCONDUCT

Misconduct, like beauty, is often in the eye of the beholder. Some forms of police behavior, however, are almost universally condemned as wrongful. Others generate heated debate and controversy as people with different perspectives judge the police action differently. We will focus on one type of police conduct which most people agree is inappropriate—corruption. In the following section we will examine some police practices about which there is considerable disagreement.

Corruption

Those who study police corruption have proposed dozens of definitions to categorize the large number and wide variety of behaviors that may be considered "corruption" by police officers. For example, is an officer corrupt if he or she

- Accepts free or discounted meals?
- Accepts payment from a motorist for not issuing a traffic citation?
- Ignores the operation of an after-hours club on the beat?
- Routinely rousts black teenagers but ignores whites?
- Checks crime records on a prospective employee for a friend?
- Steals from the scene of a crime before other police arrive?
- Commits a burglary?
- Sleeps or drinks while on duty?
- Asks a motorist for payment in return for not issuing a citation?
- Provides illegal drugs to an informant?
- Calls in sick to receive a paid day off when not ill?
- Accepts routine payments in exchange for not enforcing laws against drug sales or gambling?
- Refuses to cite a fellow officer for a traffic violation?
- Runs personal errands while on duty?
- Commits perjury in order to convict a violent offender?
- Allows his or her spouse to avoid a traffic citation or arrest for petty crime?

As these examples indicate, a variety of improper or questionable police behavior exists. Some of these behaviors are more threatening and more serious than others. Some violate laws, others violate departmental policies, and some may simply violate social norms. Whereas all of these behaviors (and more not listed) may be wrong, their categorization as police corruption depends upon the definition selected.

Police Integrity

The long-standing interest in assuring that police officers "do the right thing" has taken a turn recently. As we shall see, the traditional approach to police misconduct has been to devise ways of supervising and directing police officers through a system of rewards and punishments (mostly punishments). In 1996 a conference jointly sponsored by the National Institute of Justice and the Office of Community Policing Services focused on police integrity. In a real sense, integrity is the opposite of corruption (Gaffigan and McDonald, 1997).

This conference sought to understand the causes of police violations of the public trust and the dynamics of police integrity. In the keynote address, Stephen Vicchio (1997) observed that traditional approaches to the control of police behavior are based on the following beliefs:

1. People act correctly only out of fear of punishment.
2. There are not enough supervisors to effectively watch all police.
3. Bad cops will get media attention.

These three observations suggest that current attempts to control police misbehavior through a system of supervision and punishment are doomed to failure. Further, the cost of this failure is magnified by media coverage of those officers who do misbehave. The picture of the police painted in the media will be one of immoral and corrupt officers. Vicchio urges that the focus be on establishing and supporting police integrity instead. **Integrity,** while difficult to define, is characterized by proper behavior. Integrity is a characteristic of the individual so that officers with integrity will, on their own, refrain from misbehavior. If we could ensure integrity among police officers, police misconduct would not be an issue.

The conference was a first step in what promises to be a long-term developmental project. Among other things, those at the conference hoped to be able to identify both individual officers who have integrity and police organizations with integrity. From study of those, they hoped to derive principles to guide the selection and training of officers and the structure and administration of agencies. It will be some time before we can tell how successful they will be in this effort.

Lacking such knowledge, we will proceed in the more traditional fashion of looking at **misconduct** and **corruption** (the absence of integrity). It is instructive to note that after all this time, looking for the good is a novel approach. It suggests an important part of our perspective on police officer behavior. As Vicchio noted, we expect that most police officers, given the chance, will misbehave. Interestingly, with that expectation, we are still surprised and disappointed when we learn of police corruption.

Defining Police Corruption

Herman Goldstein (1977:188) defined *police corruption* as "the misuse of authority by a police officer in a manner designed to produce personal gain for the officer or others." This definition contains two key elements: *misuse of authority* and *personal gain.* Another definition maintains that police corruption occurs whenever an officer accepts money or money's worth either to take or fail to take an action he or she is required to take under law. The key element of this definition is *money* or *money's worth.* Finally, Lawrence Sherman (1974:30) has defined police corruption as "an illegal use of organizational power for personal gain," in which all of the terms— *illegal use, organizational power,* and *for personal gain*—are important.

Each of these definitions has its strengths and weaknesses. If we rely on the motivation of money or money's worth as a key to identifying corruption, we are not able to account for cases where the officer does not seek money—such as ignoring offenses by friends or relatives, displaying discrimination, or similar acts of misbehavior.

Goldstein's definition, which is based on the motivation of gain for self or others, is broader. It includes both money and nonmoney gains, such as satisfying prejudicial urges, strengthening ties to family and friends, and simple convenience for the officer. Sherman's use of the term *illegal* is itself broad, covering actions that are against either criminal or civil laws as well as those in violation of departmental policy. The *personal-gain* component of Sherman's definition, however, poses problems. Suppose the gain is for another—such as perjury to protect a fellow officer. Although the lying police officer may gain trust, status, and respect from her or his peers, if this is not the motivation, then the behavior is not corruption. As Klockars notes (1983:336), "Some things we might well describe as police corruption may involve *organizational* but not *personal* gain. For example, police officers may be directed by their superiors not to investigate criminal activities of politicians who are influential and supportive of the police agency. This is not an uncommon type of police corruption, and the gain involved is exclusively organizational."

One element common to all of these definitions of corruption is that they require the misuse of police authority. Whether this misuse is to invoke or threaten to invoke police powers in circumstances where they are not warranted or normally not applied, or if it is to fail to invoke police powers when they are warranted or should apply, the critical factor is such misuse of authority. A second element is intent. Most definitions skip the question of intent by specifying the motive for misuse. It is this focus on intent that muddies the definitional waters.

Corruption as Intentional Behavior

In criminal law, most offenses have two components; an *action* and an *intent* (Travis, 1998). Thus, burglary usually covers illegal entry with the intention of committing a crime therein. Neither committing a theft nor illegally entering a place, by itself, constitutes burglary. Murder is usually defined as the intentional killing of another. Simply wanting to kill someone, or killing someone accidentally, does not constitute murder.

In murder mysteries the detective is often concerned with motive. Either there are many people who have reason to kill the victim—past wrongs, lover's triangles, a desire to collect insurance or an inheritance, etc.—or there are no motives. The importance of motive is to establish intent. Even if the state can prove that someone killed his or her uncle, it cannot necessarily prove murder. The suspect can plead that the killing was an accident. By showing a motive—to collect an inheritance, for example—the state establishes that the suspect had a reason to kill and thus probably intended to kill. This intent, coupled with the act of killing, results in the crime of murder.

For our purposes, **police corruption** *is the intentional misuse of police power.* In practice, this definition means that before something can be called police corruption, two things need to be established. First, it must be shown that police powers were misused. Second, it must also be shown that the officer(s) misusing police power intended to misuse it. Whether the motive is money, personal gain, or gain for self or others is not important except insofar as it helps to show intentional misuse of power. Similarly, whether the misuse of power is technically illegal (even under Sherman's broad definition of *illegal*) is not important except to show that misuse of power occurred.

Types of Corrupt Activities

We have already seen that a host of police actions can be considered as examples of police corruption. Several observers have suggested ways to classify or categorize these individual behaviors into groups of similar misconduct. One of the first such classification schemes was presented by Ellwyn Stoddard (1979), who proposed 10 types of corrupt behavior:

1. *Mooching* is the acceptance of free or discount merchandise or services by officers.
2. *Chiseling* is the demand of police officers to be admitted to entertainment events without charge (sometimes called badging in).
3. *Favoritism,* as the term implies, is "giving breaks" to friends, family, and other officers from arrest or citation for traffic offenses.

4. *Prejudice* is the differential treatment of certain groups by the police.
5. *Shopping* refers to the practice of officers who take small items from a business without offering to pay.
6. *Extortion* is a demand by officers that citizens buy tickets, police magazines, or other things where the sales pitch is supported by a threat of citation or arrest.
7. *Bribery* is the practice of accepting cash or gifts with the understanding that the officer will overlook some violation.
8. *Shakedown* is the taking of expensive items for personal use from crime scenes.
9. *Perjury* involves lying as a witness, especially when done to cover up the misbehavior of oneself or another officer.
10. *Premeditated theft* covers those instances where police officers carry out a planned burglary or theft.

Of these types of corrupt behavior, eight clearly fit our definition of police corruption. Two, however, do not. Mooching is not necessarily an act of corruption. If merchants routinely give discounts to police officers, and if the police officers do not alter their behavior because of the discounts, then there is no misuse of authority. On the other hand, if officers patronize only a certain store or restaurant because of the discount (and thus provide that business with added police protection), then the officers may be misusing the office to obtain the free or reduced-price goods. William DeLeon-Granados and William Wells (1998) tested the impact of offering gratuities to police on levels of police patrol presence. While they found that police presence was generally low (in less than 1 percent of the times they observed 19 restaurants were police present), businesses that gave police discounts were six times more likely to be visited by police.

Premeditated theft is a crime, not an act of corruption. If the officers involved in the crime committed it because they knew, as a result of their positions, that the owners would be absent, or that the alarm was disabled, then the crime would involve police corruption. On the other hand, if an officer or group of officers commit a crime that is not an abuse of their police authority, they are criminal but not corrupt (Goldstein, 1977:189).

Other specific classes of corrupt behaviors have been identified in the literature (Barker and Roebuck, 1973). James Inciardi (1987:272) suggests that all of these can be classed into nine categories:

1. *Meals and services*—includes all forms of mooching and chiseling.
2. *Kickbacks*—receiving payment for referring clients to attorneys, bail bondsmen, automobile service stations, and the like, or providing sales leads to these and related businesses.
3. *Opportunistic theft*—includes both "shopping" and "shakedowns" as defined by Stoddard.

4. *Planned theft*—includes burglaries and other instances in which officers engage in premeditated crimes.

5. *Shakedowns*—what Stoddard called "extortion," where officers coerce civilians into buying tickets or other products, as well as officers who seek bribes.

6. *Protection*—where officers provide additional police service to those who either pay through discounts and gifts, or perhaps those whose support for the police is important.

7. *Case-fixing*—where citizens' bribes to avoid arrest or conviction are accepted by officers.

8. *Private security*—where officers hire out as bodyguards or security officers to private citizens and use their police position to make them more effective in this private role as well as, perhaps, in securing such employment to begin with.

9. *Patronage*—involves the falsification of personnel records, the assignment of officers, vacation days, and promotions for a fee within the department, and granting access to information or making recommendations about applications for various licenses for persons outside the department.

Although they may disagree about the exact number of classes and the titles to be used for each set of corrupt actions, those who have studied police corruption would agree that the opportunities for misconduct abound in policing. Given the nature of policing in America, perhaps we should be less surprised by how often and how much police corruption is disclosed than by how little and how infrequently it is discovered.

Sources of Police Corruption

Herman Goldstein (1977:197–200) listed seven factors that contribute to corruption in policing. In essence, he argues that these characteristics of policing and criminal law in America enable police to become corrupt.

1. *Unenforceable laws*—Legislators across the country have used the criminal law to regulate behaviors that many citizens do not view as criminal. Further, police officers know that they cannot enforce all the laws against all the lawbreakers all the time in our society. Thus, the overreach of the criminal law creates a situation in which citizens are willing to bribe officers to ignore a law they feel is unjust. Further, officers can rationalize their behavior by thinking, "Since most violators of the speed limit avoid detection and citation, what difference does it make if I let this person off?"

2. *Organized crime*—Illegal businesses develop to serve the interests of those citizens who wish to pursue illegal activities such as gambling,

substance abuse, and the employment of prostitutes. These criminal organizations require that the police not disrupt their operations to be profitable. As they tend to be lucrative enterprises, organized criminals have both the motive and the means to bribe police officers to ignore their illegal activities.

3. *Improper political influence*—Elected officials often attempt to influence police operations to bolster their political positions in the community. In the early days of policing in America, police positions were awarded based on political patronage. Even today, political leaders often attempt to micromanage or otherwise direct police activity. They might indicate to the police that certain behaviors should be overlooked—public drinking at community events, for example—or that other behaviors should be cracked down on, such as loitering around downtown businesses by panhandlers.

4. *Police work*—As a result of the experiences on the job, police officers are exposed to all sorts of crime and corruption. They learn that even respectable people steal by tax evasion, insurance fraud, and other means. One result of this is a cynical realization that everyone cheats sometime. To this realization of the apparently widespread existence of dishonesty is added the feeling that officers are not adequately compensated for the rigors and risks of their job. Thus, the officer can justify some forms of corruption.

5. *Prosecutors and courts*—The lack of effectiveness of the courts in controlling the behavior of offenders—plea bargaining, case dismissals, and similar practices—enables the officer to justify her or his failure to arrest or cite an offender. Certainly the officer's failure to take action seems no worse than a similar failure on the part of the courts. A flip side to this is the officer's observation that the courts are not sensitive to the reality of the streets and often allow a dangerous offender to return to the community. The desire to prevent judges or prosecutors from making such a mistake might lead an officer to justify committing perjury to ensure that an offender is convicted and punished.

6. *Police discretion*—The ability of officers to choose whether and how to intervene with citizens may very well be the most important contributing factor to police corruption. The fact is that police officers have wide powers of discretion that are little controlled by supervisors, statutes, or departmental regulations. For this reason, it is possible for police to overlook violations, whether or not their decision not to invoke the criminal law is a misuse of power. The ability of police to use their authority appropriately in a variety of ways allows them to misuse it in similar ways as well.

7. *Addictive element of corruption*—Once an officer has accepted the profits of corruption, either through monetary gain or being able to do favors for friends, it is difficult to stop. In the case of money, the officer

may come to rely on the additional income. In the case of nonmonetary motivations, the fact that the officer was able to satisfy some need or desire becomes its own incentive to do so again. To this addictive quality must be added the solidarity of police—the fact that any officer is open to charges of corruption and that what appears to be corrupt may be a proper exercise of discretion. Thus, police officers who do not engage in corrupt activity often tolerate or tacitly support those who do.

Varieties of Police Corruption

Lawrence Sherman (1974:7) identifies three levels of corruption that might be found within a police agency. These different levels are based on how widespread corrupt behavior is in the agency, how organized the corrupt practices are and where bribes originate. He describes these different levels as rotten apples and rotten pockets, pervasive unorganized corruption, and pervasive organized corruption.

Rotten apples and **rotten pockets** describes organizations in which corruption is relatively infrequent. When only a few officers engage in corruption on their own, they are "rotten apples" in an otherwise sound barrel. A "rotten pocket" is the cooperation among some corrupt officers—perhaps members of an investigative unit—in corruption. Though more threatening and more difficult to control than rotten apples, the rotten pocket is still confined to a small band of officers.

Pervasive unorganized corruption describes departments where the majority of police officers accept bribes or do other corrupt things, but they do so independently. Thus, while nearly every officer might be willing to overlook some minor violation in return for a bribe, the officers do not talk about the practice nor do they cooperate in corrupt activities.

Pervasive organized corruption is the most serious form and describes those instances where misconduct is shared among all or most members of the department including administrators. The most common example of such corruption is the "pad" as described in the story of New York detective Frank Serpico (Maas, 1973) and the report of the Knapp Commission (1972). The police officers in these cases organized to routinely collect bribe payments from citizens. The payments were then divided into standard shares, which were distributed to officers monthly. Supervisors and administrators received a higher payment each month than did patrol officers (just as they received higher city salaries). The pad developed a separate, corrupt "shadow" organization within the police department.

In addition to this classification of police corruption in organizations, individual officers also differ in terms of their styles of corruption. One basic difference is drawn between grass eaters and those known as meat eaters. **Grass eaters** are officers who take advantage of opportunities for graft that might arise. If a motorist, for example, offers a bribe, the grass

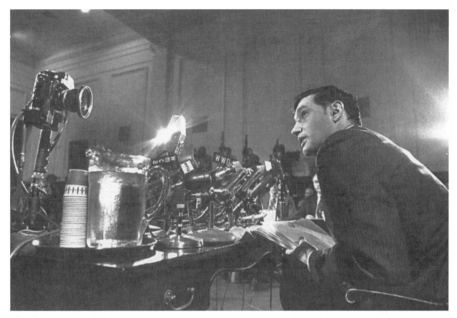

The hearings of the Knapp Commission on corruption in the New York City Police Department gained national attention for the problem of police bribery and other misconduct. (UPI/Bettmann)

eater will accept it. In contrast, **meat eaters** aggressively seek out opportunities for graft. Rather than availing themselves of opportunities that occur, meat eaters create opportunities by asking for bribes from citizens they encounter.

A second stylistic difference in police officer corruption relates to a morality of graft. Some graft is defined as harmless or clean, such as bribes from businesses to overlook parking or other code violations. Those who hold this view that some graft is acceptable argue that no one is really hurt by this dereliction of duty. Other graft is viewed as harmful or dirty, such as accepting payments from drug dealers to allow drug transactions to occur. The worst type of corrupt officer would be the meat eater who accepts dirty graft.

CONTROVERSIAL POLICE CONDUCT

Most people agree that corruption or the intentional misuse of police authority is wrong, but other forms of police conduct generate debate. Thoughtful, fair-minded people may have honest disagreements about the nature and extent of police misconduct. These disagreements tend to center on the definition of the conduct involved. What is police brutality to some is seen by others as a justified use of force. Ethnic discrimination to one observer is an example of street sense to another. As the general public reaction to the Rodney King case in Los Angeles during the early 1990s indicates, even when almost everyone says the police have misbehaved, some (like the jury in the original trial of the police officers) will disagree.

Three areas of police behavior will illustrate the controversial nature of police conduct: police discrimination, substance abuse, and brutality. These terms are themselves value laden. They all conjure negative images. The issues involved in these behaviors determine whether what the police officer did fits this negative label.

Discrimination

Donald Black (1980:13) concludes that discrimination is an unavoidable aspect of policing: "If discrimination in the application of law is taken to occur when the police systematically handle similar cases in different ways, and when this is related to the social characteristics of the people involved, it should be apparent that discrimination is ubiquitous in police work." In short, since the police handle cases differently, they discriminate.

The issue, then, is not one of discrimination *per se,* but of the reasons for discrimination. The police are expected to discriminate; indeed, they must discriminate if they hope to maintain order. Our concern is not with the fact that police discriminate in their handling of situations, but rather with the basis on which such discriminating judgments are made. If the sole justification for arresting someone is that she or he is a minority group member, or poor, or dressed wrong, then we are likely to conclude that the arrest is an example of bad discrimination. On the other hand, if the decision to arrest someone is based on seriousness of the offense, the willingness of a victim or witness to cooperate, the officer's need to gain control of the situation, or some similar reason, we might decide that it is an example of good discrimination.

The nature and extent of police discrimination, particularly bad or unjustified discrimination, is unknown. What data are available on the issue are limited and often contradictory. Do the police discriminate against minorities, males, the poor, and the young? Persons having these characteristics are disproportionately represented in the population of persons arrested by the police. On the surface, at least, it would seem that the police play favorites.

Studies that have examined the impact of ethnicity, gender, age, and other factors on police decision making have not generally concluded that police officers base their decisions solely, or even largely, on these characteristics (Weitzer, 1996; Son, Davis, and Rome, 1998). Rather, when researchers have taken offense seriousness, suspect demeanor, the wishes of the complainant, and other factors into account, discrimination does not prove to be a very powerful explanation of officer behavior (Klockars, 1985:103–104). Still, the data reveal patterns of apparent discrimination. If the police are not more lenient with women and whites, and if the police are

Are these officers racists, brutal, or both? The use of force is almost always troublesome because of difficulties in defining what is excessive and in understanding the officers' decisions to resort to force. (Christopher Smith/Impact Visuals)

not less tolerant of the poor, youth, and minorities, how is it that these groups face such different rates of arrest?

Part of the answer may be found in institutional discrimination that is rooted in society. If men drive the automobile when traveling with women, men have a greater chance of violating traffic laws and being cited or arrested for those violations. If most arrestees come from cities (where we also find most police officers), and minority group members are concentrated in urban areas, then they have a greater chance of being arrested than whites living in rural or suburban areas.

Browning et al. (1994) report a study of perceptions of being hassled by the police. They found that African Americans were significantly more likely to report that they felt they were *hassled* (unjustly bothered) by the police than were whites. Browning and her coworkers suggest that this difference in perception could be a result of African Americans being more sensitive to police actions, or to the fact that they are more likely to be the targets of differential patrol strategies (e.g., drug sweeps in urban ghetto neighborhoods [Barnes and Kingsnorth, 1996]). Even if African Americans are not the targets of different treatment by the police, their perceptions of discrimination are real and important. The specific influences of minority group status on evaluations of police actions are complex and require further research (Stroshine-Chandek, 1999), but if citizens expect the police to

behave in certain ways, these expectations are likely to influence their interpretation of events.

Recalling for a moment our earlier discussion of the reactive nature of
American policing, we can understand that the police will reflect the interests and values of the people in society. Thus, to the extent that there is discrimination based on gender, socioeconomic status, and ethnicity in society,
this discrimination will be reflected in the actions of the police (Black, 1980;
Klockars, 1985).

In recent years concern about racial discrimination by police has
grown. Part of this may be a reaction to the increased involvement of police
in order maintenance or quality of life policing. When no crime has been
committed, or no serious crime, police intervention is less likely to be
viewed as legitimate by citizens. Almost everyone, including criminals, expect police to try to arrest lawbreakers. In contrast, youth congregating on
a street corner, or panhandlers plying their trade, or homeless persons loitering in a park might not agree that their behavior is any business of the
police. When the police try to intervene, there will be questions of legitimacy
and a heightened potential for conflict. The emphasis on more aggressive
policing, coupled with the most recent "war on drugs" which has disproportionately affected minority neighborhoods, has led many to question police
motivations.

"Racial profiling" is an especially controversial and contentious topic.
In some pure form, **racial profiling** relates to the police use of race as an
indicator of criminality. Decisions about whom to stop and question, search,
arrest, or issue a traffic citation to are conditioned on the race of suspects.
As it is more broadly understood, racial profiling has become synonymous
with discriminatory policing. Fridell et al. (2001) comment on the conceptual limits of the term "racial profiling." Police narrowly define the term to
relate to interventions based solely on the race of the citizen. Citizens, in
contrast, broadly define the term to include all instances of racial bias in
policing. They write (2001:4), "These contrasting, but unspoken, definitions
lead to police defensiveness and citizen frustration." Rather than racial profiling, they chose to examine "racially biased policing."

Substance Abuse

In some ways citizens hold the police to a higher moral standard than that
which applies to civilians. Those charged with enforcing laws against public intoxication and the use, possession, or sale of controlled substances
are expected to abstain from drinking and using drugs. Obviously, police
officers are human and occasionally succumb to the temptations of mood-
altering substances (McEwen, Manili, and Connors, 1986).

Substance abuse among police officers is a topic on which little data
are available (Swanson, Territo, and Taylor, 1988). The nature and extent of

substance abuse among police officers is unknown, although it is clear that some officers do abuse alcohol and drugs (Dishlacoff, 1976; Carter and Stephens, 1988). Researchers have estimated that as many as 25 percent of officers have serious alcohol problems (Dishlacoff, 1976) and nearly one-third have used other illegal substances (Carter, 1990a).

The use of alcohol or drugs by on-duty officers is generally a violation of police department policies and hence cause for disciplinary action including termination of employment. In addition, off-duty abuse of alcohol or drugs can lead to police corruption. David Carter (1990b:89) observes that although his data did not show that all drug-using officers would become corrupt, "it appears that the progression from use to corruption is an evolutionary process, eventually affecting most drug-abusing officers to some degree." These officers divert confiscated drugs from police evidence rooms, seek and accept bribes, and sell drugs to support their own drug-use needs.

Beyond the obvious facts that substance abuse is a violation of departmental policy for on-duty officers and a motive for police corruption, the issue of police officer substance abuse is controversial. On the one hand, what someone does on her or his own time is that person's own business and need not reflect on job performance. On the other hand, as Carter (1990b:92–93) notes, the expectation of higher morality or integrity on the part of police officers makes their substance abuse a violation of public trust. The abuse of alcohol and other drugs by police officers is often seen as a character flaw (Swanson, Territo, and Taylor, 1988:258) that renders the officer unfit for duty and requires dismissal from the force. That is, the abusing officer, even when abuse is limited to off-duty hours, is viewed as a "bad cop." Because of the nature of the occupation, police officers may be held accountable on the job for their personal behavior during their leisure time.

The problems of drug enforcement and the temptations of drugs, coupled with increasing requirements that agencies maintain drug-free workplaces have heightened concern about officer drug abuse (McEwen, Manili, and Connors, 1986). Many police agencies now operate employee drug-testing programs. These programs screen officers for drug abuse through the use of urine tests for the presence of drugs. These tests will detect not only on-duty drug use but also will reveal recent drug use by the officer, whether on or off duty.

Significantly, the majority of 33 police departments responding to a survey about officer drug testing indicated that positive tests results (i.e., evidence that the officer had used illegal drugs) would generally lead to dismissal from the force. Only about one-fourth of those responding said that treatment rather than dismissal might be appropriate under certain circumstances (McEwen, Manili, and Connors, 1986:2). Barbara Webster and Jerrold Brown (1989) describe the drug-testing program of the Honolulu Police Department. Policy in this department encouraged officers who tested positive for drugs to seek and obtain drug treatment. Termination of employment for officers testing positive was reserved for those who tested

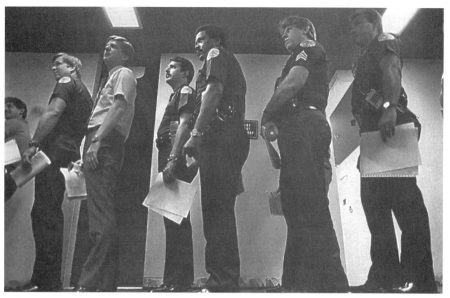

Police officers line up to give urine samples for drug testing, an increasingly common job requirement. (UPI/Bettmann)

positive on a second, later test. Yet, the officers were placed on leave and had to pay the costs of drug treatment themselves.

Brutality

As we have noted earlier, the capacity to use force is at the core of the police role in society. The use of force by police, while necessary to their function, is a controversial issue. Unnecessary force is often called **brutality,** and perceptions of police brutality have been at the heart of citizen distrust of and complaints about the police in America for nearly as long as we have had police (Johnson, 1981).

One of the greatest difficulties in understanding police brutality is that the definition of unnecessary or excessive force is ambiguous (Barker, 1978). Most statutes authorizing police to use force specify that officers may use only the degree of force necessary to effect an arrest or prevent an escape. The problem arises in determining what level of force was necessary. If an officer strikes a criminal suspect twice, is the second blow necessary or excessive? Who decides what level of force is required?

The Rodney King case in Los Angeles in 1991 provides an excellent example of the ambiguity of brutality. Several officers repeatedly struck Mr. King with batons, and one officer was seen kicking at him as he lay on the street. Even after King was knocked to the ground, officers continued to strike at him. At their state trial for assault, the officers involved main-

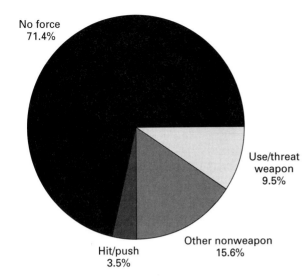

No force
71.4%

Use/threat
weapon
9.5%

Other nonweapon
15.6%

Hit/push
3.5%

CHART 16.1
Use of Force by Police in Making Arrests.
(*Source:* J. Garner, (1996) *Understanding
the Use of Force by and against the Police.*
Washington, DC: National Institute of
Justice, p. 7.)

tained that they were following their departmental training and policy in
the use of force. Their attorneys argued that Mr. King, by attempting to regain his feet, was continuing to resist arrest. Thus, the officers' use of force
was necessary to effect the arrest of Mr. King. The jury accepted this argument and acquitted all but one officer on all charges. The public, at least the
public in South Central Los Angeles, disagreed.

The use of force and the potential for police brutality is troublesome
because of the ambiguity that surrounds the decision to initiate or cease
applying force. Not only do people disagree over the amount of force to be
applied, but they also disagree about the type of force used. Force can be
thought of as a continuum ranging from threats to the use of deadly
force—shooting a suspect. Chart 16.1 shows that the actual use of physical force by police, even when making arrests, is relatively infrequent.
Along this continuum lie a number of alternatives for the officers including, where available, restraint, pushing and shoving, punching and kicking, chemical mace, electronic stun guns, batons and blackjacks, and
firearms. It is often unclear how one decides to use a baton rather than
mace, or a firearm rather than a stun gun. Deadly force employed by the
police results in an average of 373 civilian deaths each year. Of those
killed by the police, most are young and male. African Americans are disproportionately involved in police use of deadly force (Brown and Langan, 2001).

Geoffrey Alpert and Michael Smith (1999) noted the difficulty in measuring and studying the use of force by police. They recognize the variety of
circumstances in which force may be used, and the range of force that can
be employed by police. In a reanalysis of the Phoenix use of force data reported by Garner et al. (1996), Charles Crawford and Ronald Burns (1998)

found that the type or level of force used by police was correlated with officer characteristics (experience), suspect characteristics (age, gender, ethnicity, weapon possession, substance abuse, etc.), and situational characteristics (time of day, location, presence of witnesses, etc.). What this and other research indicates is that the decision to use force, and how much force to use is a product of many factors. The use of deadly force by police, for example, appears to increase in reaction to increases in criminal homicides. That is, the police exercise deadly force in response to a perception of increased risk. As violent crime, and homicide in particular, increases, so does the use of deadly force by police (MacDonald, Alpert, and Tennenbaum, 1999).

Thanks largely to popular drama, when officers do use their firearms, citizens often wonder why the police shoot to kill. After all, Western lawmen and television cops routinely shoot suspects in the leg or shoot the gun out of a suspect's hands without killing him or her. In reality, of course, few persons have such skill in aiming a firearm, especially in the emotionally charged situation of deadly combat. Further, officers are trained to shoot for the biggest target (the center of the torso) to increase the chances of hitting the target, and to immobilize the offender.

Citizen complaints about brutality are often based not on the extreme end of the continuum that includes the use of deadly force, but on the threat end instead (Reiss, 1971). The use of physical force is relatively rare in policing (Sykes and Brent, 1980; Bayley and Garafalo, 1989), and the use of deadly force is rarer still. Often, however, what is perceived as police brutality might be better understood as police *bullying*.

Garner and his colleagues (1996) studied police use of force in Phoenix, Arizona. They found that nine factors helped explain the use of force by officers, and the most important was the use of force by the suspect. Still, these nine factors explained only about one-third of the use-of-force decisions by police officers. Most use of force by police remained unexplained. When police officers did use force, however, it was typically at the low end of the severity scale. That is, force used by police in arrests usually involved weaponless tactics or applying handcuffs. The police used weapons in only 2 percent of the arrests studied.

Citizens often complain of police brutality because of what they perceive as the bullying tactics of officers. Reiss (1971:334) observes (and later researchers [Decker and Wagner, 1985] agree), "What citizens object to and call 'police brutality' is really the judgment that they have not been treated with the full rights and dignity among citizens in a democratic society. Any practice that degrades their status, that restricts their freedom, that annoys or harasses them, or that uses physical force is frequently seen as unnecessary and unwarranted."

In short, police brutality often refers to nonphysical affronts to citizens by police. We noted earlier that the most common form of police intervention with citizens involved *definitional regulation* (Sykes and Brent, 1980), where officers ask questions. The followup strategy to this appears to be *imperative regulation,* where officers issue commands. Bayley and Garafalo (1989:11) note that those officers perceived as most skillful in handling conflict situations are more confrontational and more likely to take verbally forceful actions. Thus, much of what is called brutality may actually be bullying by police officers. Citizens feel threatened and do not appreciate the officers' menacing attitude.

SOME CORRELATES OF POLICE MISCONDUCT

Police misconduct—including corruption, discrimination, substance abuse, and brutality—is a constant source of friction between the police and the citizens they are sworn to serve. A primary problem with misconduct by police is that it raises a question of control (Travis, 1998). These police are violating the expectations of the citizens. The question naturally arises, who controls the police?

The final section of this chapter addresses means of ensuring police accountability and methods of controlling police behavior. The likely effectiveness of these control measures depends on how well they reflect and effect the factors that are associated with police conduct. Thus, before turning to an examination of ways to control the police, we must identify the correlates of police misconduct. As with earlier subjects, these correlates exist at three levels; the community, the organization, and the people involved in policing.

Community

Whereas some police misconduct is probably inevitable in any community at some time, certain community characteristics are associated with varying levels of misconduct. Lawrence Sherman (1974) observes that police corruption is linked to the values and norms of a community. He suggests that corruption is likely in communities where there is more diversity in terms of both social values and economic opportunity. Further, he believes that corruption is less likely in communities where civic leaders see their jobs as serving the community rather than serving themselves or a small set of constituents.

These three characteristics of communities are related to the levels of police corruption. If some citizens have limited advantages to pursue lawful occupations, they are likely to turn to unlawful ones, like gambling. In turn, these illegitimate businesses have both the motive and the resources to bribe police officers. If there is a lack of agreement among the citizens about what is proper behavior—in some neighborhoods public drinking is tolerated and in others it is not—officers have readymade excuses for ignoring some law violations. Finally, if civic leaders are themselves corrupt, they are likely to exert a corrupting influence on the police and unlikely to oppose most forms of police corruption.

Not only corruption, but other types of police behavior and misbehavior correlate with community characteristics (Alpert and Dunham, 1988). The use of force by police is linked to community characteristics. In particular, the ethnic composition of the community (Smith, 1986; Sherman, 1980; Lester, 1982) is associated with the frequency with which police officers use coercive force, including deadly force, against citizens. The greater the proportion of nonwhites and the more heterogeneous the racial composition of residents, the more likely it is that police officers will rely on force. Also, the "reputation" of the neighborhood—high crime/low crime or dangerous/safe—is associated with use of force. Police are more likely to use force in bad neighborhoods than in good ones (Geller and Karales, 1981; Waegel, 1984). Levels of crime, and violent crime in particular, may also explain police willingness to use deadly force (MacDonald, Alpert, and Tennenbaum, 1999).

In the same ways that communities create opportunity structures for crime, they create structures for police misbehavior. Communities that have relatively high levels of drug-related crimes provide opportunities for police officers to obtain and use drugs (Carter, 1987). Communities that have relatively high rates of violence create a climate conducive to police use of force. Communities characterized by organized crime are ripe for widespread police bribery. In regard to police misbehavior as well as police behavior, communities often get the kind of policing they want or deserve.

Organization

The police organization itself correlates with different types and levels of police misconduct. The organizational influence on officer behavior occurs within both the formal organization and the informal relationships that surround the members of the police agency. In a sense, the organizational influence on police misconduct is found in the degree to which such misbehavior is tolerated or even encouraged by the organization. Liqun Cao, Xiaogang Deng, and Shannon Barton (2000) report a study of police organizational characteristics and rates of citizen complaints about excessive force. They found that the organizational characteristics of police agencies, as well as the environments (communities) in which they operated were important correlates of the rate of citizen complaints.

In regard to the use of force by police and the potential for police brutality, Friedrich (1980) notes that Wilson's (1968) conceptualization of different police department styles would tolerate different levels of force. Watchman departments would be most likely to use excessive force, while legalistic departments would be least likely. Based on this theory, he compared three departments and noted that the more professional (legalistic) had fewer instances of excessive force than the traditional (watchman) agency. The department he called *transitional* (changing from watchman to legalistic) had the lowest incidence of excessive force use. Alpert and Mac-Donald (2001) report a survey of 264 local police departments. They found that the rate of violent crime in a community was the greatest predictor of police use of force, but that organizational reporting requirements were also significantly related. Departments that required supervisors to complete use-of-force reports had significantly lower levels of force used than those that relied only on officer self-reports.

William Westley (1970) attended to the informal organization and its effects on the use of force. He concludes that a police culture exists in which officers share beliefs and values that support or even require the use of force. Officers are justified, if not expected, to use even excessive force in reaction to obviously guilty persons and to those who challenge police authority. This informal value set, while at odds with public expectations and departmental regulations, works to define situations as requiring a forceful police response.

With regard to corruption, Goldstein (1977) firmly supports the view that the police administrator, through the police organization, is able to control, and is responsible for controlling, police corruption. The administrator must be committed to eliminating police corruption and must structure training, regulations, performance appraisals, and the departmental reward structure to support zero tolerance of police corruption. Lundman (1980) similarly suggests that corruption is evidence of *organizational deviance*. His position is that police corruption cannot long exist without at least tacit acceptance by the police agency.

The existence or lack of departmental regulations and training to control corruption, discrimination, substance abuse, and brutality can be expected to affect the rates at which each of these problems occur. In some cases an organizational (formal, informal, or both) attitude may develop that supports and encourages misconduct. In other cases, a lack of effort at controlling misconduct can create a climate in which, while misconduct is not encouraged, neither is it aggressively controlled. Though not the sole correlate of misconduct, the police organization affects both types and levels of officer behavior.

People

Ultimately, police misconduct involves the actions of individual officers and citizens. It is not surprising, then, that the characteristics of the people involved in police misconduct correlate with both the types and rates of police misbehavior. Given that the misconduct occurs within a community and organizational context, the individual persons involved still exert an influence on what happens.

The traditional police response to revelations of misconduct among officers has been to attribute misbehavior to individual rotten apples (Goldstein, 1977). That is, the cause of misbehavior was assumed to be the existence of a "bad cop" or two. Indeed, much of the early research on police corruption focused on the characteristics of individual officers and sought to explain misbehavior as the product of officer characteristics.

Arthur Neiderhoffer (1967) suggests that corruption is one adaptation an officer may make to cynicism. He suggests that experience on the job leads police officers to question the value of their work and the integrity of the public, the justice system, and the police organization. In time, officers must adapt to the mismatch between their expectations of policing and the reality. One way a disaffected officer may adapt would be to become corrupt, either as a loner (rotten apple) or as part of a deviant group of officers (rotten pocket).

In this vein, discrimination is the product of bigoted police officers, brutality the product of sadistic officers, substance abuse the product of weak officers, and corruption the product of venal officers. Beginning in the 1970s, thinking about the causes of police misconduct began to shift focus to community and organizational factors. Pogrebin and Atkins (1976:10) observe that an emerging consensus supported the notion that "police corruption should be viewed as failure of the criminal justice system and not as a flaw in the character of a few wayward policemen." More recently, Paul Muscari (1984) has argued that the individual officer is still a critical factor in understanding police misconduct. The fact remains, Muscari observes, that some officers succumb to community and organizational pressures while others do not. Robin Haar (1997) examined police misconduct and found that while misconduct was fairly common in the department she studied, officer characteristics helped explain types of misconduct. Officers who were committed to the job and the organization were likely to commit acts of misconduct aimed at achieving organizational goals (arresting offenders, achieving convictions) while officers less committed to the profession or organization were more likely to engage in misconduct against the department, such as avoiding work.

The officer is not the only one influencing the likelihood of police misconduct; the citizenry sometimes plays a part. Goldstein (1977:187) notes that police corruption is often initiated by the community. For example, the impetus for bribery, or the first bribe, comes from the public and is not sought by the officer. Similarly, research on the police use of force (Westley, 1970; Reiss, 1971; Friedrich, 1980; Sherman, 1980; Smith, 1986; Bayley and Garafalo, 1989; Garner et al., 1996) indicates that force is more common when the citizen is antagonistic, under the influence of drugs or alcohol, or openly rebellious or hostile to the officers.

David Klinger (1994) has questioned the importance of citizen demeanor as an explanation for police use of force or making an arrest. He draws a distinction between the types of disrespect towards the police that are legal and those that are illegal. He contends that citizen demeanor has no effect on the likelihood of arrest. In response to this, Richard Lundman (1994) suggests that the influence of citizen demeanor is complicated by several factors, including how demeanor is measured. His analysis nevertheless suggests that citizen disrespect is associated with a greater likelihood of police taking formal action (arrest). In a later analysis, Klinger (1996) reports that there does appear to be a relationship between demeanor and arrest, but only when the citizen is extremely hostile. Finally, Worden and Shepard (1996) conclude from a reanalysis of data taken from the Police Services Study that citizen demeanor does sometimes explain police behavior.

In summary, the research indicates that citizen behavior is a correlate of police use of force or decisions to arrest. However, inconsistencies in the research indicate that we must be cautious in interpreting correlations, since each of the analyses reported was based on police–citizen encounters in a certain place at a certain time. Therefore, these studies were not able to test for the influence of other correlates such as organizational policies or community characteristics.

In short, police misconduct is the product of a balance of forces existing at the community, organizational, and individual levels. A certain mix of community, organizational, and personal characteristics makes misconduct either likely or unlikely to occur. The community affects the police organization, which in turn affects the individual officers. These factors create a context that is more or less likely to support various types of misconduct. In the end, the interaction between the officer and the citizen, occurring within this context, is the locus of police behavior.

CONTROLLING POLICE MISCONDUCT

Police misconduct illustrates the fundamental conflict of policing in a democratic society. The behaviors we have labeled "controversial" are perhaps the best examples. If the police are charged with the maintenance of public order, how does one resolve the problem of public disagreement over what constitutes order? Alpert and Dunham (1988:10–14) note that police officers tailor their practices to meet community expectations. Further, when a community lacks consensus about the definition of order, the police are often ineffective, or worse, actually in conflict with the residents.

The power of the police is awesome, including the power to use lethal force. In a democratic society, this power is granted grudgingly and with suspicion. The ability of police to use their powers is cause for concern and hence gives rise to the development of controls on police behavior.

Richard Lundman (1980:170–174) identifies four problems in community control of the police. *Coercive force* is a prerogative of the police, but one

that has few defined limits, and the ability of the police to use force can deter citizen complaints. Lundman also maintains that most police misconduct is a product of *organizational deviance,* so that what needs to be
controlled is not individual behavior but organizational climates. He identifies a third problem as *external opportunities for misconduct* in that the
causes of much police misconduct (vice laws, for example) are not controllable by the police. Finally, he notes the existence *of police opposition to citizen control* because of fears of improper interference and a concern that
civilians cannot understand the dilemmas of policing.

To this list of four problems can be added at least three more. First, the
police are often isolated from the public they serve and, by virtue of civil
service procedures and too-broad departmental regulations, insulated from
much disciplinary action. Often the allegation of misconduct rests on the
complaint of a citizen that is denied by the officer. We have seen that police
tend to work without much supervision and make low-visibility decisions
that are not amenable to review and control. Second, there is the problem
of what some have called the *blue curtain,* or police secrecy. Officers tend
neither to report the misconduct of their colleagues nor to cooperate with
investigations of misconduct (Haar, 1997). Third, it must be remembered
that the relationship between the police and the public is, in the main, adversarial (Goldstein, 1977:160–161).

Corruption, discrimination, substance abuse, brutality, and other
forms of police misconduct reduce public confidence in the police, depress
officer morale, and generate conflict. Thus, there is a need to control police
behavior. Two basic types of controls on the police exist: controls external to
the police organization and controls internal to the organization. Both types
have their strengths and weaknesses and both serve to structure and limit
the behavior of officers.

External Controls

As an agency of government, the police are ultimately accountable to the
public, and for this reason should be controlled by the public. However,
when the public is divided about the propriety of different police behaviors,
public control of the police is not feasible. In an effort to ensure that the police officers and organization do not engage in misconduct, a number of public, or external, controls on the police have developed.

If the police behave in ways that threaten citizens or fail to meet citizen expectations, the citizens must have a means of influencing the police.
Reliance on the police organization to police itself tends to be unsatisfac-
tory, and the truly deviant police organization is probably incapable of self-
control. Yet, the ability of authorities external to the police organization to
affect the behavior of officers is limited and fraught with its own set of problems. The tradition of political patronage and corruption of American police

agencies in the 1800s illustrates some of the limitations of too much external control.

If citizens do not trust the police because of what they see as police corruption or misconduct, it is unreasonable to expect them to make complaints to the police organization. Instead, some authority external to and independent of the police is necessary to engender citizen trust and cooperation. Over the years, six sources of control external to the police organization have emerged: government, the media, ombudspersons or mediators, the courts, citizen monitoring organizations, and other police agencies.

Government.   Elected officials can influence police behavior through their control over the police agency budget, tenure of the police administrator, and other oversight powers. As elected representatives of the citizens, these officials are themselves directly accountable to the public at each election. Thus, in some cases citizen complaints about police misconduct are channeled through elected officials. In recent years concern about racial profiling has spurred the passage of legislation mandating that police agencies keep statistics on the characteristics of citizens stopped for motor vehicle violations (Strom and Durose, 2000).

An alternate form of government oversight is the appointment of a city manager and/or safety director who is a nonpolice boss of the police agency. This civilian administrator is responsible for the operation and control of the police agency. In this way, the police are ultimately directed by a civilian.

The history of the police in America, however, illustrates the dangers of political/governmental micromanagement of the police. Allowing elected and civilian administration government officials to control police policy too closely runs the risk of political corruption of the police organization. Further, this effort to control the police can actually backfire and result in a politically active police department that openly supports candidates and laws supportive of the police while opposing those deemed not supportive of police interests.

Media.   News coverage of police activities can alert citizens and civic leaders to the existence of misconduct and spur appropriate reform actions. In its role as watchdog, the press can expose corruption or other misbehavior on the part of police officers and generate public support for change. However, the media itself has no power to enforce change. Another limitation of the media watchdog is the reality of the "news game." Stories or issues have a relatively short shelf life before public interest wanes and the media lose readers or viewers. Further, only those issues deemed newsworthy will receive attention.

Ombudspersons/Mediators.   Some cities have created special officers and offices for the handling of citizen complaints about police behavior.

These ombudpersons or mediators operate outside the police agency and are charged with investigating and resolving citizen complaints. They often do not have authority to enforce changes, though, and do not proactively work to prevent misconduct.

Courts.   As we increasingly become an ever more litigious society, the courts take on a more important role in all aspects of our lives. Citizens can seek court intervention to prevent or control many forms of police misconduct. Obviously, corrupt police officers can be brought to trial on charges of accepting bribes, but, in addition, citizens can file civil lawsuits against officers and agencies when police misconduct harms them. Further, courts can issue orders and injunctions to either require police agencies to perform some duty or to refrain from taking some action. The courts give citizens a means of influencing police operations by providing a forum for civil actions against them (Vaughn and Coomes, 1996).

Court intervention, however, occurs only after some wrong has transpired and thus is not well suited to preventing police misconduct. Further, with court delay being a reality in most jurisdictions, the ability of citizens to obtain speedy relief through the courts is questionable. Finally, the court process is both intimidating and expensive, so that many citizens are unable or unwilling to file suit against the police. In some cases, criminal actions can be brought against the police, such as the federal civil rights trials of the officers involved in the Rodney King assault after their acquittal in state court. When criminal charges are brought, the onus of filing suit falls on the government, and not on the individual citizen.

Citizen Monitoring Organizations.   Variously titled over the years, *citizen monitoring organizations* are official panels of citizens charged with the oversight and correction of police operations. Frequently called *civilian review boards,* these organizations receive and investigate complaints of police misconduct.

Most such review boards are understaffed, relying on the police department to conduct any needed investigations. Also, these review boards typically have no direct power or influence over the police agency, so they are unable to enforce changes or punish police misbehavior. Finally, these boards, like the courts, deal with wrongdoing that has already transpired, so they are not well suited to preventing police misconduct.

Peter Finn (2001) studied civilian oversight processes and identifies four basic types. Some communities use **civilian investigators** who report to the police organization. Others use **police investigators** whose reports are reviewed by citizens. The citizens then recommend that the police administrator accept or reject the findings of the investigators. A third type uses a **civilian appeal process** where a panel of citizens can appeal findings of the police agency to a civilian board. Finally, some jurisdictions use an **auditor** who investigates the process by which the police agency accepts and investigates complaints from citizens, and reports the audit results to both the police agency and the public.

Other Police Agencies.   When allegations of corruption, brutality, discrimination and other forms of police misconduct have been made, it is possible to ask an outside police agency to conduct an investigation. Thus, state police agencies or county sheriff's offices frequently investigate bribery and brutality charges leveled against city police departments.

If the investigation yields evidence of criminal misconduct, then this outside police agency may be able to intervene directly into the situation. Absent evidence of criminality, however, the outside police agency is not authorized to make any changes. Yet another limitation of using an outside agency is that investigators may feel constrained by their loyalty to fellow officers. In the event that the investigation clears the suspected department or officers, citizens are not likely to trust the findings of the police who investigate the police.

Internal Controls

Police organizations themselves have an interest in the control of officer misconduct. As Kelling, Wasserman, and Williams (1988:1) observe,

> Police chiefs continually worry about abuse of authority: brutality; misuse of force, especially deadly force; over-enforcement of the law; bribery; manufacture of evidence in the name of efficiency or success; failure to apply the law because of personal interests; and discrimination against particular individuals or groups. . . . Scandals associated with abuse of authority . . . jeopardize organizational stability and continuity of leadership.

Police misconduct is bad for business. The organization is at risk of outside interference, and the chief is at risk of losing his or her job. Thus, to protect both the organization and the career of the chief, administrators of police organizations are motivated to control and limit police officer misconduct and develop mechanisms within their agencies to direct officer behavior.

Two approaches to the control of officer behavior that a police agency can employ were identified by Wilson (1968) as the professional and bureaucratic models. The **professional model** works by ensuring that only the best-trained, most honest candidates are employed as police officers. The **bureaucratic model** depends on the issuance and enforcement of rules and regulations through close supervision of police officer activities.

Professionalism.   Attempting to control police misconduct through professionalism is an essentially individualistic approach to the problem. In short, professionalism suggests that good officers will behave appropriately.

Police officers resist the creation of citizen-monitoring organizations (civilian review boards), as this New York City demonstration illustrates. (Joe DeMaria/*New York Post*)

The organization ensures that the police will be professional by hiring only the best-qualified applicants and then training these officers as well as possible. Having highly qualified, well-trained personnel will, it is assumed, prevent misbehavior.

A related internal control has been sought through the publication of a law-enforcement officer *code of ethics* and the creation of training curricula for ethics in policing. Carl Klockars (1995) observes that a need for special police ethics emerges from the fact that the police are given the legitimate right to use both force and fraud to achieve just ends. For this reason, the police must develop a moral code that can regulate the use of force or deception by the police. To that end, a Center for Law Enforcement Ethics was created in 1992 (Southwestern Legal Foundation, 1997). The core notion is that adherence to a professional code of ethics will prevent most police misbehavior.

Lundman (1980:175–182) criticizes professionalism as a control on police misconduct for two main reasons. He suggests that professionalism, by focusing on the individual officer, ignores the social and organizational correlates of misconduct. In addition, professionalism is, to Lundman, an obstacle to citizen control, since the idea of a professional is one who has special knowledge and skills that the average person lacks. Rather, Lundman supports different control structures.

The tradition of American policing has been to rely heavily on a professional model of police control. Strict entry requirements and lengthy training are designed, in large measure, to insure that only qualified and well-prepared persons are appointed as police officers. Unlike the integrity model described at the start of this chapter, the current practice is aimed at *"weeding out"* unfit officers. Eric Metchik (1999) examined the "screening out" model of police selection and notes that while it has some strengths, it often fails to correctly identify persons who should not be officers, frequently screens out qualified individuals, and is not very effective at predicting which officers might later develop problems.

Bureaucracy.   The bureaucratic model suggests that police behavior can be controlled by rules and regulations enforced through the supervision of personnel. Thus, if one wishes to prevent bribery, one establishes a rule that says officers are not allowed to accept any gifts, payments, or other compensation beyond their departmental salary. If an officer is found accepting such gifts, disciplinary action follows.

The problems with this form of control are numerous. First, it is not possible to anticipate every conceivable instance of misconduct and produce a rule to prevent it. Also, there is a tendency for rules to be negative—"thou shalt not"—rather than positive, so that officers do not receive guidance on how they should behave, but only on what they cannot do. Further, the drafting of a rule creates further opportunities for officers to justify almost any behavior as being in conformance with some rule. Still, the common practice in American policing is to attempt internal control of officer misconduct through various bureaucratic structures.

Herman Goldstein argues for the use of rules and bureaucratic measures to control police behavior. He suggests, however, that police agencies take specific steps to ensure an effective control system that would be acceptable to the public. In this regard, Goldstein (1977:167–178) proposes the following guidelines:

1. *Make maximum use of positive approaches.* Rather than punishing officers who misbehave and drafting rules and regulations based on problems and failures, police administrators should emphasize what works. They should reward proper behavior and provide appropriate

role models. This way the behavior of officers can be molded to reflect the values and definitions of police work desired by the police administration. The development of guidelines to assist officers in decision making is another positive approach to controlling police behavior.

2. *View individual wrongdoing as an agency problem.* Rather than subscribing to the rotten-apple theory, police administrators should consider every instance of officer misconduct as a problem for the police agency. By taking responsibility for officer wrongdoing, the agency becomes committed to efforts to control and eliminate misconduct. Although there will probably be some instances in which the agency cannot prevent misconduct, taking responsibility for all such events encourages police leaders to alter policies, training, rules, and so forth, to reduce misconduct.

3. *Measure performance and identify patterns of wrongdoing.* Police leaders should not wait for complaints of problems and misconduct but should monitor their agencies on a continuous basis. Citizen surveys, tests (calling for police service and measuring how well the agency responds), and observational methods can be used to identify departmental strengths and weaknesses. This information can then be used to guide changes in training, policy, regulations, or other aspects of the agency's administration.

4. *Identify officers with a propensity for wrongdoing.* It is often the case that a few officers in any police agency routinely are the objects of citizen complaints. In a sense, these officers are habitual offenders in terms of police misconduct. Police leaders should institute systems whereby officer involvement in misconduct can be tracked and assessed. This information can then be used to design remedial programs—training, increased supervision, changes in assignments, and so on—to reduce the incidence of misbehavior. Walker, Alpert, and Kenney (2000) describe what are known as "**early warning systems**" in police departments. These systems involve careful selection of officers, monitoring of officer performance to identify those who might become involved in serious misconduct, and interventions (remedial training, etc.) to improve their performance. They report that about one-quarter of police agencies have these systems, and that more and more police agencies are interested in implementing early warning systems. Kim Lersch and Tom Mieczkowski (2000) report a study of complaints against police officers in one department. They found that only 11 percent of officers accounted for nearly one-half of all citizen complaints. They labeled these officers as "problem prone."

5. *Institute training specifically aimed at preventing improper conduct.* Most police training is focused on the routine or expected functions of the job. In addition, police training should include exercises and discussions aimed at examining misconduct. Similar to the national effort to get children to "just say 'no' to drugs," police leaders should

design officer training programs that will assist officers in recognizing and avoiding misconduct.

6. *Provide citizen redress.* A major problem with internal control mechanisms is that they often lack public support. When citizens complain about police misconduct such as brutality or discrimination, the typical response is an internal investigation by the police agency. When this investigation concludes that there was no wrongdoing, the citizen is dissatisfied and also cynical. Police investigation of allegations of misconduct is akin to the fox guarding the chicken coop. Police administrators should be sure to communicate the complete findings of investigations to complaining citizens and to instruct citizens on how they can appeal the police agency's decision. This extra effort will help increase citizen confidence as well as protect against the possibility of a police "whitewash" of actual misconduct.

A Comparison of Controls

External methods of control suffer from the following weaknesses:

1. They are reactive and not preventive.
2. There is potential for improper influence on police practice.
3. They lead to a lack of confidence in decisions among police officers.
4. They fail to provide direct control over police actions.

On the other hand, internal control methods suffer from other weaknesses:

1. They fail to inspire citizen confidence.
2. They are isolated from citizen input.
3. There is potential for dishonesty.

The benefits of each are similarly different. Whereas external controls ensure accountability of the police to the public, internal controls ensure that the police administration is responsible for the behavior of police officers.

The issue of controlling officer behavior remains a central dilemma in policing in America. It is likely that the best solution to the problem of controlling the police reflects a combination of both internal and external controls. The optimum mix of such control mechanisms, however, is likely to vary among communities and departments as well as within them over time. If there is widespread citizen confidence in the police, internal controls may be sufficient. Should citizen distrust of the police grow, however, external controls help convince citizens that their concerns are taken seriously.

CONCLUSION

Because of the power and role of the police in society, especially in a democratic society, control of police behavior is a pressing concern. Misconduct by police officers takes many forms, and the definition of both the nature and extent of misconduct is open to interpretation. At issue in police misconduct is the question of accountability—to whom are the police responsible—or, who controls the police?

Some types of misbehavior generate widespread agreement, such as police corruption or the intentional misuse of police powers by officers. Other types of police behavior are more controversial, such as the use of force, the possibility of discrimination, and the off-duty morality of officers. Whether and how to control such conduct are matters of debate.

In general, two models of controlling police behavior exist. One approach places the authority and responsibility to control police in the public and outside the police organization. The alternative model relies on the police organization to control the behavior of officers. External controls tend to be more trusted by civilians, distrusted by police, and lack the ability to prevent misconduct or cause change in police practices. Internal controls are more trusted by police but often fail to engender citizen confidence.

That no optimal method of controlling police behavior has emerged is a product of both the variability of police misconduct and the adversarial nature of the relationship between the police and the public in a democratic society. As has been the case for the past century and a half, we are likely to continue to wrestle with the issue of controlling the police into the future.

CHAPTER CHECKUP

1. What is *police corruption?*
2. What kinds of activities can be considered to be police corruption?
3. What factors seem to explain the existence of police corruption?
4. What varieties or different levels of corrupt behavior have been identified for both police organizations and individual officers?
5. What are three controversial (possibly corrupt) types of police officer conduct?
6. In what ways do the characteristics of the community, police organization, and police officers affect types and levels of police corruption?
7. What methods of controlling police misconduct are found outside the police organization?

8. What models can the police organization follow to control police corruption internally?
9. How do external and internal methods of controlling police misconduct compare?

REFERENCES

Ahern, J. (1972) *The police in trouble.* (New York: Hawthorn).

Alpert, G. and R. Dunham (1988) *Policing multi-ethnic neighborhoods.* (New York: Greenwood).

Alpert, G. and J. MacDonald (2001) "Police use of force: An analysis of organizational characteristics," *Justice Quarterly* 18(2):393–409.

Alpert, G. and M. Smith (1999) "Police use-of-force data: Where we are and where we should be going," *Police Quarterly* 2(1):57–78.

Barker, T. (1978) "An empirical study of police deviance other than corruption," *Journal of Police Science and Administration* 6(3):264–272.

Barker, T. and J. Roebuck (1973) *An empirical typology of police corruption: A study in organizational deviance.* (Springfield, IL: Charles C. Thomas).

Barnes, C. and R. Kingsnorth (1996) "Race, drug, and criminal sentencing: Hidden effects of the criminal law," *Journal of Criminal Justice* 24(1):39–56.

Bayley, D. and J. Garafalo (1989) "The management of violence by police patrol officers," *Criminology* 27(1):1–25.

Black, D. (1980) *The manners and customs of the police.* (New York: Academic).

Brown, J. and P. Langan (2001) *Policing and homicide, 1976–1998: Justifiable homicide by police, police officers murdered by felons.* (Washington, DC: Bureau of Justice Statistics).

Browning, S., F. Cullen, L. Cao, R. Kopache, and T. Stevenson (1994) "Race and getting hassled by the police: A research note," *Police Studies* 17(1):1–11.

Cao, L., X. Deng, and S. Barton (2000) "A test of Lundman's organizational product thesis with data on citizen complaints," *Policing: An International Journal of Police Strategies and Management* 23(3):356–373.

Carter, D. (1987) *Policy issues in police drug abuse.* (Washington, DC: Police Executive Research Forum).

Carter, D. (1990a) "Drug-related corruption of police officers: A contemporary typology," *Journal of Criminal Justice* 18(2):85–98.

Carter, D. (1990b) "An overview of drug-related misconduct of police officers: Drug abuse and narcotic corruption," in R. Weisheit (ed.) *Drugs and the criminal justice system.* (Cincinnati, OH: Anderson):79–109.

Carter, D. and D. Stephens (1988) *Drug abuse by police officers: An analysis of critical policy issues.* (Springfield, IL: Charles C. Thomas).

Crawford, C. and R. Burns (1998) "Predictors of the police use of force: The application of a continuum perspective in Phoenix," *Police Quarterly* 1(4):40–63.

Decker, S. and A. Wagner (1985) "Black and white complainants and the police: A comparison of individual and contextual characteristics," *American Journal of Criminal Justice* 10(1):105–116.

DeLeon,-Granados, W. and W. Wells (1998) "Do you want extra police coverage with those fries? An exploratory analysis of the relationship between patrol practices and the gratuity exchange principle," *Police Quarterly* 1(1):71–85.

Dishlacoff, L. (1976) "The drinking cop," *Police Chief* 43(1):32–39.

Finn, P. (2001) *Citizen review of police: Approaches and implementation.* (Washington, DC: National Institute of Justice).

Fridell, L., R. Lunney, D. Diamond and B. Kubu (2001) *Racially biased policing: A principled response.* (Washington, DC: Police Executive Research Forum).

Friedrich, R. (1980) "Police use of force: Individuals, situations, and organizations," *The Annals* 452(November):82–97.

Gaffigan, S. and P. McDonald (1997) *Police integrity: Public service with honor.* (Washington, DC: U.S. Department of Justice).

Garner, J., J. Buchanan, T. Schade, and J. Hepburn (1996) *Understanding the use of force by and against police.* (Washington, DC: National Institute of Justice).

Geller, W. and K. Karales (1981) "Shootings of and by the Chicago police: Uncommon crises—Part I: Shootings by the Chicago police," *Journal of Criminal Law and Criminology* 72(1):1813–1866.

Goldstein, H. (1977) *Policing a free society.* (Cambridge, MA: Ballinger).

Inciardi, J. (1987) *Criminal justice,* 2nd ed. (New York: Harcourt, Brace, Jovanovich).

Haar, R. (1997) " 'They're making a bad name for the department': Exploring the link between organizational commitment and police occupational deviance in a police patrol bureau," *Policing: An International Journal of Police Strategies and Management* 20(4):786–812.

Johnson, D. (1981) *American law enforcement: A history.* (St. Louis, MO: Forum Press).

Kelling, G., R. Wasserman, and H. Williams (1988) *Police accountability and community policing.* (Washington, DC: U.S. Department of Justice).

Klinger, D. (1994) "Demeanor or crime? Why 'hostile' citizens are more likely to be arrested," *Criminology* 32(3):475–493.

Klinger, D. (1996) "More on demeanor and arrest in Dade county," *Criminology* 34(1):61–82.

Klockars, C. (1985) *The idea of police,* 2nd ed. (Beverly Hills, CA: Sage).

Klockars, C. (1995) "Police ethics," in W. G. Bailey (ed.) *The encyclopedia of police science,* 2nd ed. (New York: Garland):549–553.

Klockars, C. (ed.) (1983) *Thinking about police: Contemporary readings.* (New York: McGraw-Hill).

Knapp Commission: New York City Commission to Investigate Allegations of Police Corruption and the City's Anti-Corruption Procedures (1972) *The Knapp commission report.* (New York: George Braziller).

Lersch, K. and T. Mieczkowski (2000) "An examination of the convergence and divergence of internal and external allegations of misconduct filed against police officers," *Policing: An International Journal of Police Strategies and Management* 23(1):54–68.

Lester, D. (1982) "Civilians who kill police officers and police officers who kill civilians: A comparison of American cities," *Journal of Police Science and Administration* 10(2):384–387.

Lundman, R. (1980) *Police and policing: An introduction.* (New York: Holt, Rinehart & Winston).

Lundman, R. (1994) "Demeanor or crime? The midwest city police–citizen encounters study," *Criminology* 32(4):631–656.

Maas, P. (1973) *Serpico.* (New York: Viking).

MacDonald, J., G. Alpert, and A. Tennenbaum (1999) "Justifiable homicide by police and criminal homicide: A research note," *Journal of Crime and Justice* 22(1):153–166.

McEwen, T., B. Manili, and E. Connors (1986) *Employee drug testing policies in police departments.* (Washington, DC: U.S. Department of Justice).

Metchik, E. (1999) "An analysis of the 'screening out' model of police officer selection," *Police Quarterly* 2(1):79–95.

Muscari, P. (1984) "Police corruption and organizational structures: An ethicist's view," *Journal of Criminal Justice* 12(3):235–245.

Neiderhoffer, A. (1967) *Behind the shield: The police in urban society.* (Garden City, NY: Anchor).

Pogrebin, M. and B. Atkins (1976) "Probable causes for police corruption: Some theories," *Journal of Criminal Justice* 4(1):9–16.

Reiss, A. (1971) *The police and the public.* (New Haven, CT: Yale University Press).

Sherman, L. (1978) *Scandal and reform.* (Berkeley, CA: University of California Press).

Sherman, L. (1980) "Causes of police behavior: The current state of quantitative research," *Journal of Research in Crime and Delinquency* 17(1):69–100.

Sherman, L. (ed.) (1974) *Police corruption: A sociological perspective.* (Garden City, NY: Anchor).

Smith, D. (1986) "The neighborhood context of police behavior," in A. Reiss and M. Tonry (eds.) *Crime and justice: A review of the research, Vol. 8.* Chicago: University of Chicago Press):313–341.

Son, I., M. Davis, and D. Rome (1998) "Race and its effect on police officers' perceptions of misconduct," *Journal of Criminal Justice* 26(1):21–28.

Southwestern Legal Foundation (1997) "The ethics center: A work in progress!" *Ethics Roll Call* 4(1):1.

Stoddard, E. (1979) "Organizational norms and police discretion: An observational study of police work with traffic violators," *Criminology* 17(2):159–171.

Strom, K. and M. Durose (2000) *Traffic stop data collection policies for state police, 1999.* (Washington, DC: Bureau of Justice Statistics).

Stroshine-Chandek, M. (1999) "Race, expectations and evaluations of police performance: An empirical assessment," *Policing: An International Journal of Police Strategies and Management* 22(4):675–695.

Swanson, C., L. Territo, and R. Taylor (1988) *Police administration.* (New York: Macmillan).

Sykes, R. and E. Brent (1980) "The regulation of interaction by the police," *Criminology* 18(2):182–197.

Travis, L. (1998) *Introduction to criminal justice,* 3rd ed. (Cincinnati, OH: Anderson).

Vaughn, S. and L. Coomes (1996) "Police civil liability under section 1983: When do police officers act under color of law," *Journal of Criminal Justice* 23(5):395–415.

Waegel, W. (1984) "The use of lethal force by police: The effect of statutory change," *Crime & Delinquency* 30(1):121–40.

Walker, S., G. Alpert, and D. Kenney (2000) "Early warning system for police: Concept, history, and issues," *Police Quarterly* 3(2):132–152.

Webster, B. and J. Brown (1989) *Mandatory and random drug testing in the Honolulu police department.* (Washington, DC: U.S. Department of Justice).

Weitzer, R. (1996) "Racial discrimination in the criminal justice system: Findings and problems in the literature," *Journal of Criminal Justice* 24(4):309–322.

Westley, W. (1970) *Violence and the police.* (Cambridge, MA: MIT Press).

Wilson, J. (1968) *Varieties of police behavior.* (Cambridge, MA: Harvard University Press).

Worden, R. and R. Shepard (1996) "Demeanor, crime, and police behavior: A reexamination of the police services study," *Criminology* 34(1):83–105.

chapter *17*

CURRENT TRENDS AND FUTURE ISSUES IN POLICING

CHAPTER OUTLINE

Thus far we have maintained our theme that policing in practice is a product of the balance of a variety of forces that shape the decisions and actions of police officers. In general, we have focused on three sources of influence on police practice: communities, organizations, and people. In this chapter we will identify some of the forces that appear to be emerging and predict their likely effects on policing in America in the future.

Prediction is always troublesome, with a large margin for error. Still, we study the police in hopes of understanding how it is that police do what they do. This understanding, in turn, has a pragmatic purpose—to enable us to direct, and change where desirable, the practice of policing. Based on what we know about the police in America, this chapter attempts to identify the future of American policing. More accurately, it tries to describe alternative futures that may emerge from contemporary practice and developments.

The list of current issues and crises in policing is long, including the traditional concerns over discretion, use of force, and corruption, but some issues of relatively recent vintage appear to be the most likely candidates for causing change in the future of policing. We have not, and probably cannot, eliminate the problem of police misconduct, and it will remain an important issue into the future. Several current developments, however, create the possibility of significant change for policing in America.

We saw in Chapter 11 that changes in American society are often reflected in less solidarity, or consensus, in our communities. One effect of disorganized community *is greater independence for the police from community expectations, or more proactive policing. The proactive police focus on law-enforcement functions. To the degree that this change in community life in the United States is widespread, the variation traditionally observed between police agencies in different communities will diminish. Policing in America will become increasingly homogenized. Thus our set of alternative futures involves a choice between uniformity and diversity for policing in America.*

FORCES FOR UNIFORMITY

Changes in community structure and solidarity are themselves both the cause and effect of other social forces. A large number of forces are at work that could produce greater uniformity, or homogenization, of American policing. From among many issues, we will focus on terrorism, civil liability of the police, police accreditation, the impact of technology, and the war on drugs as portents of a significant shift toward uniformity in policing in the future.

Terrorism

On the morning of September 11, 2001, four airliners were hijacked by terrorists intent on using the planes as flying bombs. Two planes crashed into the World Trade Center in New York City, one struck the Pentagon in Washington, DC, and the fourth crashed in rural Pennsylvania as passengers struggled for control of the aircraft. After the attacks, President George W. Bush declared "war" on terrorism, and a massive federal effort was mobilized. At this point, it is unclear how, if at all, the response to these attacks will change policing in America.

Despite the shock of the September 11 attack, terrorism is not a new phenomenon in the United States. Between 1987 and 1991, the Federal Bureau of Investigation recorded thirty-four terrorist incidents in the United States and Puerto Rico (Poland, 1995:768). Since 1991, major terrorist attacks in the United States have included the World Trade Center bombing in 1993, the Oklahoma City Bombing, and the September 11 attacks. What is potentially different about this most recent event is the reaction it has generated.

The Federal Bureau of Investigation has long had responsibility for preventing or investigating terrorist activity in the United States. One apparent change is the appointment of a Director of Homeland Security and attempts to improve communication and cooperation among federal law enforcement, military, and intelligence agencies. Congress has enacted legislation to reduce constraints on surveillance, including wire taps, and the use of surveillance evidence at trials. As mentioned earlier, airport police will now be federal law enforcement officers. All of these developments suggest increased centralization of anti-terror efforts, and as a result, an increased uniformity in police responses to terrorism.

At the same time, the Chief of Police in Portland, Oregon, Mark Kroeker, has received a great deal of attention because of his refusal to investigate resident aliens of Middle Eastern descent. The U.S. Department of Justice has requested that the Portland police conduct the interviews, but Oregon state law prohibits collecting information about political or religious views of persons not accused of a specific crime. On advice of the city attorney, Chief Kroeker refused to honor the federal request (Thomas and Isikoff, 2001:42).

The case of Portland illustrates the problem with predicting the ultimate impact of the war on terrorism. On the one hand, there is clearly a desire to strengthen protection against terrorist attacks. On the other, the fragmented nature of policing in America impedes efforts to centralize police authority. At this point, the war on terrorism seems to have caused a reassessment of the current balance between liberty and civility that was discussed in Chapter 1. The long term impact of the September 11 attacks will most likely depend on whether or not we change that balance.

Civil Liability

Police officers and agencies that wrongfully cause some harm to citizens can be held liable in civil court actions. In the past few decades, civil lawsuits against the police have increased greatly. Citizens filing these suits seek monetary damages for inappropriate police actions. Del Carmen (1989) has reviewed the issue of police legal liability and concludes that most suits are brought in either state civil courts or under the civil rights protections of the U.S. Code, section 1983, in federal courts.

In brief, suits are based on claims of police misconduct that is either intentional or negligent. **Intentional conduct** is when the police purposely do something harmful, such as false arrest or false imprisonment. **Negligent conduct** is behavior that is careless or insufficiently careful, like negligent operation of a motor vehicle, failure to adequately train officers, and similar acts of omission. If police officers or agencies intentionally misbehave or negligently cause harm, they may be required to compensate the injured party.

In the past, most suits brought against the police were dismissed by the courts. More recently, however, courts are increasingly willing to hold police officers and agencies liable for their behavior. As Del Carmen (1989:462) observes,

Liability lawsuits have become an occupational hazard in policing. The days are gone when the courts refused to entertain cases filed by the public against police officers and agencies. The traditional "hands off" policy by the courts is out; conversely, "hands on" is in and will be with us in the foreseeable future.

Harry Barrineau and Steven Dillingham (1984) reviewed the history and theory of the civil suit provisions of the federal Civil Rights Act. They illustrate that the rulings in federal courts have evolved in ways that make civil actions more likely to be successful. Essentially, when a government official, acting in that official capacity, infringes on the constitutional rights of a citizen, a suit can be brought. Further, over time the courts have allowed plaintiffs to sue not only the offending officer but also the agency and government unit (department and city, for example). Certain procedural rules of the federal courts, coupled with awards for attorney's fees, have made federal courts the preferred avenue for filing civil suits (Barrineau and Dillingham, 1984:131).

The use of federal courts to resolve complaints about the police raises questions of federalism and local autonomy (Barrineau and Dillingham, 1984:137–139). In short, federal court decisions have the effect of setting police policy and requiring police action. As Barrineau and Dillingham (1984:141) conclude,

> The decisions frequently mandate the redistribution of limited resources, the adoption of specified policies and practices, the imposition of burdens upon the behaviors and careers of public employees, the adherence to directives which appear antithetical to prevailing notions of federalism, and policies which are difficult to implement.

By deciding, for example, that a specific police practice violates a citizen's constitutional rights, the courts effectively ban the practice. In finding that a police department is liable for an officer's behavior because the department's policy allowed or required that action, the court mandates that the agency change its policies. The impact of these rulings can be easily seen in what has happened with police policy regarding the use of deadly force.

Tennessee v. Garner.   In 1985 the U.S. Supreme Court in *Tennessee v. Garner* held that state statutes and police policies allowing a police officer to use deadly force to effect all felony arrests were unconstitutional and therefore violated the rights of criminal suspects. Commonly called the **fleeing-felon rule,** these standards authorized police to use all necessary force to prevent the escape of a felony suspect. In *Garner,* the Supreme Court held that deadly force could only be used to prevent a dangerous felon from escaping. The use of deadly force to apprehend an apparently unarmed, nondangerous, fleeing person violated Fourth Amendment protections against unreasonable seizures.

The case began as a civil suit, *Garner v. Memphis* (Fyfe and Walker, 1990) brought by the father of a burglary suspect who was shot to death while fleeing from the scene by a Memphis police officer. The suit named the officer, the Memphis Police Department, and the city of Memphis as codefendants. In its 11-year history, the case finally revolved around the question of the reasonableness or constitutionality of the Tennessee statute that authorized a fleeing-felon rule.

With its decision in *Tennessee v. Garner,* the U.S. Supreme Court had essentially rewritten the deadly-force policies of many American police agencies. Fyfe and Walker (1990) assessed the impact of the decision on police policies. They concluded that very few states had changed their statutes or issued attorney general's advisories to police departments. Nonetheless, they write (1990:181), "Our regular exposure to a variety of police agencies suggests that the case encouraged many small and medium-sized police departments to revise their deadly-force policies or, more typically, to fill policy voids by creating such policies."

Civil Courts and Police Policy.   The effect of civil liability on police policy and practice extends to circumstances other than the use of deadly force. Goldstein (1977) notes that police administrators in several jurisdictions have developed policies regarding high-speed pursuits of criminal suspects to prevent accidents and the lawsuits that followed them. This type of police policy as reaction to liability continues (Alpert and Anderson, 1986). The National Law Enforcement and Corrections Technology Center (1996:3) notes that one of the most important reasons for the development of police pursuit policies is to "minimize agency liability." Several analyses of police liability appearing in the professional literature analyze the legal principles involved in liability cases and offer guidance to police administrators concerning how they can better protect themselves and their agencies from being found liable (Ross, 2000; Smith, 1999; Ross, 1998; Fyfe, 1998). In most cases the best advice to police administrators is that they have written policies and that officers be trained adequately.

Changes in policy on pursuits, just as those for use of deadly force, do seem to change police officer behavior. Geoffrey Alpert (1997a; 1997b) reports a national survey of police pursuit policies, observing that most recent changes in pursuit policies have been in the direction of making the policies more restrictive. He also notes that policy changes are associated with reductions in the numbers of high-speed pursuits. In contrast, Tad Hughes (2001) reports a survey of police officers in which the officers report that they do not generally consider civil liability in making policing decisions, and that they do not believe civil liability influences police practice. The issues around civil liability and police reactions to it are complex, involving balancing interests in crime control (such as capturing a fleeing suspect) against interests in protecting citizen rights (avoiding accidents and injuries). It is possible to estimate expected costs (of lawsuits) and benefits (crime prevention) of various police policies, but the issues involved do not always neatly translate into simple cost comparisons (Crew and Hart, 1999).

Similar administrative policies to guide officers' decisions in domestic-violence cases attend court decisions holding the police liable for failure to protect the victims of domestic assault. Vaughn (1994) notes that several court decisions have found the police liable for *abandonment* when police policy failed to provide for adequate protection of citizens.

In an analysis of police liability for failure to arrest drunk drivers, Kappeler and Del Carmen (1990) suggest that recent developments in case law may result in a future tendency for courts to hold police liable for failure to arrest. If this trend continues, they believe that the ultimate impact of court rulings will influence police arrest decisions in cases involving drunken driving. They write (1990:130) that these decisions "may presage a legal development that may impact significantly the way police deal with drunk-driving cases in the future."

To the degree that the police are held liable for their actions in civil courts, the courts can be said to establish police policy. When the U.S. Supreme Court, for example, decides that a particular policy or practice is unconstitutional, it sets a standard nationwide. Similarly, to the extent that civil courts agree in their findings about police liability for failure to arrest those suspected of drunken driving or domestic assault, or for negligence in pursuing offenders at high speed, they sanction police decisions and policies in general. Thus, rather than the tradition of local autonomy and substantial community influence on police practice, legal liability portends greater uniformity of police practice. Becknell, Mays, and Giever (1999) found that the presence of pursuit policies was correlated with both the number of pursuits and the use of force after pursuits. They conclude that more restrictive policies on pursuits help control officer discretion and result in fewer pursuits and lower rates of use of excessive force. Uniformity in practice and policy may also be produced by the current movement toward national accreditation of police agencies.

Law-Enforcement Accreditation

The Commission on Accreditation of Law Enforcement Agencies was created in 1979 to develop standards and an accreditation process for police agencies. The commission is supported by four professional associations of police administrators: the International Association of Chiefs of Police, the National Sheriffs Association, the National Organization of Black Law Enforcement Executives, and the Police Executive Research Forum. The commission developed a set of requirements for accreditation and a procedure for self-study, application, external review, and the grant of "accredited" status to police agencies.

Requirements for accreditation were classified into six areas: agency role and responsibility, organization/management/administration, personnel, law-enforcement and traffic-enforcement operations and support, prisoner/court services, and auxiliary/technical services. In most instances, the accreditation process requires the agency to develop or have policies for such issues as use of informants, personnel selection, and training. In some cases the commission requires that the agency have a specific policy, such as a defense-of-life policy on the use of deadly force.

Since the first police agency achieved accreditation—the Mt. Dora, Florida, police department in 1984—approximately 275 other agencies were accredited, and several times that number were in the process of preparing for accreditation by the early 1990s (Carter and Sapp, 1994). The Commission on Accreditation for Law Enforcement Agencies expects the accreditation process to benefit police departments through increased effectiveness and better relations with other agencies and the public. For many agencies, accreditation holds the promise of protection from some civil suits (McCabe and Fajardo, 2001; Cole, 1992).

The movement to accredit law-enforcement agencies is controversial (Carter and Sapp, 1994). Some question whether accreditation will produce any substantive improvements in policing. As most standards require the police to have a policy but do not speak to the specific content or enforcement of the policy, accreditation lacks the ability to affect change. Further, accreditation is voluntary, so any agency that wishes not to develop policies is not required to do so. One positive outcome of the accreditation movement, though, is likely to be the development of administrative rules to guide and control police behavior. And the requirement of written policies, with the ultimate judgment of the adequacy of such policies by an accreditation review team, may produce standardization and uniformity in policing. As the commission operates nationally, one outcome of the law enforcement accreditation process may be the development of national police standards and policies that might supercede local traditions and expectations. In an initial comparison of accredited versus nonaccredited agencies, Kimberly McCabe and Robin Fajardo (2001) found few differences. Accredited agencies required more hours of field training, and higher educational attainment of recruits, they were more likely to use drug testing of officers, and were more likely to operate specialized child abuse and drug enforcement units. To date there has been no reported research directly assessing the impact of accreditation on police practice. Still, while the current standards are purposely vague to allow local adaptation, a natural evolution to more specific requirements (such as the deadly-force standard) should be considered. As Roberg and Kuykendall (1990:459) observe, "The Commission on Accreditation for Law Enforcement Agencies, initiated in the late 1970s, wants to standardize police organizations and behavior in the United States."

Technology and the Police

Technological change affects all aspects of life in American society. Video-cassette recorders, microwave ovens, cordless telephones, and other modern conveniences were barely being planned 50 years ago. As with other areas of social life, the police must adapt to technological change. Some of these adaptations are less disruptive than others, but all require some change in police practice. One of the growing concerns in policing is the increase in computer and Internet crimes.

Perhaps the greatest change in the history of American policing came as a result of the application of technology to the police task. The combination of radio-dispatched patrol cars with widespread public access to telephones fundamentally altered the way police agencies operated (Johnson, 1981). Police officers were removed from the face-to-face contact with citizens they had enjoyed when patrolling on foot. Further, the police became increasingly reactive as the citizenry called for services and the agency dispatched motor patrol units in response to calls. It became easy for citizens to mobilize a police response.

These developments also fundamentally altered the relationship between the police officer and police supervisors. The radio allowed supervisors to maintain almost constant communication with officers. Officers

In the information age, computer-aided dispatch, priority ranking of calls for service, and other developments have changed many aspects of policing. (Glen Korengold/Stock Boston)

were required to acknowledge dispatched assignments, and to clear themselves when finished with that job. Thus, supervisors knew who was working, at what, where, and for how long. In this way, the activities of the typical patrol officer came under both increasing scrutiny and increasing control.

In more recent years a variety of technologies have been adapted to police use. Some, such as new methods of forensic analysis, improved the efficiency and accuracy of what police were already doing (Peterson, 1987). Others, such as computerization, may have changed the nature of the police job in more fundamental ways. In order to support the data needs of computerized information systems, the paperwork aspects of the police officer's job have mushroomed (Alpert and Dunham, 1992:76–77). Many of the reports required of patrol officers are prepared on coded forms so that the information can be easily entered into the computerized database. The tasks of data collection and coding have become important parts of the police officer's job (Harris, 1997). Chart 17.1 shows how police departments are using mobile computers.

Computerized management information systems have not only placed a burden on police officers to become record keepers and data collectors, but they have also increased the ability of the police supervisor and organization to control the activities of the officer. The development of 911 emergency telephone systems, coupled with computer-assisted dispatch, routinely screen citizen calls and order officer assignments. Such systems, especially in busy urban areas, account for large percentages of officer time, reducing the chance for the police officer to exercise individual initiative or engage in conversation with citizens.

Moreover, the development of large databases containing records of criminal activity, calls for assistance, and case outcomes has supported centralization of police decision making (Adderley and Musgrove, 2001). Police administrators can assign officers and rank the importance of tasks based on assessments of these data. As we discussed in Chapter 12, these information systems support administrative efforts such as directed patrol. Based on an assessment of problems on the beat conducted at headquar-

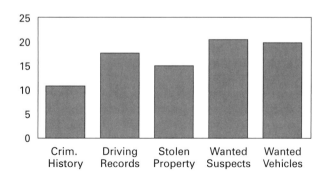

CHART 17.1
Information Available by Mobile Computers in Local Police Departments, 1999, Percentage of Departments. (*Source:* M. Hickman and B. Reaves (2001) *Local Police Departments, 1999.* Washington, DC: U.S. Bureau of Justice Statistics, p. 18.)

ters, patrol officer activities can be specifically guided by the police administration. In this fashion, the work of patrol officers can be routinized and standardized.

Similarly, an assessment of police efforts and their success in prior cases underlies the development of solvability factors and the priority ranking of detective workloads. In short, by tracing and evaluating past work, the police administration is in the position to better control and direct present and future efforts. Further, the routine collection of a range of information about officer activities enables the police administration to better evaluate and assess the performance of officers. This, in turn, places the organization in a stronger position to manipulate its reward structure in ways that encourage officers to behave in a manner that the administrator prefers.

Another product of the computerization of police information has been the implementation of geographic information systems in policing (Rich, 1995; LaVigne and Wartell, 2000). These systems link maps of the agency's jurisdiction with other police records and replace the old "pin map" with computer-generated maps of the distribution of crimes, calls for service, and police activity. While in some ways no different from having someone place pins into a map to represent each criminal complaint, the speed and flexibility of **computerized crime mapping** systems have influenced the allocation of police resources and the identification of priority targets for police action in many agencies (Rich, 1995; 1996; Davison and Smith, 2001).

Developments in communications technology have had the effect of making the police more accessible to the public and, in the early going at least, more responsive to community pressures. However, because communications between citizens and officers are mediated through the police organization, the importance of the organization in controlling officer behavior has increased. With the advent of computerized record systems and their application in management information systems, the control over the allocation of resources, including officer time, and the ability of the central police administration to both monitor and control officer behavior has been strengthened. One long-term effect of these changes may be that police officers increasingly become interchangeable cogs in the police department machine.

The War on Drugs

"There is little question that the recent public outcry to get tough on drugs—particularly street-level drug activity—has taken its toll on law-enforcement workloads" (Manili and Connors, 1988:3). Responding to a needs assessment of police and sheriff departments, over 90 percent reported a need for training and assistance in controlling drug crimes. Nearly two-thirds noted that drug arrests were responsible for an increase in their agency workloads (Manili and Connors, 1988:3). A later survey (McEwen, 1995) revealed that these problems were continuing.

The pressures to do something about drugs have spurred most police agencies to take some action, from crackdowns to the creation of special units to combat drug offenses. Moore and Kleiman (1989) describe a number of strategies that police agencies have employed in response to drug-related crime. They (1989:6) note that most "departments generally have 'comprehensive' approaches to the problem. Departments differ, however, in the overall level of activities they sustain and in the relative emphasis they give to each." Some police agencies focus their attention on high-level drug suppliers, others on street pushers and users, and still others on educational and preventive programs in the community.

What is important about the war on drugs in recent years is its impact on policing in general. In many ways the drug offenders or, more simply, drugs, serve as a common enemy for the majority of American police departments. The presence of a common enemy has encouraged local police agencies to enter into cooperative agreements and formal inter-jurisdictional enforcement efforts with state and federal law enforcement agencies, as well as with other local police departments (Jeffris et al., 1998). The cooperation among agencies and the focus on a particular form of crime and enforcement effort have brought these diverse agencies

The war on drugs has encouraged many police organizations to create specialized drug enforcement units that are distinct from uniformed patrol. (Piet van Lier/Impact Visuals)

closer together, thereby generating some level of cooperation, at least in terms of drug enforcement. Bradley Smith, Kenneth Novak, and their colleagues (Smith et al., 2000) report that such task forces do not appear to be related to differences in drug enforcement outcomes (e.g., numbers of arrests), but do account for differences in how police perceive and present their drug enforcement efforts. That is, task forces might not help police agencies achieve increased drug enforcement, but might serve process goals.

This effect of the drug war is an extension of the general impact of federal intervention in local policing that has been ongoing for at least the past quarter-century. The availability of federal assistance (typically in the form of financial support) to local agencies for drug-related law-enforcement efforts has helped encourage police administrators and local political leaders to focus attention on drug-related crime. Dunworth, Haynes, and Saiger (1997) report that such task forces have received four times as much federal funding between 1989 and 1994 than any other criminal justice effort, as shown in Chart 17.2. In fiscal year 2000, the federal government allocated over one billion dollars for the Byrne Memorial and Local Law Enforcement Block Grant programs (Bureau of Justice Assistance, 2000a; 2000b). Relatedly, the federal statute on civil forfeiture of the assets of drug offenders provides for sharing proceeds with local law enforcement. Under forfeiture, the proceeds of an illegal enterprise, such as drug sales, can be confiscated

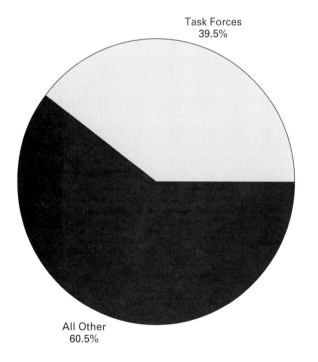

Task Forces
39.5%

All Other
60.5%

CHART 17.2
Byrne Program Funding of Multijurisdictional Drug Task Forces, 1989–1994. (*Source:* T. Dunworth, P. Haynes, and A. Saiger (1997) *National Assessment of the Byrne Formula Grant Program.* Washington, DC: National Institute of Justice, p. 5.)

by the government. In some states, and under the federal statute, the seized assets are given to law enforcement agencies, creating a "profit motive" for drug enforcement. Critics of this practice have been successful in several states in preventing police agencies from getting a share of forfeited property. Under the federal statute, however, local police can still share in the proceeds. John Worrall (2001) studied the forfeiture process and concludes that at least some police agencies may be "addicted" to the drug war. He suggests that some local police cooperate with federal drug law enforcement (rather than local prosecution) precisely so that they may obtain some of the seized property. Here again, local police are enforcing (or supporting enforcement of) federal laws, rather than local priorities and laws.

Federal financial support usually also carries federal requirements. To the extent that independent local police agencies are required to adopt or follow specific procedures and policies established by federal authority, these agencies become more similar to each other in practice. Despite relatively broad definitions of how the funds can be used, the federal programs do identify program areas/activities eligible for support, and prohibit using the funds for other activities/purchases. The acceptance or reliance on federal grant money for local policing activities represents an increase in vertical relations in the communities applying for such support. While they are still local police, at least some of their policies and procedures are established federally, leading to greater uniformity in local policing.

FORCES FOR DIVERSITY

The forces just described exert pressure toward greater uniformity in policing in America, but two opposing forces promise greater diversity in police practice. First, the contemporary movement to community-oriented policing, at least in theory, supports decentralization of police authority and more variation in police practice on the street. Even within specific jurisdictions, proponents of community policing encourage experimentation and variety. Thus, a large local police department would be organized into a number of semiautonomous community or neighborhood police units. Within a single city, then, one might find the police in different neighborhoods adopting different styles of policing. Second, the existence of many different police employee unions supports diversity and may block attempts at creating more uniformity in policing.

Community Policing

As we discussed in Chapter 15, during most of the twentieth century, the police in America (at least in large cities), became increasingly bureaucratic, impersonal, and focused on crime-control issues. The traditional relationship between the police and public was severed. Increasingly, the police and the public failed to understand or trust each other, yet the fact remains that they are codependent—they need each other. Community policing emerged as an attempt to bridge the gap between the police and the citizenry by increasing communication and understanding and by making policing more responsive (and relevant) to the public.

Jerome Skolnick and David Bayley (1986:214–220) summarize the primary elements of community policing as police-community reciprocity, areal decentralization of police command, reorientation of police patrol, and civilianization of the police. These elements, they contend, will produce a police agency more responsive to the people it polices and better able to control crime and maintain order.

By *reciprocity,* they mean that the police and community must share responsibility and authority for policy-making and for identifying and setting police priorities. *Areal decentralization* refers to the need for police agencies to organize in a way that allows reciprocity. Since community needs and concerns (not to mention policing) vary by community characteristics (Smith, 1986), police officers must be free to respond to different concerns. Police command, then, should be decentralized to those areas that constitute communities or neighborhoods within the city. The *reorientation of patrol* entails a shift from anonymous automobile patrol and response to citizen calls to a more personal and pervasive foot patrol. The purpose of patrol is less to respond to calls for assistance than it is to prevent the need for those calls. Finally, *civilianization* is designed to reduce policing costs and to free sworn officers from administrative and support duties to increase their ability to work with the community.

Hans Toch (1997) refers to this movement as evidence of a *democratization* of policing. He suggests that the decentralization of police command required by community policing has the effect of increasing diversity of police practice and strengthening ties between the police and the neighborhoods they serve.

Critics of community policing (Klockars, 1991; Mastrofski, 1991; Williams and Wagoner, 1992) observe that although there appears to be a broad-based movement toward community policing that, on the surface, shows remarkable similarity in American police agencies, the product of this change may be increased variation. If successful, policing will vary not only from jurisdiction to jurisdiction, but from neighborhood to neighborhood within a jurisdiction. Alpert and Dunham (1988) have observed that within the city of Miami, Florida, neighborhoods differed in what was desired of the police and what police actions were deemed acceptable by residents.

It is entirely likely, should the theory of community policing be realized in practice, that a more democratic and reactive mode of policing will result. In turn, police practice will vary depending on neighborhood priorities and policies. Thus, substantial differences in policing will be observable within police agencies. As police agencies attempt to follow the desires of citizens, neighborhood differences in the definition of police problems and appropriate police

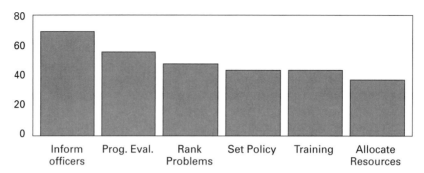

CHART 17.3 Uses of Citizen Survey Data by Local Police Departments, 1999. *(Source: M. Hickman and B. Reaves (2001) Local Police Departments, 1999. Washington, DC: U.S. Bureau of Justice Statistics, p. 12.)*

actions may create substantial differences in policing across neighborhoods. To understand citizen interests, police agencies have adopted the use of public surveys (Chart 17.3) (Beck, Boni, and Packer, 1999; Travis et al., 2000). Still, citizen participation in community policing appears to be greatest among "activist" citizens, and not among the average citizen (O'Shea, 2000). If more citizens can be encouraged to participate in planning and policy development, the likely outcome will be less clear directions for the police.

Interestingly, if community policing were fully realized, at the organizational level there would be less diversity. Unlike Wilson (1968), who was able to categorize police agencies, future researchers would be less able to identify any particular agency style. Rather, precincts or districts within agencies might develop unique styles that, when summed for the entire department, cancel each other out. The net result could be an enormous increase in the number of police organizations. Instead of some 13,500 local police agencies there could be a much larger number of decentralized police units functioning as independent organizations.

Police Unions

A second force that supports diversity and may effectively block attempts at creating greater uniformity in policing is police employee unions. Hoover and DeLord (1995) observe that police officers belong to many different labor and fraternal organizations. The sheer multitude of these groups prevents the establishment of any single national police union. Thus, the concerns and activities of most police employee labor organizations have a decidedly local flavor (Alpert and Dunham, 1992:98). As Hoover and DeLord (1995:792) put it, "The police labor movement today is best described by the term *fragmented.*"

Without a national organization representing police employees, there is no police officer voice for the creation of national standards, policies, and

Supporters of community-oriented policing hope to improve police–community cooperation for the future. (John Moore/The Image Works)

procedures. National standards, such as law-enforcement accreditation, would support and encourage more uniformity in policing. But because of the local orientation of most police unions, these groups can thwart the spread of national movements and prevent or delay the adoption of policies and procedures at the local level. As Guyot and Martensen (1991:445) observe, "Police unions seldom initiate change but often use government institutions to delay or overturn police management decisions." Pfuhl (1995) claims that unionization, including the right to strike, serves to reduce the barrier between the police and citizens by defining the police as *labor*, rather than as crime-fighting specialists. He writes (1995:749), "To the degree that it promotes democratization, the effort to promote police unionization in our society, including the right to strike, may prove to be one of the healthiest contributions to the future of the American system of policing."

The vital and public nature of policing prevented early unionization of police employees. The most celebrated early attempt at the creation of a police union in the United States was the Boston police strike of 1919 (Russell, 1975), in which striking Boston police officers were fired and the union's attempt at organizing officers was blocked. Later efforts to unionize police also failed until the 1960s.

In this era, several conditions had developed favorable to the creation of police unions. Pfuhl (1995) notes that at this time a variety of public employees, including teachers, firefighters, and sanitation workers, were unionized. This was also a time of great public concern about community safety, and there was some sympathy for the plight of police officers. This, too, was a period in which the importance of the police and the reality that police positions went unfilled and were generally underpaid were being acknowledged. Thus, there was considerable social and political support for unionization, and few openly opposed the movement.

As with most employee organizations, the goals of police unions focus on the topics of job security, job conditions, and compensation. Nowicki, Sykes, and Eisenberg (1991:302) identify union objectives as union recognition and security, improvements in wages and benefits, clear disciplinary procedures, and job conditions. The last two items, however, allow unions to effect efforts to change the police.

The disciplinary process is critical to the ability of the police administrator to control and direct the activities of officers. To the extent that the police union can delay or deny disciplinary action, the police administration is hindered in its efforts to direct or change the work of officers. The job conditions under which officers are employed provide several permissive issues for negotiation. Most public collective bargaining is guided by statutes that define topics of negotiation as **mandatory issues** (salary, benefits, hours, etc.) that must be negotiated, **excluded issues** (typically called *management rights,* to determine staffing, training, recruitment, etc.) that cannot be negotiated, and **permissive issues** that may be negotiated.

A recent survey of police chiefs in Ohio (Travis, 2000) revealed that chiefs did not see unions as a major obstacle to management. The chiefs generally reported that they felt they could take necessary managerial actions. What the chiefs did report, however, was that the presence of a union influenced how, and how quickly, they could act. Thus, while unions might be a constraint on managerial freedom, the Ohio chiefs did not see unions as a permanent obstacle. A national survey of police and police union leaders revealed similar findings. Colleen Kadleck (2001) reports that actual conflicts between unions and police administrators are relatively rare. The majority of police chiefs responding to this survey did not report that they felt the union represented a major obstacle to effective management.

The local nature of most police unions and the ability of the unions to negotiate changes in the definition or requirements of jobs support continued diversity in the daily practice of policing in America. For example, an attempt to remove patrol officers from automobiles and place them on foot patrol can be seen as a management prerogative—determining the best allocation of agency resources. On the other hand, the officer who must now brave the elements on foot and who no longer enjoys the mobility and safety of the automobile can define this change as an alteration of job conditions. To the extent that this or similar changes must be negotiated, the ability of the police administrator to change aspects of policing is constrained. To the degree that these and similar negotiations occur and are resolved at the local level, diversity will remain.

BALANCING FORCES

Whether the future of policing in America holds greater uniformity or diversity remains to be seen. On the one hand, the influence of court intervention, accreditation, technology, the drug war, and other common demoninators may create a more national or global style of policing in America's municipalities. On the other, the gathering momentum of the community policing reform movement may break down the police bureaucracy and produce a greater number of smaller, more local police organizations.

A third alternative also exists—the absence of significant change. In many ways this third alternative seems most probable. Given that police practice is caught between two powerful forces pushing in opposite directions, it is entirely likely that policing in America will remain static, wedged in place as it were, between these two forces. Several commentators see this future as likely, if not preferable.

In discussing the current efforts to change the police to a more proactive role in crime control and prevention, especially through the adoption of community policing strategies, Williams and Wagoner discount the chance of significant change. They write (1992:4), "While some change is undoubtedly taking place among police and in police organizations, the current situation leads us to believe that nothing substantial is taking place except, perhaps, in idiosyncratic circumstances and isolated instances." They predict (1992:5), "Through the end of this decade, there will be minor modifications in policing, all of which will be strongly resisted, or modified to such an extent that the consequences will be minimal or nonexistent." As we enter the twenty-first century, Willard Oliver (2000) foresees a coming failure to the movement towards community policing. The influx of federal money and the imposition of some federal direction on community policing development changed the nature of the movement. Further, Oliver foresees the end of federal financial support as a harbinger of the end of the community policing movement.

Against all of the forces for change to either more uniformity or more variation are aligned the forces of tradition and simple inertia. Malcolm Sparrow (1988) examined the resistance to change that exists within police organizations. He concludes that while change is possible, those seeking to reform police organizations must be sensitive to the powerful resistance to change that is grounded in the police culture and the bureaucracy of large police agencies.

Still, change is possible, at least in the case of an individual police department. In a study of the police of Troy, New York, Dorothy Guyot (1991) describes a steady alteration of one police agency from a bad department into a good one that values the people involved, both officers and citizens, and that strives to achieve both fairness and effectiveness in its operations. Guyot (1991:299) concludes,

> However, departments still exist in which much work is needed to bring up the level of fairness. The importance of the Troy experience is that

a police manager who has clear ideas about where policing ought to be going undertook a dozen years of work with a department that was far behind the times. That department has grown into an organization that supports the professional commitment of officers.

Guyot's general optimism about the prospects for the future of policing in America must be tempered, however, by our recognition of the forces that shape it. The difference between a "bad" and "good" police department is often the difference in perspective of those who observe it. No doubt there are some in Troy, New York, who long for the "good old days." Further, if this department is no longer "far behind the times," we can be certain that there are at least several hundred more in America that remain so. Moreover, there is always the possibility that the next manager of the police department will spend a dozen years working to change it into something quite different yet.

What remains to be seen is whether policing in America, in general, is amenable to change in the future. Variety is a hallmark of American police. Although they are all the same on some level, every local police department is different. Each police agency is the product of a unique balance of forces within its jurisdiction. It is this balance that explains the difference between police agencies.

THE BIG PICTURE

This book began by asking the reader to visualize the police. We expect that there have been almost as many different mental pictures of the police as there have been readers. The interesting thing about this variety is that each of those pictures is both an accurate portrayal of the police in America and at the same time an incomplete, biased stereotype. It is time now to look at the "big picture" of policing in America.

In many ways this is a speculative venture, theoretical in nature and open to empirical test. The "big picture" is a product of what we know, and think we know, about the police in America. It is a description of how the various forces that affect policing interact. The specific forces at work in any instance of policing, as well as their unique combination, enable us to understand how policing works in that instance.

Policing occurs on at least two levels: the street level, or the practice of policing where officers interact with civilians to ensure social control, and the social level, where the characteristics of the community and its members shape the nature of the police task and the role and structure of the police organization. Though interrelated, each of these levels can be examined alone.

Where We Are

We began in Chapter 1 with a description of Lundman's (1980) model explaining types of policing. He suggests that the type of policing that

emerges in a community is a product of the interaction of the pattern of social organization, the interests of the elites, and various conceptions of the
crime problem (rates and images of crime). Applying this model both cross-
culturally and historically to the American experience, we concluded that it
was generally supported by available data.

We then examined the police industry in America and described the
number, organization, and types of police agencies in our society, including
federal, state, municipal, private, and special-purpose organizations. Next
we described the characteristics of police officers, police organizations, and
communities—the correlates of policing. These included various mixes of
officer/supervisor types, organizational types, and community types. This
was followed by an examination of police functions in America as crime
control, service delivery, and order maintenance. We followed this with a
description of issues in policing, including an application of our correlates
model to the development of community policing in America.

One difficulty in studying the police is that because it is such a broad
topic, a variety of people have conducted and reported research on police issues, and a range of language has been used to describe important concepts
and observations. In the rest of this chapter, we will attempt to reduce these
differences in language while summarizing how the forces that affect policing may come into balance, first on the social, or community, level and then
on the street, or practice, level.

Policing American Societies

The explanatory model proposed by Lundman (1980) suggests that the way
a society is organized is related to the type of policing it displays. Clearly,
contemporary American society has an organic pattern of solidarity with a
division of labor and moral diversity. Social solidarity, however, is a continuum from wholly mechanical to wholly organic. Some communities have
less conflict and more shared understandings of acceptable and unacceptable behavior than others. Because American police are organized locally
for the most part, our police agencies reflect differing levels of organic solidarity in our communities.

These differences in the degree of organic solidarity are reflected in
both (1) the rates and images of crime that exist in a community and (2) the
development of elite interests. Where members of the community have
more agreement on what is or is not acceptable behavior, the police tend to
enforce these shared norms. This is the pattern described in Chapter 11 as
the solidary community. Where there is conflict between community members, some group may emerge as more powerful—an elite.

If there is an elite in a conflicted community, its conceptions of order
and crime will dominate and define the function of the police. In such circumstances, the police are likely to enforce the laws and maintain the order that is deemed important by the elite. Nonelite members of the

community, naturally, will disagree with these priorities and may resent the police.

It may also occur that no single group dominates so that there is not any elite with sufficient power to direct the police. In those cases, either the community government will be strong enough to direct the police, resulting in a formal, legalistic style of policing, or the police will direct themselves. When there is conflict in the community and no strong nonpartisan or government direction, the police agency is more or less free to govern its own affairs.

Thus, at a social or community level, the degree of conflict over values that exists, as well as the relative distribution of power among members of the community, shape the structure and function of the police organization. Consistent with a democratic orientation, the police organization will reflect the dominant interests in the community. If no set of interests dominates (either by consensus or power), the police organization is unlikely to intervene much in community life.

Policing American Streets

Moving beyond an explanation for types of policing or police organizations, we now examine policing in practice. As mentioned earlier, policing typically occurs on the streets of our communities as an interaction between police officers and citizens. Both sets of actors reflect the values of the community in their own right. When there is conflict in the community over values, there is likely to be conflict in police–citizen interactions.

Further, the actions of the individual officers are influenced by the training, rules, regulations, and reward structure of the police organization. Thus, the decisions of officers in their dealings with citizens are related to the culture of the organization in which they work. So, too, the expectations of citizens are colored by their experiences with the police organization.

Beyond the influence of social and organizational factors, policing on the street is also related to the individual characteristics of those involved. The attitudes, beliefs, and values of police officers color their decisions. The same sort of characteristics affect the citizens. Policing on the street therefore represents an interaction of community, organizational, and individual correlates of police decisions.

An arrest may be the product of a departmental norm—officers are expected to arrest whenever possible, are trained to do so, and are rewarded based on the numbers of arrests they make. It may also be the product of an officer's style. An aggressive or confrontational officer may provoke citizen resistance. So, too, it may be a product of citizen initiative. A more powerful citizen (say, a storekeeper) may demand that police arrest a less powerful citizen (say, a youth who has been caught shoplifting). Similarly, a citizen who is very sensitive to personal affronts may perceive an officer's request as an insult or challenge, and cause the interaction to escalate to the point of arrest.

Our purpose, the reader will recall, was to develop an understanding of policing on the street. Over a long, and often not very straight, path, we have identified most of the important correlates of policing. This does not mean, of course, that we can accurately predict the outcome of every police–citizen encounter. Rather, by identifying the forces at work in these interactions, and by determining how these forces are balanced, we hope to be able to explain most of them. From this level of understanding, we may be able to develop workable techniques with which to change those aspects of policing that trouble us.

CONCLUSION

Returning to the question of the future of policing in America, perhaps the only thing which can be said with any certainty is that American policing has a future. We may be wrong here, too, but barring some revolutionary social development, the police in America will be here for many years to come. One product of our study of the police is a recognition of the complexity of the topic and the variety of factors and outcomes involved. This very complexity hinders prediction, because at the same time, everything (including nothing) is possible. Before seeking to know the future, it is perhaps better that we try to understand the present.

CHAPTER CHECKUP

1. Which of the three possible futures of policing in America do you believe is most likely, and why?
2. Identify four forces working to create uniformity across jurisdictions in American policing.
3. How is it that the current movement to community policing may create even greater diversity in American policing?
4. What forces affect policing at the community or social level?
5. How do community characteristics affect policing at the street or practice level?

REFERENCES

Adderley, R. and P. Musgrove (2001) "Police crime recording and investigation systems: A user's view," *Policing: An International Journal of Police Strategies and Management* 24(1):100–114.

Alpert, G. (1997a) *Police pursuit: Policies and training.* (Washington, DC: National Institute of Justice).

Alpert, G. (1997b) "Pursuit driving: Planning policies and action from agency, officer, and public information," *Police Forum* 7(1):1–12.

Alpert, G. and P. Anderson (1986) "The most deadly force: Police pursuits," *Justice Quarterly* 3(1):1–14.

Alpert, G. and R. Dunham (1988) *Policing multi-ethnic neighborhoods* (New York: Greenwood).

Alpert, G. and R. Dunham (1992) *Policing urban America,* 2nd ed. (Prospect Heights, IL: Waveland).

Barrineau, H. and S. Dillingham (1984) "Section 1983 litigation: An effective remedy to police misconduct or an insidious federalism?" *Southern Journal of Criminal Justice* 8(2):126–145.

Beck, K., N. Boni, and J. Packer (1999) "The use of public attitude surveys: What can they tell police managers?" *Policing: An International Journal of Police Strategies and Management* 22(2):191–213.

Becknell, C., G. Mays, and D. Giever (1999) "Policy restrictiveness and police pursuits," *Policing: An International Journal of Police Strategies and Management* 22(1):93–110.

Bureau of Justice Assistance (2000a) *Edward Byrne memorial state and local law enforcement assistance.* (Washington, DC: Bureau of Justice Assistance).

Bureau of Justice Assistance (2000b) *FY 2000 local law enforcement block grants program.* (Washington, DC: Bureau of Justice Assistance).

Carter, D. and A. Sapp (1994) "Issues and perspectives of law enforcement accreditation: A national study of police chiefs," *Journal of Criminal Justice* 22(3):195–204.

Cole, G. (1992) *The American system of criminal justice,* 6th ed. (Pacific Grove, CA: Brooks/Cole).

Crew, R. and R. Hart (1999) "Assessing the value of police pursuit," *Policing: An International Journal of Police Strategies and Management* 22(1):58–73.

Davison, E. and W. Smith (2001) "Informing community policing initiatives with GIS assisted multi-source data and micro-level analysis," *Journal of Crime and Justice* 24(1):85–108.

Del Carmen, R. (1989) "Police legal liabilities," in W. G. Bailey (ed.) *The encyclopedia of police science.* (New York: Garland):453–463.

Dunworth, T., P. Haynes, and A. Saiger (1997) *National assessment of the Byrne formula grant program.* (Washington, DC: National Institute of Justice).

Fyfe, J. (1998) "Good judgment: Defending police against civil suits," *Police Quarterly* 1(1):91–117.

Fyfe, J. and J. Walker (1990) "Garner plus five years: An examination of Supreme Court intervention into police discretion and legislative prerogatives," *American Journal of Criminal Justice* 14(2):167–188.

Goldstein, H. (1977) *Policing a free society.* (Cambridge, MA: Ballinger).

Guyot, D. (1991) *Policing as though people matter.* (Philadelphia, PA: Temple University Press).

Guyot, D. and K. Martensen (1991) "The governmental setting," in W. Geller (ed.) *Local government police management.* (Washington, DC: International City Management Association):431–462.

Harris, K. (1997) *Law enforcement mobile computing: Armed with information.* BJA Technical Bulletin. (Washington, DC: Bureau of Justice Assistance).

Hoover, L. and R. DeLord (1995) "Unionization," in W. G. Bailey (Ed.) *The encyclopedia of police science,* 2nd ed. (New York: Garland):792–796.

Hughes, T. (2001) "Police officers and civil liability: 'The ties that bind'," *Policing: An International Journal of Police Strategies and Management* 24(2):240–262.

Jeffris, E., J. Frank, B. Smith, K. Novak, and L. Travis (1998) "An examination of the productivity and perceived effectiveness of drug task forces," *Police Quarterly* 1(3):85–197.

Johnson, D. (1981) *American law enforcement: A history.* (St. Louis, MO: Forum Press).

Kadleck, C. (2001) "Police unions: An empirical examination." (Unpublished doctoral dissertation, University of Cincinnati).

Kappeler, V. and Del Carmen, R. (1990) "Police civil liability for failure to arrest intoxicated drivers," *Journal of Criminal Justice* 18(2):117–131.

Klockars, C. (1991) "The rhetoric of community policing," in C. Klockars and S. Mastrofski (eds.) *Thinking about police: Contemporary readings.* (New York: McGraw-Hill):530–542.

LaVigne, N. and J. Wartell (2000) *Crime mapping case studies: Successes in the field, Vol. 2.* (Washington, DC: National Institute of Justice).

Lundman, R. (1980) *Police and policing: An introduction.* (New York: Holt, Rinehart & Winston).

Manili, B. and E. Connors (1988) *Police chiefs and sheriffs rank their criminal justice needs.* (Washington, DC: U.S. Department of Justice).

Mastrofski, S. (1991) "Community policing as reform: A cautionary tale," in C. Klockars and S. Mastrofski (eds.) *Thinking about police: Contemporary readings.* (New York: McGraw-Hill):515–530.

McCabe, K. and R. Fajordo (2001) "Law enforcement accreditation: A national comparison of accredited vs. nonaccredited agencies," *Journal of Criminal Justice* 29(2):127–131.

McEwen, T. (1995) *National assessment program: 1994 survey results.* (Washington, DC: National Institute of Justice).

Moore, M. and M. Kleiman (1989) *The police and drugs.* (Washington, DC: U.S. Department of Justice).

National Law Enforcement and Corrections Technology Center (1996) *High-speed pursuit: New technologies around the corner.* (Washington, DC: U.S. Department of Justice).

Nowicki, D., G. Sykes, and T. Eisenberg (1991) "Human resource management," In W. Geller (ed.) *Local government police management.* (Washington, DC: International City Management Association):272–307.

Oliver, W. (2000) "The third generation of community policing: Moving through innovation, diffusion and institutionalization," *Police Quarterly* 3(4):367–388.

O'Shea, T. (2000) "The political dimension of community policing: Belief congruence between police and citizens," *Police Quarterly* 3(4):389–412.

Peterson, J. (1987) *Use of forensic evidence by the police and courts.* (Washington, DC: U.S. Department of Justice).

Poland, J. (1995) "Terrorism in the United States," in W. Bailey (ed.) *The encyclopedia of police science,* 2nd ed. (New York: Garland Publishing):767–771.

Pfuhl, E. (1995) "Strikes and job actions," in W. G. Bailey (ed.) *The encyclopedia of police science,* 2nd ed. (New York: Garland):744–749.

Rich, T. (1995) *The use of computerized mapping in crime control and prevention programs.* (Washington, DC: National Institute of Justice).

Rich, T. (1996) *The Chicago Police Department's information collection for automated mapping (ICAM) program.* (Washington, DC: National Institute of Justice).

Roberg, R. and J. Kyukendall (1990) *Police organization and management: Behavior, theory and processes.* (Pacific Grove, CA: Brooks/Cole).

Ross, D. (1998) "Examining liability factors of sudden wrongful deaths in police custody," *Police Quarterly* 1(4):65–91.

Ross, D. (2000) "Emerging trends in police failure to train liability," *Policing: An International Journal of Police Strategies and Management* 23(2):169–193.

Russell, F. (1975) *A city in terror—1919—The Boston police strike.* (New York: Viking).

Skolnick, J. and D. Bayley (1986) *The new blue line: Police innovation in six American cities.* (New York: Free Press).

Smith, B., K. Novak, J. Frank, and L. Travis (2000) "Multijurisdictional drug task forces: An analysis of impacts," *Journal of Criminal Justice* 28(6):543–556.

Smith, D. (1986) "The neighborhood context of police behavior," in A. Reiss and M. Tonry (eds.) *Communities and crime.* (Chicago: University of Chicago Press):313–341.

Smith, M. (1999) "Police pursuits: The legal and policy implications of County of Sacramento v. Lewis," *Police Quarterly* 2(3):262–282.

Sparrow, M. (1988) *Implementing community policing.* (Washington, DC: U.S. Department of Justice).

Thomas, E. and M. Isikoff (2001) "Justice kept in the dark," *Newsweek* (December 10, 2001):37–43.

Travis, L. (2000) *Managerial freedom and collective bargaining in Ohio municipal police agencies: The current state of the art.* (Columbus, OH: Ohio Law Enforcement Foundation).

Travis, L., K. Novak, C. Winston, and D. Hurley (2000) "Cops at the door: The impact of citizen surveys by police on public attitudes," *Police Quarterly* 3(1):85–104.

Toch, H. (1997) "The democratization of policing in the United States: 1895–1973," *Police Forum* 7(2):1–8.

Vaughn, M. (1994) "Police civil liability for abandonment in high-crime areas and other high risk situations," *Journal of Criminal Justice* 22(5):407–424.

Williams, F. and C. Wagoner (1992) "Making the police proactive: An impossible task for improbable reasons," *Police Forum* 2(2):1–5.

Wilson, J. (1968) *Varieties of police behavior: The management of law and order in eight communities.* (Cambridge, MA: Harvard University Press).

Worrall, J. (2001) "Addicted to the drug war: The role of civil asset forfeiture as a budgetary necessity in contemporary law enforcement," *Journal of Criminal Justice* 29(3):171–187.

AUTHOR INDEX

SUBJECT INDEX